Introduction to Environmental Management

This book is directly aligned to the NEBOSH Certificate in Environmental Management, which is a qualification aimed primarily at those in business who influence the environmental performance of their organisation by the decisions that they make as managers or the actions that they take as operators. This book aims to provide an introduction to the main areas of concern and how the challenges can be addressed.

This new edition takes account of recent changes in international guidance and legislation and the recent update of the International Standard in Environmental Management ISO 14001. The contents are important for businesses that wish to stay within the law and avoid adverse publicity. It explains how the concept of sustainability can be achieved in practice and what benefits – especially financial – can accrue. Recent developments in the definitions of sustainability and the growing interest in the circular economy are introduced. It pays to be ahead of the game because decisions made now need to reflect an awareness of the coming pressures and there are opportunities available that can bring other benefits.

This book is intended for candidates for the NEBOSH qualification, but it will also be useful to anyone who wishes to understand the problems and how they can be tackled within their own organisations, be they industry, public service, voluntary bodies, or even as individuals.

Brian Waters spent 15 years in the water supply industry, principally with Severn Trent Water, covering water treatment and supply, waste water and industrial waste treatment and the management of river and groundwater quality. He then joined the National Rivers Authority and subsequently the Environment Agency, gaining experience and senior management responsibility for all of the different functions across the Midlands. Since retiring from the Environment Agency he has worked in consultancy and training and as an examiner for NEBOSH.

Introduction TO Environmental Management

For the NEBOSH Certificate in Environmental Management

Second Edition

Brian Waters

LONDON AND NEW YORK

Second edition published 2020
by Routledge
2 Park Square, Milton Park, Abingdon, Oxon, OX14 4RN

and by Routledge
52 Vanderbilt Avenue, New York, NY 10017

Routledge is an imprint of the Taylor & Francis Group, an informa business

First edition published by Routledge 2013

British Library Cataloguing-in-Publication Data
A catalogue record for this book is available from the British Library

Library of Congress Cataloging-in-Publication Data
Names: Waters, Brian, 1946- author.
Title: Introduction to environmental management : for the NEBOSH certificate in environmental management / Brian Waters.
Description: Second Edition. | New York : Routledge, 2020. | Revised edition of the author's Introduction to environmental management, 2013. | Includes bibliographical references and index.
Identifiers: LCCN 2019054407 (print) | LCCN 2019054408 (ebook) | ISBN 9781138098084 (hardback) | ISBN 9781138098107 (paperback) | ISBN 9781315104522 (ebook)
Subjects: LCSH: Environmental management.
Classification: LCC GE300 .W38 2020 (print) | LCC GE300 (ebook) | DDC 363.7/05—dc23
LC record available at https://lccn.loc.gov/2019054407
LC ebook record available at https://lccn.loc.gov/2019054408

ISBN: 978-1-138-09808-4 (hbk)
ISBN: 978-1-138-09810-7 (pbk)
ISBN: 978-1-315-10452-2 (ebk)

Typeset in Helvetica Neue LT Std
by Swales & Willis, Exeter, Devon, UK

Contents

LIST OF FIGURES vi
LIST OF TABLES vii
LIST OF BOXES viii
ABBREVIATIONS ix
PREFACE x
ABOUT THE AUTHOR xii
ACKNOWLEDGEMENTS xiii
INTRODUCTION xiv

1 FOUNDATIONS IN ENVIRONMENTAL MANAGEMENT 1

2 ENVIRONMENTAL MANAGEMENT SYSTEMS 21

3 ENVIRONMENTAL IMPACT ASSESSMENTS 47

4 CONTROL OF EMISSIONS TO AIR 65

5 CONTROL OF CONTAMINATION OF WATER SOURCES 77

6 CONTROL OF WASTE AND LAND USE 99

7 SOURCES AND USE OF ENERGY AND ENERGY EFFICIENCY 121

8 CONTROL OF ENVIRONMENTAL NOISE 143

9 PLANNING FOR AND DEALING WITH ENVIRONMENTAL EMERGENCIES 151

10 THE EXAMINATION FOR THE NEBOSH CERTIFICATE IN ENVIRONMENTAL MANAGEMENT 167

APPENDIX 1 UNITS OF MEASUREMENT USED IN ENVIRONMENTAL MANAGEMENT 171

APPENDIX 2 BACKGROUND BRIEFING TO SOME OF THE SCIENTIFIC TERMINOLOGY USED IN THE TEXT 173

APPENDIX 3 SOME EXAMPLES OF ENVIRONMENTAL MANAGEMENT IN PRACTICE 177

APPENDIX 4 FURTHER DETAILS ON CHAPTER REFERENCES TO UK LAW 179

BIBLIOGRAPHY 195

INDEX 199

Figures

1.1 Environment 3
1.2 The greenhouse effect 5
1.3 An illustration of sustainability 12
2.1 PDCA model 25
3.1 Simplified life cycle analysis for a washing machine 50
4.1 Cyclone separator 73
4.2 Bag filter 74
4.3 Electrostatic precipitator 74
4.4 Water scrubber 75
4.5 Activated carbon adsorber 75
5.1 Surface water features 79
5.2 Groundwater recharge 79
5.3 Pollution threats to surface water 82
5.4 How not to stack drums 90
5.5 How not to stack drums 2 90
5.6 Colour coded drains 90
5.7 Section through an oil interceptor 91
5.8 Where is the bund? 92
5.9 Section through a bunded storage tank 92
5.10 Using a boom on a stream 93
5.11 Using a larger boom on a river 93
5.12 Screening and solids separation 94
5.13 Horizontal and vertical sedimentation 95
5.14 Flotation for solids separation 95
5.15 Biological oxidation by activated sludge and biological filter 95
5.16 Activated sludge by surface aeration 96
5.17 Open water filter 96
6.1 The waste hierarchy 104
6.2 Clear signage is essential 110
6.3 A tidy site with multiple containers 110
6.4 Cutting oil escaping from a waste metal skip 110
6.5 Landfill operations in progress 114
6.6 Main requirement of a closed landfill site 114
6.7 Contaminated land 118
7.1 Coal fired power station 123
7.2 Flat plate solar heater 125
7.3 Solar heat panels on the roof of a Spanish hotel 125
7.4 Photovoltaic cells on the roof of social housing 126
7.5 Medium sized wind turbines on the edge of a business park 126
7.6 Illustration of two designs of wind turbine 127
7.7 Nobody likes power lines 127
7.8 Illustration of hydropower 128
7.9 Illustration of geothermal power 129
7.10 Principles of a nuclear fission power reactor 130
7.11 Principle of combined heat and power 133
7.12 A bus fuelled with natural gas in Barcelona 139
8.1 Frequency and noise 145
8.2 Some methods to control noise 146
9.1 Planning for an environmental emergency 155
9.2 Surface aerators used to reoxygenate a lake 162

Tables

1.1 Water used in the production of some common products 8
1.2 United Nations 2030 Goals for Sustainable Development 12
2.1 Internal or external audit? 42
3.1 Example of extraction of ore to produce one tonne of metal 51

Boxes

1.1 Examples of companies that have generated bad publicity for environmental incidents 10
1.2 Environmental law in England (Appendix 4) 179
1.3 Examples of international agreements to protect the environment 15
1.4 The institutions of the European Union involved in legislation 16
1.5 European Commission definition of Best Available Technique 17
1.6 Environmental enforcement agencies in the UK (Appendix 4) 181
1.7 Role of the Environment Agency in England (Appendix 4) 181
1.8 Facilities requiring a permit (Appendix 4) 182
1.9 The roles of the Agency and local authorities in permitting under IPPC (Appendix 4) 182
1.10 Powers available to inspectors from the Environment Agency (Appendix 4) 183
1.11 Enforcement powers available to the Environment Agency (Appendix 4) 183
1.12 Designated statuses by Natural England for protected sites (Appendix 4) 184
2.1 The phased approach to developing an EMS based on BS 8555:2016 24
2.2 An example of a simple draft environmental policy 27
2.3 Ecosystems and food chains 28
2.4 Aspects and impacts 28
2.5 SMART objectives 31
2.6 Topics for a training programme for the EMS 32
2.7 A simple plan for waste reduction 36
2.8 Suggested performance indicators to monitor progress with the EMS 40
3.1 Projects that require an environmental impact assessment within the EU (Appendix 4) 185
3.2 UK requirements for environmental impact assessment (Appendix 4) 185
3.3 Using life cycle analysis to compare products 52
3.4 Biodiversity, bioaccumulation and food chains 56
3.5 UK government and related sources of information 59
3.6 Examples of the source, pathway, receptor model 61
4.1 Inversion 67
4.2 Air pollution control in the UK (Appendix 4) 187
5.1 The regulation of water in the UK (Appendix 4) 188
5.2 Legionella 80
5.3 Pollution Prevention Guidelines issued by the environmental agencies in the UK (Appendix 4) 188
5.4 Some examples of applying the control hierarchy to reducing water pollution 85
5.5 Measures of organic carbon in water 86
5.6 List I and List II substances as the basis for priority control 87
6.1 What is waste? 101
6.2 The EU waste catalogue 102
6.3 Defining hazardous waste 102
6.4 Waste definitions in the UK (Appendix 4) 190
6.5 UK waste production (Appendix 4) 190
6.6 Benefits of waste recycling 107
6.7 Limitations to waste recycling 107
6.8 Regulation of waste in the UK (Appendix 4) 191
6.9 Duty of care as respects waste (Appendix 4) 191
6.10 EU Landfill Directive targets 114
7.1 Heat exchangers 125
7.2 The basis of nuclear fission 129
7.3 Storing nuclear waste 130
7.4 Principles of operation of a fuel cell 134
8.1 Noise in the work place (Appendix 4) 192
8.2 Noise as nuisance 145
9.1 Some common causes of pollution of the environment 152
9.2 Legal requirements for emergency plans in the UK (Appendix 4) 193
9.3 Suggested services required in an emergency centre 157
9.4 Key information required in an emergency centre 158

Abbreviations

This list contains abbreviations that recur. Some abbreviations that appear only once are defined in the text.

BAT	best available technology
BOD	biochemical oxygen demand
BPEO	best practicable environmental option
BS	British Standard
BSI	British Standards Institute
CFC	chlorofluorocarbon
CHP	combined heat and power
CNG	compressed natural gas
COD	chemical oxygen demand
COMAH	Control of Major Accident Hazards
COSHH	Control of Substances Hazardous to Health
CSR	corporate social responsibility
DECC	Department of Energy and Climate Change (UK)
DEFRA	Department for Environment, Food and Rural Affairs (UK)
EC	European Commission
EIA	environmental impact assessment
EEPA	European Environment Protection Agency
EMAS	Eco-Management and Audit Scheme
EMS	environmental management system
EPA	Environment Protection Agency (US)
ESG	Environmental, Social and Governance
EU	European Union
HSE	Health and Safety Executive (UK)
ISO	International Standards Organisation
IPPC	integrated pollution prevention and control
LEV	local exhaust ventilation
LCA	life cycle analysis
LPG	liquefied petroleum gas
MSDS	material safety data sheet
NIEA	Northern Ireland Environment Agency
NOX	oxides of nitrogen
PPE	personal protective equipment
PPG	Pollution Prevention Guidelines (UK)
SEPA	Scottish Environment Protection Agency
SI	System International
SOX	oxides of sulphur
TOD	total oxygen demand
UKCIP	United Kingdom Climate Impact programme
UN	United Nations
UNEP	United Nations Environment Programme
UNFAO	United Nations Food and Agriculture Organisation
UV	ultraviolet light
VOC	volatile organic compound
WHO	World Health Organisation

Preface to the second edition

Hardly a day goes by without some reference to the environment in the news. Usually the news is bad – a serious pollution incident, political disputes over climate change, the growing demand for and cost of energy, damage to the rain forests, flooding, drought, food shortages. The list seems endless. Occasionally there is a good news story about otters returning to rivers or renewable energy sources coming on stream but it is not surprising that the bad news dominates and that the public look for someone to blame.

This quotation is from the Preface to the first edition of this book written in 2012. If anything has altered it is that the profile and pressure for change has increased. In 2018 a Swedish teenager, Greta Thunberg, captured the public imagination, especially of the young, in her protest about a climate emergency by striking outside her school. Her enthusiasm took her all the way to a speech at the United Nations. The movement has become more activist as 'Extinction Rebellion' which is campaigning to get governments world-wide to respond more rapidly to climate change and ecological damage. These two issues are widely recognised as the consequences of unconstrained growth, consumption of resources and energy, damaging emissions to the environment and exploitation of the biosphere. The causes are multiple and are as much the fault of the wider population as the businesses that get caught in the spotlight.

But business is a major cause and industries such as oil exploration, power generation or chemical manufacture are obvious culprits. They operate in such a way and on such a scale that the potential risks of environmental damage are high and fairly obvious. However, looking more closely at other businesses, the use of energy, demand for raw materials, production of waste and emissions to air and water apply to virtually every activity and drive the activities of those bigger companies. Many things that we do at work have an impact on the environment in one way or another even if we are in a service industry or a public or charitable body. The same can be said of our home and leisure activities. At any level these may be small but collectively they add up such that the potential availability of some raw materials is becoming a global problem, prices are rising and we are making changes to the atmosphere and seas that could be irreversible and have the potential to cause severe problems for future generations. This vague concept is our children and grandchildren.

Responsibility to deal with these issues falls to everyone but business is often seen as an easy target. Business exists to meet the demands of its customers and they must be the ultimate culprits by demanding the latest gadget or faster cars or the right to wander about at home in a tee shirt in freezing weather. Of course the public is fickle: it demands the goods and services that make life easier and more enjoyable but castigates the businesses that meet those needs. Tackling that broader issue is not the purpose of this book but businesses in the firing line can do a lot to mitigate the damage that these demands place so that they can continue to operate, supply their customers and reduce the impact that they have. This book is aimed at that purpose.

Public perception is important for most businesses and the rise of social media such as Twitter and Facebook can spread information (and misinformation) rapidly around the globe. Public concerns are reinforced in many cases by the law. Stricter controls are applied to reflect those concerns and understanding and complying with the requirements is becoming more difficult and costly. Larger companies often employ environmental managers to advise and monitor what is happening but compliance is usually down to those who carry out the tasks that are at the heart of the problem. Whether or not an environmental manager is in post, it is those who specify, design, produce, purchase, market, deliver and dispose who have the potential impact. They are overseen by directors, managers and supervisors who influence what they do.

So the easily identifiable risks to business are reputation, legal liability and cost escalation. There is also the risk to the environment which most responsible people at any level in a business would wish to avoid. However, the risk is not always easy to spot and may be even harder to manage. The broader relationships among environmental risk, economic development, social progress and resource issues are now rolled into the concept of sustainability. Sustainability can apply at the global level down to the very local and is becoming the overarching measure for the future. The sustainability of their business may be what matters most to the business manager and that can be at risk from bad behaviour but there are business opportunities through reduced costs and new markets for those that are prepared to rise to the challenge. Understanding and then responding to the issues that apply at the business level will help to meet the global objectives for a sustainable world more able to deliver a brighter outlook for future generations.

The National Examination Board in Occupational Safety and Health (NEBOSH) operates an Environmental Certificate for those who are not environmental managers (for whom there is a Diploma) as a means to gain the knowledge and a recognised qualification to help their businesses meet these challenges. It has proved popular outside of the UK and the syllabus reflects the international reach of the qualification. The syllabus contents are relevant in some way or other to every level in a business from the most senior manager to the cleaner. They may not wish to take the examination but the principles are still relevant. This book is structured around the syllabus of the Certificate. It is split into logical sections to make for easy identification of the key issues. It is not aimed at the specialist; the target audience is those who can make a difference on the ground as outlined above. It assumes limited background knowledge and there are explanations of technical content where this is necessary to understand the detail.

About the author

Qualifications: BSc Chemistry, PhD Biochemistry, both from Birmingham University.

Fellow Royal Society of Chemistry, Fellow Chartered Institution of Water and Environmental Management, Member Institute of Environmental Management, Member Chartered Management Institute. Brian spent 15 years in the water supply industry, principally with Severn Trent Water, covering water treatment and supply, waste water and industrial waste treatment and the management of river and groundwater quality. He then joined the National Rivers Authority and subsequently the Environment Agency, gaining experience and senior management responsibility for all of the different functions across the Midlands. Since retiring from the Environment Agency he has worked in consultancy and training and as an examiner for NEBOSH.

Acknowledgements

The preparation of this book would not have been possible without the help and support of many people. Staff at NEBOSH have been helpful over many years in my involvement with the syllabus and as an examiner. Steve Simmons provided images 6.2, 6.3, 6.4 and 6.5 on waste practice and UKCIP provided the image 1.2 to illustrate the greenhouse effect. Matthew Ranscombe and colleagues at Routledge, the publisher, have supported me through the production process. A big debt is also owed to the many colleagues that I really enjoyed working with over many years and who taught me a lot. I must also acknowledge the few environmental rogues that crossed my path who added even more to my education.

A special mention is due to my wife, Janet, who has put up with my isolation and deferred projects at home to enable this book to be completed.

All of the content is my responsibility. I apologise for any errors but would be pleased to be told about them and to receive any other comments.

Introduction

This book is in support of the NEBOSH Certificate in Environmental Management. It is targeted at those in business who can influence the impact that an organisation can have on the environment by virtue of the decisions made and the activities carried out. The objective is to improve the understanding of the issues involved and the ways that businesses can avoid damaging the environment and their own reputations in the process. In many cases there are opportunities to reduce costs and other benefits may also accrue. There are also legal issues in environmental management of which the people in the organisations need to be aware. Most countries now have laws to protect the environment with penalties of fines and prison terms for breaches. Some of the fines may be on companies but they can also be on individuals as would terms of imprisonment.

The book is not aimed at environmental professionals such as an Environmental Manager. NEBOSH offers a Diploma qualification for them with more legal and technical information. At Certificate level the technical content is more restricted and some of that and additional background information is presented in red boxes. However, to understand the content it has been necessary to assume some background knowledge. The main areas considered necessary are:

- The ability to recognise and use some units of measurement not in general use.
- Limited understanding of some sciences, mainly chemistry and biology, and particularly the terms used.

It is recognised that some readers may have not formally studied these subjects or that it was such a time ago that it is a distant memory. If you have a recent relevant GCSE or above from a UK school or an equivalent in these subjects you certainly ought to be able to manage. There are two appendices to help those with limited background knowledge or who need some revision. Do not be put off – it is not difficult.

The other hurdle for some examination candidates is that English is not their first language. However English is the language of international trade and many candidates for NEBOSH examinations see them as a way to help gain employment in companies that have international businesses. The language has been kept as simple as possible and the book should help them recognise any terms that are not in general use and that they will need to know.

Other readers of this book may be studying for their own satisfaction or to seek ideas for improving the environmental performance of the organisation that they work for. There is no obligation to take the examination. If you do, the NEBOSH Certificate is examined in two stages: a written examination on the content of the syllabus and a practical application in the workplace or other suitable location. The purpose of the practical application is for the candidate to demonstrate the ability to apply the knowledge gained in the first stage. This should present no problem to a candidate who has studied and understood the syllabus.

As this book is aimed at an international audience the legal requirements are stated in general terms in the main text with examples taken from UK practice. Unfortunately for the reader, UK legal requirements are not consistent as there are differences among the different administrations of England, Northern Ireland, Scotland and Wales. Reference to UK regulations and practices is mostly for England and Wales and may not be valid in the other devolved administrations in all cases but the principles should still be constant, as they are mainly derived from EU directives, and they should also be applicable more generally. Recognising that many readers will be from the UK or from countries with a similar legal system, there are boxes with the more detail on some of the legal points. These are mainly for information and are distinguished by being with a blue background and collected together in Appendix 4. More detailed legal information can also be obtained by consulting the relevant legislation or the various compendiums and summaries provided commercially in book form or electronically. Nothing in the text should be taken as legal advice!

Although the legal content has been limited, there may be a need for a reader to follow something up to help understanding or to deal with a local problem. The text includes references to international, European and UK legislation and guidance. Do bear in mind that these are subject to change and if you want to be up to date check that you are consulting the latest version. Change is constant, particularly in the UK where every change of government seems to result in a shuffling of responsibilities among different departments and a switch in emphasis on what is important. At the time of writing (late 2019) the UK is in a long, tortured process of planning to leave the EU and at the start of an election. If it does leave much existing legislation will initially be carried forward but it is probable, depending on the final terms agreed, that there will be changes over time. It is fortunate now that most of the necessary information can be found on the internet and downloaded in pdf format. The web site references given were valid in November 2019 but they do have a habit of moving to archives or some other site. A search engine should track them down if the reference given is no longer valid. References for European Directives are from the Official Journal but these can also be found on the internet.

The structure of the book is based on the NEBOSH syllabus. The chapter numbers and titles refer to the elements within the syllabus and the sections are similarly structured. The headings for the sections are mainly the same but a few headings or their order have been changed to help the flow of the text. The content does go beyond the limits of the syllabus as the aim of the book is to help candidates understand the environment and use that knowledge to protect and enhance it. In addition they will help their organisation to keep within the law, save money and protect its reputation. The aim is not just to gain a piece of paper.

As with most textbooks you do not have to read it in the order presented. However, environmental topics are inter-related: air quality can affect both water and land quality, for example, in a number of ways. This means that there are cross-references between chapters to avoid too much repetition.

The final chapter is about the examination for the certificate and includes some example questions. Even if you are not contemplating taking the examination, you may find them useful to check your understanding.

Hopefully you will not see this as the end of your learning. It is important to keep up to date with changes in the topics in the chapters but, more importantly, to be aware of developments in technology if you maintain an interest in the subject. Developments in lighting technology have already reduced energy demand dramatically; vehicle design is changing and will be very different in a few years' time; sources of energy, their supply and use will continue to change; the production and management of waste has to be tackled to avoid more environmental damage; everyone will have to play a part in minimising climate change but also in adapting to some which will be inevitable.

CHAPTER

1

Foundations in environmental management

After this chapter you should be able to:

1. Outline the scope and nature of environmental management
2. Explain the ethical, legal and financial reasons for maintaining and promoting environmental management
3. Outline the importance of sustainability and its relationship with corporate social responsibility
4. Explain the role of national governments and international bodies in formulating a framework for the regulation of environmental management.

Chapter Contents

Introduction **2**

1.1 The scope and nature of environmental management **2**

1.2 The reasons for maintaining and promoting environmental management **9**

1.3 The importance of sustainability **12**

1.4 The role of national governments and international bodies in formulating a framework for the regulation of environmental management **15**

INTRODUCTION

This chapter starts with a look at what we mean by 'the environment' and 'environmental management' and why it is considered important to protect it. This leads into a study of the meaning of 'sustainability' and how this fits into the way that a business may project its image to the wider world, followed by the legal issues surrounding environmental protection. This background forms the basis for the following chapters which elaborate on some of the topics raised here, giving more detail on why they are important and how they can be managed.

1.1 The scope and nature of environmental management

1.1.1 Some definitions

Words like environment and pollution are in everyday use and may mean slightly different things to different people. They will most likely reflect their immediate surroundings or their concerns. For example, my immediate environment may be my home or work site but could extend to include the garden or cultivated area, village, adjacent park, surrounding farmland or forest, coastal stretch or the sea (see Figure 1.1). However, increasingly people are concerned about the wider environment such as the change in climate or the impact of plastic waste on wildlife. This is usually in response to publicity by pressure groups or in response to documentaries on television illustrating the events in dramatic ways.

Generally, though, our concerns about what affects our environment are usually those which we can easily sense and have a direct impact: noise, smell, visual intrusion, flooding, perhaps something that is causing illness or killing our plants. We may refer to these as pollution. To complicate it further our response to these effects is coloured by our expectations. We expect some factories to be noisy but escape to the quiet of the countryside. The smell of animal manure being spread will be less acceptable to the local villagers than to the farmer. The sea environment is important to fishermen who rely on it for their livelihood. Before reading further think about the environments that you come across regularly and what you like about them and if there are things happening around them that concern you; why do they concern you?

In order to study our subject comprehensively we need to adopt standard definitions because, as we shall see, decisions we make have implications that can extend well beyond our immediate surroundings. A good example concerns waste disposal: as societies develop they produce more waste. In a simple society the waste may be just food and crop residues which can be dealt with locally by composting along with limited quantities of other materials such as paper or metal, many of which will be put to other uses. In an urbanised industrial society households and businesses produce larger volumes of waste which cannot be dealt with on site and have to be removed and dealt with elsewhere. This involves transferring a potential local problem to somewhere else and in a globalised economy that could be another country. Our waste does not affect our environment directly but can affect somewhere far away.

In Chapter 2 we shall learn about environmental management systems and the international standard ISO 14001. This has a definition of **environment** which should suit our purposes well:

> **The surroundings in which an organization operates including air, water, land, natural resources, flora, fauna, humans, and their interrelation.**

ISO 14001 adds a note that surroundings in this context extend from within an organisation to the global system (ISO 2015).

There is no similar internationally accepted definition of **pollution**. If you try to look it up in a dictionary you will find several definitions, some specific to a medium such as water, air or land, or related to specific issues such as noise or visual intrusion. We also need a definition that does not become too restrictive. Consider water: it evaporates from the earth, leaving any contaminants behind and condenses in clouds as 'pure' water. When it falls as rain it dissolves gases and picks up solids from the air before dissolving more contaminants in its passage over and through the land. But for most people it is not considered polluted unless it has been exposed to

Figure 1.1 Environment

something toxic; indeed, many may drink it directly. For our purposes we need a general definition that does not include benign contaminants. One such is:

> **The presence of substances or objects in the environment which may cause adverse effects on the natural environment or on life.**

This definition brings in the requirement to cause harm. It should be recognised that the substances referred to could arise from natural causes (such as a volcano) as much as from man's activities and that objects can extend to large structures causing visual pollution. The effects on life may be confined to human beings but could extend to any form of life.

1.1.2 The meaning of environmental management

Man has tried to manage the environment for millennia. The earliest humans were hunter-gatherers who took small quantities of food relative to the amounts available. They then progressively cleared forest for farming and extracted timber and stone for building materials. As societies developed technology, they extracted minerals and converted them to useful materials such as bricks and metals. Modern man does all of this on a much bigger scale and builds roads and bridges, flies aircraft round the world, wages war on a more destructive scale and consumes energy at an ever increasing rate. These activities all have an impact on the environment and most are humanity trying to exploit it for some direct benefit such as feeding the population and providing goods and services. Our reason for studying environmental management though is understand the consequences of this and to minimise the damage to the environment from human activities. As the population grows and expects a higher standard of living, the degree of exploitation is causing extensive environmental damage on a scale not seen before – as we shall see in the next section.

The challenge is becoming more complex. Small-scale operations such as farming or mining or waste disposal did not present a problem to earlier communities. Such damage as was caused was of a similar small scale and confined to the immediate neighbourhood. If one person upset his neighbours by his activities they would likely take direct action. Now food is grown on an industrial scale. Resources such as oil or minerals are extracted far from the centres of demand. The link between the demand and the damage caused is remote. Further, in an increasingly complex society, a simple action can have multiple consequences: inappropriate disposal of waste can cause pollution of water, land and air. Washing contaminants out of an air stream converts it to water pollution; removing that pollution can result in solid waste that may pollute land. The nature of environmental management has become more complex, involving the basic sciences of chemistry, biology and physics and the more complicated sciences such as climate modelling, toxicology and hydrology. It has become multi-disciplinary. This book will not go into the detail of all of these but rather give an overview of the main issues. However, it is important to bear in mind the inter-relationships that can be involved in the media (air, water and land) and the processes taking place in the environment.

The factors that complicate environmental management do not stop there. Within any business or community there are competing and conflicting demands. People want adequate food, housing and energy as a minimum. In many societies this is still a wish whereas in others there is excessive use and wastage. Few people give thought to where their food, energy or raw materials come from – it is likely to be well away from their immediate environment. As population and prosperity grow, demand increases, people expect more and competition for resources increases. Energy and minerals are extracted from increasingly remote and difficult places, often in sites previously considered wild or in their natural state and even on another continent. The consumers may not be aware or just ignore this, but if there is a proposal to develop a quarry near to their homes they are likely to use every available avenue to object. We want the benefits that technology can bring but not the problems, especially if they impinge on our quality of life.

As prosperity grows not only do we consume more but we waste more. Items that were routinely repaired a generation ago are now replaced if they fail. Some items are too complex to be economically repaired (such as electronic goods), others are replaced often because there is a new model available (consider mobile phones), demand for new clothes is driven by fashion rather than need. The effects of these decisions are remote from the decision maker and even if they are aware of the consequences they take a simplistic view that their lone action cannot have major implications. It is the individual decisions of 7.7 billion people that have compounding consequences.

1.1.3 The scale of the environmental problems

We have already realised that some pollution problems remain local. Noise does not travel far and odour rarely goes far before dilution or some chemical change renders it harmless. Light pollution is the presence of light from street lights and urban areas that causes background illumination which can disturb wildlife and hide the night sky. (See http://apod.nasa.gov/apod/ap020810.html for an image from space of the sources of light – set aside some time if you get lost exploring the many other images available from satellites!). It is an increasing problem changing animal behaviour and interfering with other people's lives – not just astronomers. Contamination of land by waste, chemicals, oil, etc. is also likely to be mainly local although some contaminants can migrate to cause pollution further away as we shall see in later chapters. Many of these local effects can be mitigated at source or dealt with locally in other ways but even if little is done the impact remains local. It is the pollution that arises locally but has an impact over a wide area, even globally, that is more of a problem. Some important examples are considered below.

1.1.4 Population growth

The world population grew slowly up until the beginning of the last century. In 1972 it was estimated at 3.85 billion and had grown to 6.1 billion by mid-2000. It is estimated to grow to over 9 billion by 2050 but the rate of growth is slowing down so that the population will reach nearly 11 billion by 2100 (Our world in data 2019). These additional people will drive up consumption of materials and the production of pollution, especially as all peoples aspire to a higher standard of living.

In 2019 the United Nations Food and Agriculture Organisation estimated that over 800 million people were undernourished (UNFAO 2019), just over 10 per cent of the total population. Increasing food production to meet the current shortfall, let alone the projected increase in population, will require more land, fertiliser and energy as well as the lesser items such as pesticides, farming implements, etc. These hungry people also consume little else of the world's resources at the moment.

1.1.5 Sourcing raw materials

Phosphate fertilizer to help meet the need for food is extracted from large mines and quarries. But we get many of our other raw materials in the same way: ores for the production of metals such as iron and copper, gold, building stone, sand and gravel, limestone for various uses such as cement production and clay for making bricks and pottery. Production of the end products has moved from craft industries to major industrial factories requiring large volumes of raw materials to keep them supplied. To make matters worse, the quality of metal ores is declining; those with a high metal content are becoming exhausted or too expensive to extract and poorer quality ores are being used. The lower content means that more material has to be extracted for the same yield of finished metal (if the metal content falls from 2 per cent to 1 per cent then twice as much is required). This is more expensive in transport and processing costs, uses more energy to process, produces more waste material (the residue after the metal has been extracted) and damages more of the environment adjacent to the mine. There is more on this in Chapter 3. There are often land rights issues associated with increasing food production or the exploitation of minerals in remote areas as forest is cleared or land is exploited that has a history of traditional rights for indigenous peoples such as being the source of food and timber.

1.1.6 Energy supplies

Similar issues apply to sourcing energy. Although some developing rural communities still burn wood for cooking, this can be difficult to find. Urban and industrialised communities rely mainly on coal, gas and oil products as primary sources of energy. Electricity has traditionally been produced from these primary sources although, as we shall see, this is changing. Burning these primary sources causes air pollution in ways described below but finding them is becoming a problem as well. Coal is available in some countries in plentiful supply but is potentially the most polluting. In some cases it is close to the surface and can be open-cast mined. Elsewhere it involves deep underground tunnels which are often dangerous due to poor health and safety practice. Either way coal extraction also produces large quantities of solid waste known as 'spoil'. Coal is being replaced by gas as a fuel for generation of electricity in many countries, although China and India are still building new coal-fired stations on a large scale. Oil and gas are extracted from wells into suitable geological strata. Reserves are declining or becoming uneconomic in many of the existing sources as consumption continues to rise. There is a school of thought that we are close to the point where rising consumption will overtake falling supply, especially of oil – referred to as 'peak oil'. Political risk associated with some supply countries also drives a search for new supplies and so wells are being drilled in more remote areas such as the Arctic and in deeper seas such as off the Falkland Islands. Extracting gas by fracturing the rocks holding it (hydraulic fracturing or fracking) is also being used. All of these have potential problems, particularly those of pollution and these are detailed further in Chapter 7. The point here is that energy demand increases as standards of living rise and the demand for energy has similar consequences as that for other raw materials.

1.1.7 Emissions of carbon dioxide and climate change

Burning fossil fuels such as coal, oil and gas produces carbon dioxide – the carbon in the fuel is combined with oxygen in the atmosphere, releasing heat at the same time. It has been known for over 100 years that the concentration of carbon dioxide in the atmosphere was responsible for maintaining the earth's climate within a range suitable to support life. The concentration has varied significantly over geological time but in more recent centuries it has been constant at about 260 to 280ppmv (see Appendix 1 on units and Section 4.1.1 for measurements in air). It started to rise with the industrial revolution in the 19th century as coal was used to fire the factories and produce energy. The concentration is now about 410ppmv, over 50 per cent higher (Mauna Loa Observatory 2019) and has risen in parallel with the increasing use of fossil fuels to produce energy.

The carbon dioxide is evenly dispersed throughout the atmosphere by atmospheric currents and plays a key role in absorbing infra-red radiation from the sun. The detailed mechanism by which this occurs is complex (Houghton 2009) but can be briefly summarised as follows. As the radiation falls on the earth it is absorbed by the carbon dioxide and by water vapour. This warms up the atmosphere and the earth's surface. Some of the radiation is radiated back into space again but the carbon dioxide absorbs this as well. If the carbon dioxide concentration is constant a balance is achieved such that the temperature remains fairly constant but as the concentration rises, less is radiated back into space and the temperature rises. A simplified illustration is given in Figure 1.2.

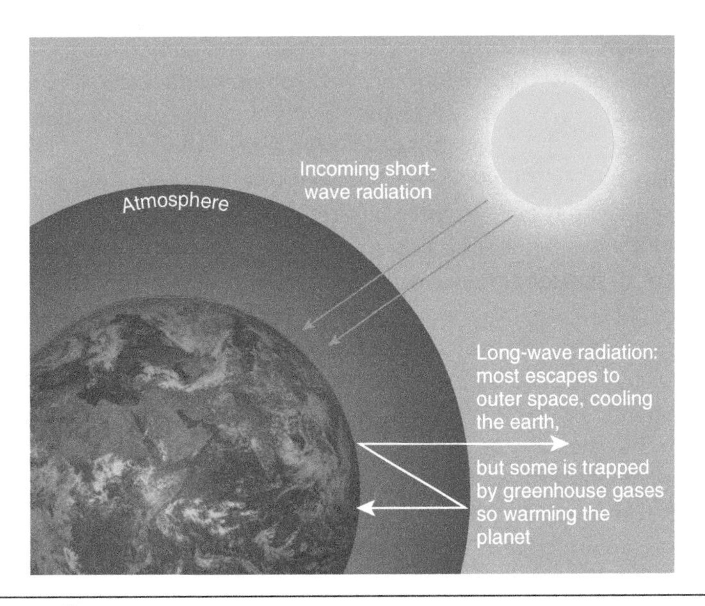

Figure 1.2 The greenhouse effect

This effect is known as the greenhouse effect as it is similar to what happens inside a greenhouse where the glass prevents some of the heat being radiated away. Water vapour, methane and other industrial emissions of gases such as some chlorofluorohydrocarbons, nitrous oxide and sulphur hexafluoride also act as greenhouse gases. Methane is a more powerful greenhouse gas but carbon dioxide dominates the effect because of its relative concentration. Both of these have increased in concentration in the last hundred years or so due to industrialisation: carbon dioxide from burning fossil fuels and methane from various sources such as releases of natural gas, emissions from anaerobic conditions (lack of oxygen: see Appendix 2) in waste tips and releases by ruminants.

The term 'global warming' has been used for this effect and it has a number of consequences, some of them serious, especially if the effect continues unabated. The first of these is an increase in global temperature. For some communities this could seem welcome: if the British climate became more Mediterranean there would be few complaints. However southern Europe could become as hot as North Africa, which could be less welcome. Globally many areas could become too hot for comfort and severe droughts could become more frequent in developing countries which already suffer from this. A further consequence is that water expands on heating so that, over time, sea levels will rise. Higher temperatures cause glaciers and polar ice to melt also adding to sea level rise. It is estimated that average sea level has risen by between 13 and 20cm in the 20th century (Smithsonian Institute 2019) and that the rate of increase is itself increasing, so coastal areas and low-lying islands are increasingly vulnerable to flooding.

A warmer atmosphere contains more energy and can hold more water vapour and this will affect rainfall patterns and the risk of storms. Increased rainfall may lead to inland flooding and reduced rainfall may lead to drought. If drought becomes more severe or frequent there are consequences for nature and agriculture and for fire risk in forests and moorland. Changing temperatures and water availability will also change wildlife distribution: some species may disappear altogether, especially trees and plants which are slow to migrate or animals which can find no suitable habitat nearby. Spring now arrives about one week earlier in the northern hemisphere than 20 years ago, meaning that bird breeding cycles are getting out of synchronisation with insect breeding patterns which they rely on to feed their chicks. Another well-publicised case is that of polar bears which are dependent on sea ice in the Arctic for their survival. The area of sea ice is decreasing and the melt is occurring earlier in the year so that the bears cannot spend as long at sea on the ice flows searching for food with their young.

There is plenty of evidence that the climate is changing, with changes in flood risk, frequencies of droughts, storm damage, changes to the growing season and loss of wildlife. These events happen irregularly for other reasons such as sun spot activity or random weather patterns. Consequently no one event can be attributed to global warming but a changing pattern over time confirms that it is likely. Average global temperatures have increased by about 0.9°C in the last 100 years (NASA 2019) which is seen as evidence of global warming. There are projections for the rest of the 21st century based on models of the climate which predict further increases in temperature. The amount of increase depends on many factors including what assumptions are made about the concentrations of greenhouse gases in the atmosphere. The range is from about 2° to 6°C by 2100. Neither of these seems large but they are enough to make a radical difference to future climate. More worrying is that, even if there were a dramatic decrease in CO_2 emissions today, there is sufficient inertia in the climate system to ensure that the temperature is likely to rise by a couple of degrees anyway.

Changing climate also brings changes to health risks and pests. Diseases such as food poisoning or malaria may become more common and pests able to survive in new territories may become a problem. In recent severe heat spells in Europe the death rate has increased especially among the elderly and vulnerable. A secondary effect of changes in food availability and disease risk is likely to be an increase in the numbers of refugees.

Another impact from carbon dioxide emissions that is not related to climate change is that it causes acidification in seas. That is, the carbon dioxide dissolves in sea water to make it slightly more acidic. This is believed to be causing damage to corals and other sensitive organisms.

It would be fair to point out that not everyone accepts that recent changes in climate have been caused by the greenhouse effect. The climate has changed significantly over millennia with wider swings than in recent decades or those predicted for the next few decades. However, the consensus among the scientists most concerned about the issue is that it is real and we should be doing something to mitigate it by reducing emissions of greenhouse gases as well as adapting to the current and projected changes in climate to minimise the potential damage – for example, in raising the levels of flood defences and managing water resources more efficiently. Adopting a precautionary approach would seem sensible given the potential risks. There have been moves to reach agreement on reducing emissions of greenhouse gases, particularly CO_2. These started with the Kyoto summit in 1997 (United Nations 1998). The protocol adopted sought to get countries to agree to a 5 per cent reduction of greenhouse gases by 2012 compared with 1990. Although some progress was made by some countries, the global CO_2 concentration continued to increase (Mauna Loa Observatory 2019). The most recent proposal, the Paris Agreement, adopted in 2016, provided an opportunity for countries to strengthen the global response to the threat of climate change by keeping a global temperature rise this century well below 2°C and to pursue efforts to limit the temperature increase even further to 1.5°C. The report of the Intergovernmental Panel on Climate Change (IPCC 2018), subsequently approved by governments, reinforced the argument that the increase should be limited to 1.5°C in order to minimise the risks. Attempts to reach agreement have proved difficult so far as countries are worried about the impact on their economies or development. China produces the highest total CO_2 emissions – not surprising as it has the highest population. Some smaller countries have high emissions per capita but, of the larger countries, the United States, Australia and Canada are also high; all have some of the highest standards of living. Signatories to the Paris Agreement are recorded by the UN (United Nations 2019), and most have signed but little is happening sufficiently fast to meet the commitments. Public pressure increased in 2019 as younger people started to demonstrate against the lack of progress, declaring a 'climate emergency'. This has raised the profile of climate change world-wide. However, an equitable solution that allows the aspiring nations to develop and the developed nations to moderate their demands on the earth's resources will need statesmanship well beyond that currently being demonstrated.

That is not to say that some countries have not made commitments: the EU has agreed a target of 40 per cent reduction by 2030 (compared with 1990) with talks of further cuts and the UK has a new target of 100 per cent reduction by 2050. However, some politicians and other campaigners are trying to water these down. Overall, if the projections about future climate and its impact are correct, the prospects for the next generation of children may be bleak.

A method used by organisations and individuals to monitor and manage their own emissions is known as 'carbon footprinting'. This refers to measuring the emissions of greenhouse gases from everyday activities such as energy use and transport and then devising ways to reduce the size of the footprint. There are calculators available online to help calculate the footprint from energy use, km travelled, etc. and the calculations can go into enormous detail, for example taking account of the footprint arising from food production and products manufactured or bought. As a start it would make more sense for most organisations or individuals to look at the big producers of carbon dioxide such as heating, the use of electricity or driving vehicles. Options for reducing emissions of carbon dioxide on a wider scale are mentioned again in Chapters 4 and 7.

1.1.8 Air pollution and damage to the ozone layer

The ozone layer is a part of the upper atmosphere that contains a higher concentration of ozone than elsewhere. It is caused by the interaction of the ultra-violet (UV) light range of the spectrum of the sun's rays causing the oxygen molecules in the atmosphere (O_2) to split and then recombine to form ozone (O_3). (The UV part of the spectrum is different from the infra-red part that is associated with global warming.) This layer adsorbs further ultra-violet light and prevents it reaching the earth's surface where it would be damaging to organisms; for example, it is these rays that cause sun burn and skin cancer in humans. Note that ozone up here is beneficial; ozone at ground level is not: it causes damage to the lungs and contributes to smog, as we shall see later in Chapter 4. Damage to the ozone layer was first discovered in the 1970s and eventually traced to ozone depleting substances that cause ozone to break down again to oxygen. The source of the damage was traced to halogen containing compounds (halogens is a collective term for chlorine, bromine, iodine and fluorine: see Appendix 2 for an explanation of terminology) such as chlorofluorocarbons – abbreviated to CFCs. These are a range of chemicals with slightly differing compositions some of which also contribute to the greenhouse effect. These broke down in the atmosphere to release halogen atoms such as chlorine and bromine. CFCs were used as refrigerant gases, propellants in aerosols, in fire extinguishers and as solvents as they were relatively inert and stable. It was their stability that was part of the problem as they survived in the lower atmosphere and slowly dispersed to the higher levels where they caused the damage. Their use has been progressively reduced by the Montreal Protocol (UNEP 2009). This international agreement has been successful as there are other substances available for the original purposes. Some of these still contain halogens but are less stable at lower atmospheric levels and so have less of an impact at higher levels.

Nitrous oxide is another ozone-depleting compound that occurs naturally and arising from agricultural activities but is more difficult to control. It is also a greenhouse gas.

1.1.9 Water resources

Water is essential for life: people can survive without food for weeks but without water they die within days. Our food production also depends on water and irrigation is essential to ensure germination and reliable yields, especially in hot, dry climates. Water is subject to the same strains as minerals and energy resources: an increasing population and higher expectations of hygiene, coupled with increasing demands from industry mean that safe drinking water is scarce in many parts of the world. Sea water is plentiful but is not suitable for most purposes unless treated at great expense to remove most of the salts present.

Man only needs one to two litres of water a day to survive but uses a lot more for other purposes. In developed economies household use is often 150 litres per head, as flushing toilets, washing machines and showers take their toll. This is only part of the story as the food we eat and the other goods that we use are also dependent on water. Table 1.1 shows the water traditionally used to produce some common goods. The effect of this is that in many countries true water consumption per head is significantly greater than that which is supplied directly. This is known as 'virtual water' and implies usage that is not apparent to the consumer. Some of this may be local such as food crops that are irrigated, washed and prepared nearby. Water may be part of the product (e.g. in drinks or canned food) or used in production of other finished products (e.g. leather or cotton), often from parts of the world which can least afford to use it. Food and flowers grown in Africa as cash crops for use in Europe denies the food and water used in its production to the local population; so there is a trade-off. The local direct benefits of the trade such as cash income and jobs for some may be at the expense of the wider local population who find their water supplies dwindling. Estimates vary depending on the source of the information and the country of origin of the materials and such factors as climate. These are estimates that give an indication of the relevant scale of water use to produce different products and show some interesting comparisons. Note that it takes nearly eight litres of water to produce the plastic bottle holding one litre of water.

Table 1.1 Water used in the production of some common products

Product	Water used in litres
Apple 150g	125
Apple juice 1litre	1,140
Beef 1kg	15,400
Bioethanol from sugar cane 1litre	2,107
Bottle (plastic) to hold bottled water	7.8
Car	165,000
Cement 1tonne	5,760
Chocolate 1kg	20,000
Cotton 1kg	10,000
Cup of coffee	130
Jeans 1 pair	7,620
Leather 1kg	17,000
Lettuce 1kg	240
Maize 1kg	1,220
Milk 1kg	1,020
Rice (milled) 1kg	2,500
Steel 1tonne	263,000

Sources: For agricultural products; M. M. Mekonnen and A. Y. Hoekstra, *The Green, Blue and Grey Water Footprint of Crops and Derived Crop Products*, Value of Water Research Report Series 47, UNESCO-IHE, Delft, the Netherlands: UNESCO, 2010, extracted from www.waterfootprint.org dated 2012, under http://creativecommons.org/licenses/by-sa/3.0/.

Other information from the Water Footprint Network at https://waterfootprint.org/en/resources/interactive-tools/product-gallery/.

The cost of water is increasing for much the same reasons as outlined above for other resources. It has to be sought further afield and may require additional treatment as less pure sources are developed. Much may be lost in transmission and to evaporation, especially irrigation water in hot countries. Efficient use is becoming the way to ensure supply by adopting the hierarchy that is applied to other aspects of managing resources as described in Chapter 5. Severe droughts have resulted in famine, especially in parts of Africa, and climate change is expected to result in less rainfall in many of the most vulnerable regions. A continuing need for relief, large numbers of refugees and potential for conflict are possible consequences.

1.1.10 Land issues

The damage resulting from human activities is not confined to the air and water. We have already learned of the damage to land from mineral exploration and extraction. The removal of large quantities of soil and rocks causes direct damage. The surrounding area is also usually affected by the spoil that is left behind and possibly from leaching of minerals from the spoil heaps that can render the soil useless for crops or can percolate into groundwater or run off into streams. Disposal of the wastes from mineral processing along with all the other waste arising from man's activities has the potential to cause further damage as we shall see in Chapter 6. Recent concerns have included the problems associated with plastic waste, which usually degrades slowly, if at all. The export of waste for disposal (usually dumping) in third world countries is being stopped by the recipient countries and international agreements but it remains a problem.

Deforestation is another way that land can be damaged. It may be caused by the removal of trees for firewood or construction or by clearance to provide more land to grow food crops or raise animals as well as to find mineral resources. The current controversy over forest clearance to grow crops such as sugar or palm oil as alternative sources of fuel and food is a good case in point. Often the vegetation is burned to clear it and after use for a few years the land has lost its fertility and is allowed to grow back wild. Unfortunately, forests take many years to develop and the recovery of abused land is likely to be slow if it occurs at all.

Forests are important in many ways. As the trees and other plants grow they remove carbon dioxide from the atmosphere, using it to produce carbohydrates for cell structure and releasing oxygen by a process known as photosynthesis. In this way they help to remove some of the CO_2 that is causing global warming. Forests also affect the weather more directly as the leaves transpire water from the soil into the atmosphere causing local rainfall and contributing to the high humidity associated with tropical rainforests. Because of this forests are usually very diverse in the range of animals and plants that they support; from rare trees to dependent populations of plants, insects and animals. Gibbons in the forests of Borneo are a typical threatened species. The recycling of leaves as they fall helps to return nutrients and organic matter to the soil and maintain the fertility. The soils are often fragile and the removal of the trees results in damage to the soil structure, with much being washed away by rainfall. The tree roots and the soils hold back water and water pollution and deforestation is partially blamed for increased local flood risk downstream of the Himalayas. All of these effects result in direct and indirect damage to the local economy such as loss of tourism or jobs, to the detriment of the local population who often gain little or no benefit from the causes.

1.1.11 Loss of biodiversity

Biodiversity refers to the presence of a wide range of species of plants, animals and other forms of life. A healthy natural environment has good biodiversity meaning that there is a lot of different species present in that habitat. They form an ecosystem (a term used in Chapter 2 and expanded in Box 2.3) in which species are interdependent. Loss of one species can have a knock-on effect in that other species are lost because they depend on the first for food or some other reason. The previous sections have highlighted some effects on wildlife due to causes such as climate change or loss of water or forest cover. As species are lost from a habitat or an area (or become extinct world-wide) there is a loss of biodiversity and the implications are that the habitat or area is degraded: it is not as varied as it was and the ecosystem has been disrupted. Loss of habitat such as complete deforestation means that a whole range of species is lost and the biodiversity is severely affected.

1.2 The reasons for maintaining and promoting environmental management

The previous section has outlined some of the serious global issues that are affecting the environment. In later sections we shall see that there are many more matters of concern. These may also cover a wide geographical area but many will be confined to the immediate vicinity of the source. In either case why should we be concerned? If the global impact is from a multitude of sources, as would be the case for CO_2, what effect will one person or business have by reducing their emissions? If people choose to live by a factory should they complain if it affects their quality of life or, even worse, makes them ill? In short, why should anyone be bothered to manage their impact on the environment?

There are, of course, many reasons and they can be categorised into three main groups: ethical, legal and financial. The following definitions are based on the Oxford Dictionary online:

ethical: morally good or correct and avoiding activities or organizations that do harm to people or the environment

legal: relating to the law and appointed or required by the law

financial: relating to the finances or financial situation of an organization or individual.

These overlap to some extent and we shall look at these in the wider contexts of how they interact with society, the law and the business.

1.2.1 People and society

It is understandable that people expect to live their lives as they wish and not to be disadvantaged by the activities of others. These expectations have built up over many years; the pollution in a Victorian town in the UK was a lot worse than now and people accepted it as their lot. At that time it was the price to be paid for a developing industrial society and they were glad of the work. However, life expectancy was short, not least in part due to the toxicity and disease associated with their surroundings. As they became wealthier they demanded a better quality of life. The same thing is happening in China, India and other developing countries now. Rapid industrialisation has produced pollution that would not be tolerated in many parts of the world today and now the local population is looking to the state to do something about it. Of course the problems in China are partially in response to the production of goods for sale in the wealthier parts of the world. In the same way, our demand for minerals, timber, food or fuel is damaging the environments where they are extracted and it is the poorer indigenous peoples who suffer the consequences. A business which wishes to be seen as ethical would not want to be associated with causing harm to people or their environment. Large corporations often get taken to task for just these reasons. Some examples of major incidents which have caused environmental damage and, in some cases, affected the health and safety of people are given in Box 1.1.

The assault on the reputation of companies that are seen as unethical can start locally but may well be taken up by pressure groups with a global reach such as Greenpeace or Friends of the Earth. Their approach may start off low key but escalate if they feel that their concerns are not being taken seriously. This can include activities that draw attention to the company through high-profile publicity stunts that get on to television and radio and may result in adverse publicity world-wide.

Box 1.1 Examples of companies that have generated bad publicity for environmental incidents

Year	Company/Operator	Location	Issues
2019	Vale	Brumadinho Dam, Brazil	Collapse of tailings dam
2015	Samarco/Vale/BHP Bilton	Mariana Dam, Brazil	Collapse of tailings dam
2011	Tokyo Electric Power Company	Fukushima, Japan	Damage to nuclear power installation following an earthquake and tsunami
2010	BP	Gulf of Mexico, USA	Major oil spill from Deep Horizon drilling rig
2005	Site owned by TOTAL and Texaco	Buncefield, UK	Explosion and fire at storage depot
2004	Coca Cola	Kerala, India	Water abstraction and impact on local supplies
1989	Royal Dutch Shell	River Mersey, UK	Oil pipeline leak
1989	Esso	Prince William Sound, Alaska, US	Oil spill from carrier *Exxon Valdez*
1984	Union Carbide Corporation	Bhopal, India	Gas leak from chemical plant
1976	ICMESA	Seveso, Italy	Dioxin release
1974	Nypro UK	Flixborough UK	Explosion at chemical works

These are examples of some of the most widely publicised events but many companies including such well-known brands as Apple, Nike and Adidas have found themselves at the centre of controversy due to alleged pollution by their suppliers. The sites are often in China, India, Pakistan or Bangladesh or other developing countries. A more extensive list can be found at https://en.wikipedia.org/wiki/List_of_environmental_disasters.

Customers who have high ethical standards may go elsewhere and companies in the supply chain who buy from or sell to the errant company may decide not to do business in case their reputation also becomes tarnished. A good example is builders or their suppliers of timber that will only deal with logging companies and sawmills that can demonstrate that their trees are managed in sustainable ways (defined in more detail in Section 1.3).

The company's relationship with its employees may also be affected. Those working in a poor environment will protest, strike or leave. Recruiting new staff may be difficult, especially those with specialist skills or whose services are in demand, as they apply their own moral standards to potential employers – not just about their working conditions but about the reputation of the company.

If you cause harm to people or the environment then, according to our definition, you are not being ethical. The harm may cause health problems to people or their animals, damage their crops or their wider environment (as defined in Section 1.1.1 above) or have less serious consequences such as nuisance caused by noise or smell. The implications go wider than that though as most countries have now developed laws to regulate such harm and to offer protection to their citizens with rights of redress.

1.2.2 Bringing in the law

The response of governments under pressure from its citizens is eventually to bring in laws to protect them, although they are generally slow to respond. Pressure groups representing business or companies themselves may lobby to avoid regulation citing cost and competitive disadvantage and governments have to balance the needs of everyone. Ethical companies may support regulation in principle because it coincides with their outlook but want it to be applied evenly to ensure that no one gets a competitive advantage by working to lower standards – the so-called 'level playing field'.

The law can come into play in two main ways. A set of statutes (laws passed by the legislature) sets down a framework and a range of standards that have to be met. The standards may be for environmental quality (air, water or land), emissions that cause pollution, noise or any other source of environmental impact. They will usually specify how the standards are to be enforced and any penalties for non-compliance. Penalties may include fines and even terms of imprisonment. The alternative legal route is a system that allows those affected to seek remedies. These could take the form of directions to the company to stop whatever activity or emission is causing the problem and to pay for remediation and compensation to the affected parties.

Many countries have a similar approach to establishing the legal frameworks but they differ significantly in the detail. Countries that are members of the European Union have agreed to harmonise standards in order to have a 'level playing field' across the Union. In other words, no member state can gain an advantage by adopting lower standards of environmental protection. In this case standards are agreed in Directives which member states are then obliged to translate into national legislation. This is where the detail may differ as different member states may develop and enforce standards at a national or a regional level and the mechanisms by which they do this may also differ. An example for the UK is given in Box 1.2 in Appendix 4.

Not every country has a full set of legislation to cover all eventualities and some may have very little at all. A further problem is that there may be little enforcement either through lack of resources such as trained personnel or money to pay them. The countries with the weakest laws and enforcement tend to be the developing countries which may also be vulnerable to corruption or political and military intervention to control valuable resources and it is these countries which often have large deposits of minerals or oil. Some African states such as the Democratic Republic of the Congo or Zimbabwe have endured decades of conflict.

The usual penalty for causing pollution is a fine. In the UK fines for less serious cases taken to magistrates courts may be up to £50,000 although they are usually a lot less. Serious cases that go to the Crown Court face potentially unlimited fines and there are many cases of fines in the hundreds of thousands of pounds and even millions of pounds. It is also possible to serve a term of imprisonment if there has been a severe incident that results in a criminal prosecution. Those at risk are generally the person who acted irresponsibly or the directors or senior managers if they have acted to cause the problem themselves or have condoned or ignored others not following good practice.

Defences available will depend on the jurisdiction and the relevant legislation but in general terms someone would have to demonstrate that they took reasonable care or that a manager or director had put in place all reasonable precautions such as appropriate systems of work, supervision and training.

1.2.3 The financial case for environmental management

Legal cases incur costs. There may be legal costs for lawyers and fines and compensation may be payable to those affected. In the case of the BP pollution in the Gulf of Mexico, compensation of several billion dollars has already been paid to fishermen and other local businesses and individuals who have suffered a loss. The final bill, including fines and compensation, is over $60 billion. This is an extreme example involving a large company, a major incident and the United States which has an aggressive litigious approach to such matters. However, fines of a few thousand pounds with compensation also in thousands of pounds could bankrupt a small company. These are known as direct costs, that is, costs directly attributable to an incident.

Direct costs of an incident may not stop with fines and compensation. Loss of production is very likely as processes are shut down to avoid causing further damage. If pollution was caused by a plant malfunction, this will have to be put right before production can restart. Dealing with the incident will absorb staff time in many ways: managing the process on the ground; dealing with media and other enquires; cleaning up any site and off-site contamination; repairing whatever caused the damage; dealing with lawyers, the courts and litigants; sorting out the issues arising with regulatory agencies; site investigations to avoid a repetition; any relevant staff issues such as disciplinary hearings. Much of this will be senior staff time as well as expensive external fees (e.g. lawyers) and charges (e.g. contractors involved in the clean-up). Cost could be incurred for several months, or even years, if litigation is involved.

There is also the potential for indirect costs; that is costs not directly attributable to the incident but which may arise because it happened. We have already touched on these in general terms above but not considered that there is a potential cost. Loss of reputation may result in loss of customers and business. Poor publicity may require fences to be mended with the local population as well as wider range of stakeholders. Insurance premiums may rise even if there has not been a claim. Share values may fall in anticipation of loss of profits and investors or banks may be more wary of buying shares or investing in the company. There is a group of 'ethical investors' who do not invest in some companies on principle.

All of the above have happened to BP and the fallout in that case will last for many years. It is possible to insure against some of these costs but there are limitations. It is possible to insure against most direct losses but not legal fines. Insurance often has constraints such as not paying out in the event of negligence: this can be insured at higher cost. A poor record could result in loss of insurance cover altogether.

A further financial reason for practising good environmental management is that money can be saved by reducing the use of raw materials and energy and reducing the amount of waste produced or requiring disposal. Chapter 2 goes into this in more detail and later chapters expand further on some specific cases. Many organisations have reduced some of their costs by 20 to 30 per cent with little effort. At this point we can say that money not saved when it could be is money wasted and that reinforces the financial case.

1.3 The importance of sustainability

So far we have tended to focus on the problems; environmental damage from a range of sources and the costs and penalties for failure to comply with environmental legislation. It would be far better if personal and business decisions were made in such a way that these could be avoided. An approach towards this end was developed in the 1980s and first came to a wider audience with the publication of the Brundtland Report (UN World Commission on Environment and Development (1987). This discussed the need to balance social, economic and environmental issues in order to produce the best outcome for society, not just now but also in the long term – referred to as sustainable development. The Commission came up with the definition of sustainable development as 'those paths of social, economic and political progress that meet the needs of the present without compromising the ability of future generations to meet their own needs'. Further discussion took place at the United Nations Conference on Environment and Development in Rio de Janeiro in 1992 (also known as the Rio Summit or the Earth Summit). The outcome was the Rio Declaration on Environment and Development (United Nations 1992) which contained 27 principles that were adopted to guide what had become to be known as sustainable development. Principle 3 stated that: 'the right to development must be fulfilled so as to equitably meet developmental and environmental needs of present and future generations'. This has been adapted as a definition of sustainability as **development that meets the needs of the present generation without compromising the ability of future generations to meet their needs.**

To get further up to date, in 2015 the global community adopted the 2030 Agenda for Sustainable Development and its 17 Sustainable Development Goals (United Nations 2019). The Goals, listed in Table 1.2, cover a wide range of topics and there are 169 targets associated with them. Thinking about the contents of this chapter already described above, it should be clear that environmental issues will be important factors in achieving many of the goals and targets. The detail is beyond the purpose of this book but many companies are starting to adopt these as part of their reporting procedures and to demonstrate their commitment to a wider audience.

Table 1.2 United Nations 2030 Goals for Sustainable Development

1.	No poverty
2.	Zero hunger
3.	Good health and wellbeing
4.	Quality education
5.	Gender equality
6.	Clean water and sanitation
7.	Affordable and clean energy
8.	Decent work and economic growth
9.	Industry, innovation and infrastructure
10.	Reduced inequalities
11.	Sustainable cities and communities
12.	Responsible production and consumption
13.	Climate action
14.	Life below water
15.	Life on land
16.	Peace, justice and strong institutions
17.	Partnerships for the goals

1.3.1 Sustainability in practice

There are conflicting demands that the present generation faces which will compromise future generations. All governments want to promote economic development to provide jobs and improve the standards of living of their populations. But, as we have seen, this can cause environmental damage and has social consequences for health, well-being and human rights. A traditional model of sustainable development is about trying to get the balance among these three legs (economic, social and environmental) to get the best overall outcome. The common way of illustrating this uses circles as shown in Figure 1.3

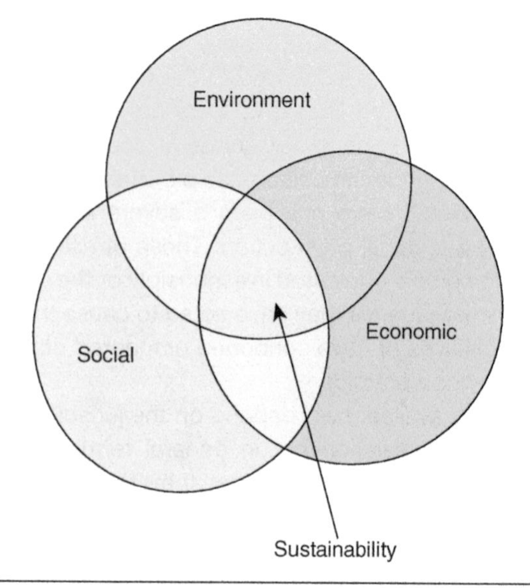

Figure 1.3 An illustration of sustainability

where the three circles overlap in the centre representing sustainability. Some models add a fourth leg or circle concerning the demand for and conservation of resources separate from the environmental leg.

A growing population, with increasing expectations about their future prosperity and dependency on energy and raw materials to meet these, is resulting in diminishing resources of these materials, environmental damage and loss of rights, especially by the indigenous peoples and the most vulnerable. If this generation achieves these expectations unconstrained, the outlook for future generations is bleak. In order to have some hope of leaving them with a reasonable future we need to tackle these issues in a consistent and constructive way.

1.3.2 Balancing the three legs of sustainability

Effective protection of the environment

We have already seen that the growing population and demand for resources is damaging the environment in many different ways. In order to achieve sustainable development we should not be leaving a legacy for future generations of polluted air, water and land, damaged climate and landscapes and insufficient raw materials or energy sources for their needs. Avoiding pollution will feature in later chapters describing the potential sources and how to reduce the risk from pollutants. However, a good start is to reduce the need for resources and the production of pollution in the first place. This is known as resource efficiency; maximising the utilisation of the resources taken. On reflection this ought to be obvious but it is only relatively recently that it has been taken seriously by many of the larger companies. Metals and energy are expensive to mine and refine. Some materials such as glass, concrete and plastics are also expensive to manufacture because they use a lot of energy or depend on expensive raw materials. So there is a good financial incentive to use resources efficiently before one takes account of the continuing needs of future generations. Resource efficiency can be achieved in several ways.

Choice of materials of manufacture or construction

Often there is a choice of material from which to construct a project or manufacture a product. Construction projects can utilise concrete, timber, metals and natural materials such as sand or clay. A dam for a reservoir could be made in a number of ways and the choice will depend on the availability of suitable materials and cost but where there is a choice the environmental impact ought to be a consideration as well. Drinks containers can be made from aluminium, tinned steel, glass, plastic or plastic lined cardboard; which is the most appropriate and would it be suitable in all cases?

Minimising the use of material in products

Design and choose suitable materials to ensure the product is fit for purpose but is not over designed or engineered (consider how aluminium drinks cans and plastic and glass bottles have become thinner and lighter in recent years).

Design to avoid waste in production

Avoid off cuts by planning the size and layout of parts such that waste is minimised and ordering materials of an appropriate size (think how the pages of this book could be cut from different sizes and shapes of paper and the boxes that they are packed in).

Designing and building products for a longer life

Built-in obsolescence was a feature of many products in the past and many domestic goods such as kettles and toasters are designed and manufactured such that they cannot be repaired. There is now more consideration of ensuring that the product will last for longer and that simple parts can be replaced. Another notable example is private cars. They now last longer as the mechanical parts are more reliable and the bodies less prone to corrosion compared with models of just a few years ago.

Ensuring products or their components at the end of their life can be reused or recycled

Some products may be able to be repaired or refurbished and used again but if not then the materials of construction should be easy to separate and be reused or recycled as raw material for new parts (this happens now with metals and plastics in vehicles).

Designing products to minimise the use of energy in production and use

Lighter products require less energy to make, transport and use (another benefit of the changes in drinks containers referred to above and the use of plastics and lighter components through specialist materials and design in the latest generation of aircraft). Efficient engines in vehicles reduce the demand for fuel.

Avoiding pollution

A further way of protecting the environment which goes beyond resource efficiency is avoiding causing environmental damage in other ways. This is principally about avoiding causing pollution at all stages of any process although, as we saw earlier, efficient use of resources will help to minimise any pollution resulting from the abstraction and processing of raw materials. Hence operating manufacturing processes to avoid emissions likely to cause harm to the environment and designing products such that their operation is also not a cause of problems must be seen as part of this process. We shall investigate much of this later but at this stage it is worth looking at car manufacture to identify a suitable example of both.

For many years car bodies were sprayed with solvent-based paints to protect the bodywork. These released solvents into the atmosphere which were in part responsible for some of the air pollution problems referred to earlier. Modern cars are dipped into or sprayed with water-based paints which do not cause air pollution. Fuel efficiency of cars has also improved significantly for a number of reasons (cost of fuel being one) but the result is that emissions of carbon dioxide and other air pollutants are reduced for vehicles travelling the same distance. You could identify other aspects of car manufacture and design which meet these expectations or apply similar thinking to other products in everyday use.

We should not ignore how our decisions as consumers have an impact on sustainability. In the developed world we want the social and financial benefits of economic growth: well-paid and secure jobs, adequate housing, a wide choice of foodstuffs at reasonable prices, the goods that make our lives easier or more enjoyable, leisure activities and toys to keep us amused. We do not want the problems associated with environmental damage or pollution on our doorsteps. Yet we often act as though our children and grandchildren are not representatives of the future generations at risk. So, if we can afford it (even sometimes if we can't), we buy the bigger, faster car, take several foreign holidays, replace perfectly serviceable products with the latest model, overheat our houses and waste water as though supplies are limitless. Although as individuals we are unlikely to be in court for causing pollution directly, our collective decisions are one of the drivers that result in the supply chain causing pollution. Think of other examples where actions in our private lives (at home or in our leisure activities) may be unsustainable and then consider how behaviour and actions may be changed to reduce their impact.

The circular economy

The sections above show how the traditional model of consumption – extract resources, manufacture products, use them, and discard as waste (take, make, waste) – is damaging the planet. A rethink is required and this has resulted in the concept of the circular economy promoted by the Ellen MacArthur Foundation. The principles are to design out waste and pollution, keep products and materials in use and regenerate natural systems. In Chapter 3 there is discussion about life cycle analysis and there are synergies with that. But the circular economy takes it further by incorporating the practices of biological cycles of materials into technical cycles: repair, reuse, remanufacture or recycle. Many companies are adopting this approach although it requires changes to their business. They see it as a long-term benefit, consistent with their corporate social responsibility (see below) but with other benefits as well, such as providing goods for rental rather than outright purchase. The circular economy is certain to feature more in the future: more information can be found on the website of the Ellen MacArthur Foundation (www.ellenmacarthurfoundation.org/).

The other aspects of sustainability

The other two legs of sustainability: economic and social, are just as important but deserve books on their own. For our purposes, we need to see how they fit into the model of sustainability and the relationship with the environmental leg but to keep it simple.

Economic development

People want economic development to improve their lives and governments want it to generate tax revenues (also, without it they will probably not get re-elected). Economic growth and the generation of profit pays for environmental and social improvements so avoiding it is not really an option. We have already seen that economic development has damaged the environment and has had social impacts in the past: what sustainable development requires is economic growth without the damage. The prudent use of resources and the application of some of the principles already presented in this section can help to achieve this. We shall explore more of this in later chapters showing how resource use and the reuse and recycling of waste materials can contribute.

Social progress

We have seen examples where economic development has been responsible for disruption to societies by displacing them and damaging their environments. This needs to be avoided, or at least minimised, if we are to achieve sustainable development. Avoiding the environmental impacts is part of the subject of this book but it goes further than that. The provision of well-paid local jobs supported by education and training can improve societies. Equity in economic development, avoiding plundering but promoting the sharing of resources and helping the developing nations to achieve their full potential, may be idealistic but without them sustainable development will not be achieved.

1.3.3 The link with corporate social responsibility

For many businesses the application of the principles of sustainability is embodied in the wider concept of corporate social responsibility (CSR). These principles take a broader view of the position of the business in the community, not just whether it is causing a nuisance or pollution but how the business interacts with all levels. So, for example, it may support local community projects such as fund raising or a new sports facility. It could get involved with local schools by providing materials or equipment or providing work experience opportunities. It could offer business expertise such as accountancy skills to local clubs. It could offer its own facilities such as printing or a meeting place to a community project. In developing countries some businesses actually provide education or health services to their employees and their families.

On a wider scale, the growing outsourcing of activities to developing countries has raised additional concerns. It is not just the impacts on the environments but such issues as the health and safety of workers or neighbours, working hours, working conditions and rates of pay that are important. These should also be a central part of CSR.

These are just a few examples where the business gets involved and the community benefits. However, the reasons for doing it may not be entirely altruistic. It gets access to an educated and healthy workforce; the local community thrives and offers more business opportunities; the image of the business is improved and people will want to work there and will defend the company against outside criticism; if there is a problem then the reaction is likely to be more muted than if it had totally ignored the local population. This is not being cynical – there are mutual benefits and most businesses would gain from building relationships with neighbouring businesses and residents. Larger businesses may practise CSR on a wider scale getting involved as sponsors of national projects or sports events (the 2012 Olympics Games in London were sponsored by several international and UK-based companies), working with universities or supporting national charities.

The reporting of CSR is being superseded in some organisations by Environmental, Social and Governance (ESG) reporting. This includes most of the elements of CSR but extends them to include issues such as slave labour, pay, diversity of labour force, animal welfare, the avoidance of corruption and the principles of investment. Large companies, especially including those that manage investment in other companies, are taking this stance. This is relevant to some of the points made earlier and elsewhere in this text of the risks to an organisation for causing environmental harm.

1.4 The role of national governments and international bodies in formulating a framework for the regulation of environmental management

There have already been some references to the international conferences such as Rio on sustainability and other conventions and the role of the EU in developing legislation. These are examples of how countries work together to tackle some of the bigger issues referred to earlier in this chapter.

1.4.1 International law governing the environment

There is no international parliament or senate setting down international laws to protect the environment. Instead there has to be a consensus amongst all interested states to find a solution to deal with an international problem. These then result in some form of agreement which may be known as a convention or a protocol. Some, such as those for managing climate change, are mentioned in this chapter. There are more examples in Box 1.3 where international

Box 1.3 Examples of international agreements to protect the environment

The OSPAR Convention (The Convention for the Protection of the Marine Environment of the North-East Atlantic) was adopted in 1992 in Paris. It replaced the **Oslo Convention** of 1972 (Convention for the Prevention of Marine Pollution by Dumping from Ships and Aircraft) and the **Paris Convention of 1974** (Convention for the Prevention of Marine Pollution from Land-Based Sources). These were agreed to limit the pollution being caused in the North East Atlantic by various discharges from countries bordering the sea and from transport discharging wastes. More information can be found at http://jncc.defra.gov.uk/page-1370.

The Montreal Protocol of 1989 on substances that deplete the ozone layer was agreed to phase out the use of ozone-depleting substances as described in the main text. More information can be found at www.unido.org/our-focus/safeguarding-environment/implementation-multilateral-environmental-agreements/montreal-protocol.

The Basel Convention on the Control of Transboundary Movements of Hazardous Wastes and their Disposal of 1989 was agreed to protect human health and the environment by controlling the transport of hazardous wastes around the world. More information can be found at www.basel.int/.

The Ramsar Convention on Wetlands of International Importance, especially as Waterfowl Habitat of 1971 was agreed to maintain the ecological character of the Wetlands of International Importance and to plan for their sustainable use. There is an internationally agreed list of wetlands to be protected. More details can be found at www.ramsar.org.

agreements have been reached to tackle some of the problems that have a global or regional impact. A selected (but still long) list of international environmental agreements can be found at www.cia.gov/LIBRARY/publications/the-world-factbook/appendix/appendix-c.html. Most of these are sponsored by the United Nations or one of the associated bodies such as the United Nations Environment Programme (UNEP) as Multilateral Environmental Agreements. The RAMSAR Convention in Box 1.3 was not agreed under the auspices of the UN but was an independently agreed international treaty. Not every country is a signatory to all the agreements. Some are regional and so of limited application but sometimes a country does not sign because it has no interest in the subject or because it may feel that it will be against their national interest.

Once the texts of the international agreements are signed by the participating governments there is usually a period of several years whilst they are ratified by each signatory. This is done now by the EU on behalf of its members and by the other national legislations according to their own statutory procedures. Implementation may require changes in legislation or there may be political difficulties in gaining further agreement within the individual states. The Kyoto Protocol and subsequent conventions on climate change have been examples where ratification has not gone smoothly due to local political opposition in countries like the United States. Within the EU, any legal changes would take the form of a Directive but countries outside of the EU may have to change an existing law or adopt a new law or issue standards under existing legislation.

1.4.2 The role of the European Union in harmonising environmental standards

EU directives are aimed at harmonising standards across the member states. These are decided within the European Parliament based on proposals from the European Commission (see Box 1.4 for an explanation of the relationships between the European institutions). The text of directives usually contains an introduction that explains why it is being adopted (a series of statements starting with 'whereas') and then goes on to set out the rules or standards to be put into place by the member states. There are deadlines by which national legislation is expected to be in place and for the subsequent compliance with rules and standards. Different countries have different ways of doing this. Within the UK most recent environmental legislation has been written in such a way that it allows the Secretary of State to issue or change detailed Regulations without further reference to Parliament (see Box 1.2 in Appendix 4).

The EU has established a European Environment Agency with the role of providing advice to the Commission on environmental matters and collecting environmental information from member states. The Commission may require information about implementation or compliance with directives to be provided on a regular basis. Failure to meet deadlines or standards can result in penalties. As a recent example, the UK has paid fines for failing to deal with air pollution in London.

1.4.3 Knowing and complying with local legislation

The result of the European approach to environmental legislation is that companies that operate across member states should expect to find the same rules, regulations and standards. However, some European countries which are preparing for EU membership (known as accession countries) have longer deadlines for compliance so there may be differences in the interim period. Across the rest of Europe and the rest of the world the laws and standards are likely to be different. Most countries now have an environmental organisation that administers the laws passed by their legislature and enforces standards. This may be an arm of a national or regional government department or an independent agency. The details for any country may be found on the internet although obtaining copies of the laws or regulations may not be so easy for some countries. It is important though that this is done if an organisation is planning to operate in other countries.

Box 1.4 The institutions of the European Union involved in legislation

The **European Union** refers to the collection of 27 member states.

The **European Commission** is the body that prepares legislation and ensures that it is applied once it has been passed. It also is responsible for the administration of the EU and is similar to the UK civil service.

The **European Parliament** is made up from elected members from all states and it has the final say on approving legislation.

The **European Environment Agency** has no statutory role but advises those developing, adopting, implementing and evaluating environmental policy and provides a coordinating role with other centres that collect and interpret data.

The need to know the local legislation is obvious if setting up a factory that may produce emissions to air or water. Importing and exporting goods, such as timber or timber products or especially waste, may seem less obvious at first but many countries now have laws to protect their natural resources or wildlife or put restrictions on what can be imported. For example, from 2012 it is illegal to import timber to the EU that has been harvested in breach of national laws in the country of origin.

There are reasons beyond the law for complying with local legislation even if it poorly enforced. The organisation's reputation locally will be damaged if it is causing pollution and there is also the risk of international reputational damage if it becomes more widely known. In the absence of national legislation a reputable company will adopt suitable regulations from its home country or elsewhere.

1.4.4 Terminology used in legislation

The various directives and national legislation use two terms that may be unfamiliar but which are a feature of gaining a permit for some operations.

Best practicable environmental option

For some decisions there may be more than one alternative way to deal with a problem. They have different implications not just for the environmental impact of any chosen route but on the wider issues such as the economy and some social issues such as employment. In determining which has the least environmental impact, the concept of Best Practicable Environmental Option (BPEO) was introduced by a UK Royal Commission initially in a report in 1976 but subsequently refined in a later report (Royal Commission on Environmental Pollution 1988). Their definition was:

> the outcome of a systematic consultative and decision making procedure which emphasises the protection and conservation of the environment across land, air and water. The BPEO procedure establishes for a given set of objectives, the option that provides the most benefits or the least damage to the environment, as a whole, at acceptable cost, in the long term as well as in the short term.

BPEO is widely used in deciding the best approach to waste management, especially at the planning stage. Decisions about the location of sites can be influenced not just by the environmental impact of the site itself but by other factors such as the environmental impact of transporting the waste. The use of BPEO balances the various alternatives to find the one with the least environmental impact overall. The final decision may still be influenced by financial factors such as cost or social factors such as employment: the result of a broader sustainability appraisal. The same principles can be applied to other environmental decisions.

Best available technique

BPEO is applied at the strategic level. Best Available Technique (BAT) is applied at the operational level for the protection of the environment. It derives from a European Directive first applied in 1996 concerned with Integrated Pollution Prevention and Control. This has been updated in 2010 (European Commission 2010) from which the definition in Box 1.5 has been taken. In the use of BAT, 'best' means the most effective, which should be fairly easy to decide in most cases, and 'available' means that the choice should from anything available world-wide. The word 'technique' has a meaning beyond the conventional one: it is applied not just to the process or equipment

Box 1.5 European Commission definition of Best Available Technique

'Best available techniques' means the most effective and advanced stage in the development of activities and their methods of operation which indicates the practical suitability of particular techniques for providing the basis for emission limit values and other permit conditions designed to prevent and, where that is not practicable, to reduce emissions and the impact on the environment as a whole:

(a) 'techniques' includes both the technology used and the way in which the installation is designed, built, maintained, operated and decommissioned;
(b) 'available techniques' means those developed on a scale which allows implementation in the relevant industrial sector, under economically and technically viable conditions, taking into consideration the costs and advantages, whether or not the techniques are used or produced inside the Member State in question, as long as they are reasonably accessible to the operator;
(c) 'best' means most effective in achieving a high general level of protection of the environment as a whole.

(European Commission 2010)

used to limit pollution but extends to the design, maintenance and operation. So the phrase BAT implies, in theory, the adoption of the best possible solution but that is not regardless of cost. The definition of 'available' does allow for costs to be taken into account and compared with the advantages gained. BAT is used in permitting major industrial sites as described in the next section.

1.4.5 The role of regulatory agencies

The national legislatures put laws in place to protect the environment but implementation and enforcement is usually by a separate agency. Sometimes the regulatory powers are split among more than one agency but the principles are likely to be the same. These agencies may be a part of national or regional government or entirely independent. This section will elaborate on this using the UK as a current example, which is similar to the models in many other countries although the range of powers may be different. Even within the UK there are differences, as explained in Box 1.6 in Appendix 4, which demonstrate how complicated it can be. To add to the confusion, the UK is currently deciding if and under what terms to leave the EU (known as Brexit) and this is likely to change these arrangements. This is likely to include a new UK regulatory body to replace the oversight currently carried out by the EU.

The UK Parliament makes laws for the whole of the UK but some aspects, including much of the environmental legislation, are devolved to the relevant national legislatures. The English arm of government has two departments that share most of the responsibilities for environmental protection: the Department for Environment, Food and Rural Affairs (DEFRA) and the Department of Energy and Climate Change (DECC). Elsewhere in the UK it is the responsibility of a department in the respective national governments. These in turn have regulatory bodies as shown in Box 1.6. In each case the national bodies responsible for environmental regulation and for nature conservation are given. These two bodies deal with most of the environmental issues. The Environment Agency, which has been used as the common model in this book, covers just England. The Welsh Assembly merged the environmental regulator for Wales with the Countryside Council for Wales and the Forestry Commission for Wales into a new multipurpose body – Natural Resources Wales – on 1 April 2013.

In addition, the local authorities (County Councils and District Councils and their equivalents in some parts of England and in the other countries) have roles in environmental protection. These are included below.

In each case the regulators have various responsibilities. Unfortunately these differ again across the four countries of the UK but the finer details can be found on the web sites of the different bodies. The examples used below for the next two sections are based on England but note that there are regular updates to the legislation and Guidance notes for all regions which mean that, if the information is critical for a specific issue, it is essential to check for the latest updates. This will be especially true as Brexit proceeds.

1.4.6 The Environment Agency

The Environment Agency has a range of responsibilities shown in Box 1.7 in Appendix 4. Some of these such as managing flood risk are operational as well as regulatory. Most of the content of this book is confined to water resources, and environmental protection of air, water and land.

The establishment of the Agency was the result of the Environment Act 1995 which incorporated and replaced much legislation from previous Acts that were concerned with environmental protection. The 1995 Act set out the aims and objectives for the Agency. In fulfilling its principal aim of protecting and enhancing the environment it had to make a contribution towards the broader objective of attaining sustainable development. The Agency meets its primary objective mainly through discharging its regulatory and advisory functions. The main ones are:

- Issuing permits, authorisations, licences and consents for discharges to air, land and water and for some other activities involving fisheries and flood protection;
- Monitoring compliance with the terms of the permits, authorisations, etc.;
- Taking enforcement action for non-compliance or causing pollution;
- Regulation of waste management and hazardous waste;
- Regulation of some activities involving radioactive substances;
- Regulation of contaminated land designated as special sites;
- Monitoring and reporting on the state of the environment: air quality, water quality and land quality;
- Regulatory activities concerned with packaging regulations and carbon emissions;
- Providing advice and guidance to government and business on environmental matters.

Note that the regulation of smaller processes with emissions to air, smaller IPPC installations (see Box 1.8 in Appendix 4) and designation of and most aspects of contaminated land are dealt with by the relevant local authority. The Agency gets involved if the contaminated land is causing water pollution and these are designated as special sites.

Issuing permits and other forms of authorisation

Activities that have the potential to cause environmental damage such as discharges to air and water or the transport and disposal of waste are controlled by issuing a permit under the Environmental Permitting Regulations

(England and Wales) 2016. Previously these were known as authorisations, consents or licences depending on which Act they were issued under and what type of activity was being regulated. The Government has issued core guidance (DEFRA 2010a) on the general principles of permitting and a series of guidance notes on specific types of permits. The aim is to simplify the permitting process such that permits cover all types of activities shown in Box 1.9 in Appendix 4 unless they are registered as exempt (some activities are exempt by virtue of their size or reduced potential for environmental damage). Sites now generally have one permit covering all activities on that site if the operator and regulator are the same. The Agency is also trying to simplify the process further by adopting standard permits for many activities and focusing its main effort on high-risk operations.

The permit will cover a range of issues. It will start with the administrative information such as business name and address, etc. but the key piece of information is the details of the activity or activities that it refers to. The permit is only valid for these and the conditions that applied when the application for a permit was made. So if the permit is for a discharge to a watercourse from an effluent treatment plant for a particular process at a specific volume, the permit holder cannot increase the volume or change any of the other details without consultation with the Agency or submitting a new application.

The key content for most organisations is the setting of limits on emissions and any other operational limitations (e.g. times of operation) that may be applied. Permits can also include content such as requirements for the competences of the operators, requirements for maintenance and requirements to return sites to their original condition after operations cease and a permit is surrendered. Permits issued under IPPC may require evidence that BAT has been applied and set conditions relating to the promotion of energy efficiency, the prevention of waste production and the avoidance of the effects of accidents.

Organisations have rights of appeal to the Secretary of State against refusals to issue permits or the conditions that are applied.

Inspection and sampling

The regulatory bodies ensure compliance with permits by regular site inspections and sampling of discharges. Inspections will include checking compliance with any conditions in the permit and looking at site records, operator training and the presence of other potential risks such as the storage of liquids or waste. Compliance with documentation for waste disposal may also be checked. Samples will be taken of emissions to air or water and data collected by the operator or from continuous monitors will also be examined. Samples may also be taken from the air or watercourse receiving discharges.

The frequency of inspection and sampling will depend on conditions that may be in the permit, requirements in legislation and a risk assessment by the regulator. If a site has a persistently good reputation it will be inspected less frequently but if problems are found the frequency may be increased.

Enforcement procedures

If there is non-compliance with the conditions in the permit or the site is being operated without the necessary permit the regulator may take enforcement action. Enforcement action will also result if there is a pollution event that causes damage to the environment or harm to human health or a site is being managed in a way that could cause pollution (e.g. poor storage conditions for chemicals and waste). Regulators can visit sites at any time if there is a report of pollution or in order to fulfil their regulatory roles. The extensive powers available to the Environment Agency in investigating pollution are shown in Box 1.10 in Appendix 4.

The Agency inspectors collect evidence as required and can use it to decide on what enforcement action is appropriate. Some form of action is likely if there is pollution or harm to the environment or significant or persistent failure to comply with a permit. The Agency's stated approach is to favour preventative measures, resorting to laws as a last resort. It has a series of responses set out in its enforcement and sanctions policy (available on the UK Government's web pages for the Agency). These start with issuing advice and guidance, followed by a warning which may resolve the matter. Failure to comply with a written warning or in other circumstances can result in statutory notices, changes to permits, injunctions, carrying out remedial works, the use of civil sanctions, issuing formal cautions or prosecution. A summary of the enforcement powers available to the Agency is in Box 1.11 in Appendix 4. (Note that Local Authorities can issue abatement notices for nuisances as described in Box 1.2 and remediation notices in respect of contaminated land.)

An additional enforcement procedure is available under the Environmental Damage Prevention and Remediation Regulations 2015 which implement the European Directive on Environmental Liability. This legislation applies only to the most serious environmental damage and requires the polluter to pay for remedial measures (the 'polluter pays' principle). The specific types of damage are to an SSSI (see below under Natural England), a protected species, or to surface, marine or groundwater causing a deterioration in its status under the EU Water Framework Directive or from contaminated land with a significant risk of adverse effects on human health. The enforcing authority may be the Agency but can include the local authorities and Natural England depending on who the regulatory body is for the specific case.

Responding to enforcement action

Operators that have caused pollution or are in breach of the other environmental legislation are advised to cooperate with any investigation and to help in mitigating the effects and putting remedial actions in place. Failure to do so may be a criminal offence (e.g. not cooperating with the powers available as listed in Box 1.10 or the statutory notices in Box 1.11). The attitude of the offender and the degree of cooperation afforded often influence the response from the regulator. Self-reporting and prompt responses to stop any problem at source and initiate clean-up procedures may head off enforcement action or at least be used in mitigation in any subsequent proceedings.

The consequences of prosecutions and their impact on the organisation are taken further in the Chapter 2.

Public information

Most of the information collected and used by the Environment Agency and other regulatory bodies is in the public domain. This includes information about permits, analytical results, compliance or incidents and any sanctions taken as well as information about the local environment (quality of rivers, etc.). Summary information is available through the Agency's web site and more details are available on request. It is widely used by local residents and pressure groups as well as business competitors and suppliers of goods and services. There are few circumstances where it can be withheld.

1.4.7 Natural England

The other regulatory body involved in protecting the environment in England is Natural England (the equivalent organisations in the other jurisdictions are included in Box 1.6). Its primary purpose is to conserve and enhance the natural environment – land, flora, fauna, geology, soils and water environments. This is achieved for many sites by giving them protected status with the details of the main ones in Box 1.12 in Appendix 4. Natural England also operates some national nature reserves; other reserves are operated by county-based wildlife trusts.

Damage to protected sites is an offence and can result in various sanctions including prosecution. The protection of the protected sites and nature reserves is one aspect that can be taken into account in determining permit conditions. For example, a discharge to a watercourse that was also an SSSI may mean that additional conditions are imposed. It is important therefore that organisations are aware of any special sites near to their operations on which they may have an impact and take precautions to avoid damaging them.

It is beyond the scope of this book to go into more detail. Further information, including the location of sites can be found at Natural England's web site www.gov.uk/government/organisations/natural-england.

CHAPTER

2

Environmental management systems

After this chapter you should be able to:

1. Identify the reasons for implementing an environmental management system (EMS)
2. Describe the key features and appropriate content of an effective EMS, i.e. ISO 14001:2015
3. Identify the benefits and limitations of introducing a formal EMS such as ISO 14001/BS 8555/EMAS into the workplace
4. Identify key members of the ISO 14000 family of standards and their purpose.

Chapter Contents

Introduction 22
2.1 Reasons for implementing an environmental management system 22
2.2 The key features of an effective EMS 24
2.3 Benefits and limitations of introducing a formal EMS into the workplace 45
2.4 Key members of the ISO 14000 family of standards and their purpose 46

INTRODUCTION

The previous chapter set out some of the problems that have arisen in the environment as a result of human activities and gave reasons for promoting environmental management. This chapter takes this further by introducing the concept of an environmental management system (EMS) that can be adopted by organisations to minimise their impacts, explaining why a structured approach is appropriate and how environmental management can be implemented in a structured way.

The term 'environmental management system' is a bit of a misnomer. We are not setting out to manage the environment but to manage our impact on the environment. The previous chapter outlined some of the issues that need to be managed: use of resources, emissions to air and water, waste production and energy use. Different businesses will have different combinations of these and it would be easy to adopt a scatter-gun approach, picking off the topics that seem simple to deal with or that are currently flavour of the month. However, this brings the risk of missing the more important issues that are having the most environmental effect or getting the priorities wrong. Equally important, if a system is to be introduced, it is important that it is one that is recognised by outside parties (customers, public, etc.) so that it gives them confidence that the company is doing its best to behave in a responsible way and it is worth doing business with them. We saw in Chapter 1 that it is easy to damage a company's reputation and that other responsible companies may avoid it if there is a risk of damaging their own reputations.

The International Organisation for Standardisation (ISO) exists to develop and agree standards for a very wide range of topics on a basis that is accepted worldwide. This avoids the problem of each country developing its own standards and then finding that other countries will not accept them for some reason or other. More information about ISO can be found on their web site at www.iso.org/iso/home.htm. The standards for environmental issues are known as the ISO 14000 series and ISO 14001 is for environmental management systems. ISO 14001:2004 was the standard for many years but this has been updated to *ISO 14001:2015 Environmental management systems – Requirements with guidance for use* (ISO 2015) with 2018 set as the final year for transition. The changes were not substantial but strengthened some of the clauses and put more onus on top management to get more involved and to support the process, incorporating it more closely into overall business processes.

It is necessary at this stage to introduce two definitions from ISO 14001 that are fundamental to the standard and will be used frequently throughout this book. They are words that we have met already: aspects and impacts. The formal definitions are:

> **Environmental aspects: element of an organisation's activities, products or services that can interact with the environment.**
>
> **Environmental impacts: any change to the environment whether adverse or beneficial, wholly or partially resulting from an organisation's environmental aspects.**

Remember we met the ISO 14001 definition of environment in Chapter 1: 'the surroundings in which an organisation operates including air, water, land, natural resources, flora, fauna, humans, and their interrelationships'.

These definitions are important as they are the basis on which an EMS is developed and are wider than would perhaps be expected from the normal usage of the words. They are also fundamental to other topics that will come up later.

So, in defining the context and scope of an EMS, we need to include the **aspects** – the activities, products and services of the business – and consider their **impacts** on the **environment**.

2.1 Reasons for implementing an environmental management system

There are several reasons for implementing an EMS, some related to the organisation and others the external perceptions of the organisation. Most organisations will wish to see some form of business case if they are being asked to commit time and money to a project. The argument for an EMS needs to quantify the cost savings, risks and business opportunities in sufficient detail to be comparable with the cost of implementation that will also have to be identified. This will require an initial review of the aspects and impacts and the risks as described below.

2.1.1 Demonstration of management commitment

The directors or senior managers (for example, chief executive or managing director) have the overall responsibility for running an organisation. For this discussion, an organisation can be a company, a public sector body such as a local authority, a utility, school or university department, charity, golf club; in fact any organisation that has the potential to affect the environment. They also need to secure the reputation of the organisation with all the stakeholders and ensure that its relationships with suppliers and customers are maintained so that it can thrive. (Stakeholders are the individuals and organisations that affect or are affected by the decisions of another organisation; other examples are employees, neighbours or pressure groups.) Implementing an EMS will take up the time of employees and cost money so it is important that the senior management recognise the importance of doing it and commit their support and provide the necessary resources. Even if outside support from consultants is sought to help with the process, they are not cheap and they will still require input from the people that know the business best.

So there need to be compelling reasons to commit resources and the management of stakeholder relationships is a good starting point. Many businesses now insist that suppliers meet relevant ISO standards and can require compliance with ISO 14001 in order to be able to tender for business. Outside pressure groups, neighbours, clients and regulatory bodies are other examples of stakeholders that may be best convinced of the business's environmental credentials if it has a recognised standard for environmental management. The adoption of an EMS would be a commitment by the management to maintain and improve the environment and a good demonstration of corporate social responsibility referred to in Chapter 1. Businesses subject to environmental regulation may find that it is a requirement of the regulator for example in issuing a permit or for sites that are subject to COMAH regulations (covered in Chapter 9). It should at least lead to a better relationship based on confidence of performance with consequential reduced inspection visits. For example, the operational risk appraisal (OPRA) used by the UK Environment Agency took account of any EMS in place when determining the targeting of its resources for sampling and inspecting a regulated business, although in 2019 this was undergoing changes.

2.1.2 Potential to reduce costs

The potential for cost savings always attracts management attention and a good EMS will seek to minimise the utilisation of raw materials and energy and the reduction of the costs of disposal of waste and emissions and the risk of legal costs and penalties. Charges for environmental permits and taxes for waste or energy use and raw material extraction are usually based on some measure of quantity and also offer some potential for saving costs. In the early stages it may be difficult to quantify all of these but looking at the achievements of other similar organisations that have already been down this route can give an indication. Most companies make their environmental reports freely available as publications or online. Savings in energy use, water use, CO_2 emissions and waste disposal costs are common to most organisations and a conservative range can be selected and applied to this potential case. It is usual for savings to be built up over a period: an initial delay may be necessary to invest in monitoring or improved control equipment and then the subsequent savings may arise quite quickly until a time is reached when all the easy pickings have been made and progress then becomes more gradual. So it is important to look at several years of reports if trying to draw comparisons with peer organisations.

2.1.3 Gain new customers

Managers are always looking for business opportunities and we have already identified the supply chain pressure from other organisations to adopt an EMS in order to do business. Retaining current clients and customers and finding new ones should bring improved productivity and profits.

2.1.4 Protection of the environment and reduction of risk

Another issue that excites senior managers is risk management. All organisations are subject to risks which can take various forms – fire or other hazard, power cuts, on-site emergencies causing health, safety or pollution problems, supply chain disruption, financial risks, reputational damage and loss of production to name a few. An EMS offers the opportunity to improve the management of some of these and, indeed, to recognise opportunities. In particular there are the risks of environmental pollution caused by business activities and the consequential legal costs, fines and compensation; the loss of production caused by such an emergency and the damage to reputation that may result.

2.1.5 Integration with other management systems

Good business practice also requires the adoption of other standards relevant to the activities or the products or services of a business and the ISO series are structured in similar ways so that it is possible to integrate aspects of different standards or use common information in putting the relevant documents together. ISO 14001 shares some characteristics with ISO 9001 for quality management, ISO 45001 for occupational health and safety management systems and ISO 50001 for energy management (these can be found on the ISO web site www.iso.org). If a business wishes to adopt standards for EMS, quality, health and safety, and energy management systems, then adopting these four rather than a mix of national and international standards will make the process simpler.

Whereas the ISO standards are developed by international collaboration and are adopted by many countries world-wide, the European Union has adopted a voluntary scheme known as EMAS – Eco-management and Audit Scheme. This has been adapted in recent years to make it more compatible with ISO 14001. The differences are modest and need not concern us here (see https://ec.europa.eu/environment/emas/pdf/factsheets/EMAS_revised_annexes.pdf).

2.1.6 Corporate Social Responsibility

The adoption of an EMS ties in with another pressure on business to adopt Corporate Social Responsibility (CSR) and Environmental, Social and Governance (ESG) across the whole business. This was covered in Section 1.3.3.

2.2 The key features of an effective EMS

The adoption of an EMS can be a daunting task especially for smaller or medium-sized enterprises (SMEs) who may not have the resources in terms of staff time and expertise to make good progress. Within the UK a British standard, BS 8555: 2016 *Guide to the phased implementation of an environmental management system,* has been published to help achieve EMAS or ISO 14001 by an approach that is built up in five phases (see Box 2.1). Each of these can be tackled on its own, in sequence and at a pace to suit the business. There is guidance and standard formats available to make each step easier. Training programmes and certification are available through a range of organisations and training providers that can be found through the internet.

Box 2.1 The phased approach to developing an EMS based on BS 8555:2016

BS 8555 is referred to in the main text as an aid to implementing ISO 14001. This abbreviated content needs to be read in association with that text.

Phase 1 Gaining commitment and establishing a baseline

Gain management commitment and leadership; define organisational context and scope; initial baseline assessment; initial draft of environmental policy; communicate awareness and initial training programme; establish data requirements and draw up overall plan to prepare EMS.

Phase 2 Identifying and ensuring compliance with legal and other requirements

The requirements other than legal include, for example, contractual requirements, relevant codes of practice, standards (company, industry wide or ISO). Checking actual compliance and the effectiveness of controls in ensuring compliance; dealing with any non-compliances.

Phase 3 Plan and develop the EMS

This phase is where the core to the EMS is developed. Finalising the management structure with clear definitions of roles and responsibilities; refining the aspects and impacts and determining their significance. Are there risks or opportunities? Finalising the policy in the light of these refinements. Developing objectives and targets for improving performance, in particular in line with the policy and dealing with the significant aspects and impacts. Preparing programmes for improvement with responsibilities and resources allocated. Monitor environmental performance.

Phase 4 Implementation of the EMS

This phase takes the previous work, which is necessary preparation for an EMS, to the stage where it can be adapted to the specific requirements of ISO 14001 or EMAS or any other alternative. Developing operational control procedures, emergency response procedures and supporting training programmes; implementing document control procedures; testing emergency response procedures.

Phase 5 Checking, audit and review

Both ISO 14001 and EMAS require auditing of the EMS and a regular management review. At this stage formal procedures can be put in place to establish an audit programme and train the necessary auditors. The findings need to be communicated and reviewed by management and any necessary corrective or preventative actions put in place. The management review also looks at the achievement of objectives and targets and the continued suitability of the EMS with a focus on continuous improvement of both performance and targets.

ISO has also produced some additional help in two standards: ISO 14004 provides additional guidelines for the development of an effective environmental management system (ISO 2016) and ISO 14005 (ISO 2019) provides guidance for SMEs to develop, implement, maintain and continuously improve an EMS. ISO has also published *ISO 14001: Environmental management systems – A practical guide for SMEs* (ISO 2017).

This section describes the key features required of an EMS to meet ISO 14001, EMAS and BS 8555. These features would be common to any EMS that is fit for the purpose of enabling a business to manage its environmental impact and improve its performance. Managing the impacts on the environment is obviously the primary purpose of introducing an EMS but a concept common to most systems is striving to go beyond mere compliance with the law or environmental standards and introducing the idea of continual improvement on achievement of targets for environmental performance (note that 'continual improvement' and 'continuous improvement' are used interchangeably in various documents).

The ISO management standards are developed around the 'Plan, Do, Check, Act' (PDCA) model as illustrated in Figure 2.1 although it has been adapted to be relevant to this particular standard:

- **Plan:** establish the objectives and processes necessary to deliver results in accordance with the organization's environmental policy;
- **Do:** implement the processes as planned;
- **Check:** monitor and measure processes against the environmental policy, including its commitments, environmental objectives and operating criteria, and report the results;
- **Act:** take actions to continually improve.

Following the review the cycle starts again with a new plan.

This section, which broadly follows the sequence of ISO 14001, is structured around the PDCA model as set out in the Standard. The Plan phase is within the following Sections 2.2.1 to 2.2.3, the Do phase within Section 2.2.4, the Check phase within Sections 2.2.5 to 2.2.6 and Act from Section 2.2.7 to 2.2.9, although there is some overlap if the terms are interpreted strictly.

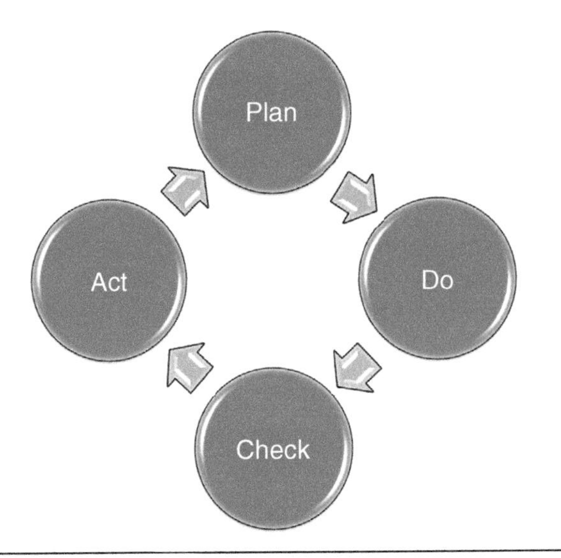

Figure 2.1 PDCA model

2.2.1 Context of the organisation

Before getting stuck into the detail it is important to draw some boundaries around the planned EMS. The first is to understand the context in which the EMS is being developed. There are the external and internal issues that need to be taken into account, either because they are relevant to the process or they affect the ability to achieve the intended outcomes. They would include:

- The requirements or expectations of interested parties (stakeholders) such as suppliers, customers, regulators or pressure groups.

These could in turn affect:

- The environmental conditions affected by or capable of affecting the organisation – such as the local atmosphere or water courses;
- The nature of the surroundings – are there areas requiring special protection?
- The presence of neighbours – business or residential;
- The nature of the materials used or the products – are they toxic?, etc.;
- The nature and relative consumption of materials or energy;
- The constraints applied by planning or operating licence conditions;
- Any legal or contractual requirements.

Some of these can have a positive effect rather than the expected negative. They all can affect the decision about what processes or which sites are to be considered for an EMS: the more constraints, the more likely an EMS will be useful. The constraints are returned to in more detail in later sections.

Scope of the EMS

Within the context of the organisation, it is also necessary to think about the scope. A large organisation with many processes on multiple sites would not want to include the whole business at once. A decision is required as to whether the EMS is confined to the whole organisation (including global to local), just one factory or branch office or one production line and which activities, products and services and which inputs (just energy or all, such as water, raw materials, etc.) are included. The initial scope, even a pilot, may be confined to one process or one site in order to manage the work, control the cost and learn what to do. However, a small office, shop or workshop could consider the whole business in one go especially if there are overlaps in the way that the business is run or managed.

The scope ought to be determined by the relevant significance and the influence that the organisation has over that aspect or impact. There is no point in listing minor noise if you are located next to an airport. The inclusion of the pay office on a steel works is unlikely to be justified – at least at the start. Equally it is important to look off site. The sources of raw material and transport of supplies and products can be significant indirect aspects. The presence of potentially hazardous materials or contaminated land needs also be taken into account. The main aspects are likely to include emissions to air and water, use of raw materials and resources including water and energy, waste management, contamination onto land, noise, transport, dust and similar sources of potential impact on the community or local environment. Also bear in mind that some things such as the weather cannot be controlled but the impacts may be managed in order to reduce the risk from flooding, for example.

2.2.2 Leadership

With most aspects of change to the way that a business operates, leadership is crucial to success. A senior director or manager needs to own the process, set the direction and monitor progress. They may well have different priorities and pressures, especially if the initiative has not started at the top. They will need to be convinced of the benefits in terms that they will understand so they will be able to recognise it as essential to securing the organisation and taking it forward. We have already seen why the implementation of an EMS is recommended but they need to be presented in a convincing way that fits in with the wider objectives of the business.

Detailed implementation may be delegated to someone else but senior management commitment is essential to ensure that the final EMS is owned by the business and deployed across the whole or relevant parts of the organisation. In addition, the management ensures that the necessary resources of peoples' time, money or consultants are available. In the end top management is accountable for the effectiveness of the EMS. Leadership is demonstrated by:

- Ensuring that the environmental policy and objectives are established and compatible with the strategic direction and the context of the organisation;
- Ensuring the integration of the EMS into the business;
- Providing the resources;
- Communicating the importance of effective environmental management;
- Ensuring that the intended outcomes are achieved;
- Assigning and communicating the responsibilities and authority to the relevant roles;
- Directing and supporting all those that contribute to the effectiveness of the EMS;
- Promoting continual improvement.

If some of the action is delegated to another person, that person needs to have their authority and responsibilities recognised throughout the organisation and they then become responsible for ensuring that the EMS conforms to ISO 14001. In addition, they will be expected to report to senior management on progress, costs, compliance issues, conformity with ISO 14001, etc.

Environmental policy

An early stage in developing an EMS is to prepare a policy for the organisation. The final policy is a demonstration of the commitment by the organisation and its senior management but it is also a statement of intent which is the basis for the implementation of the EMS (i.e. what is done to improve environmental performance). The initial draft may not be the final version but it drives the development of the EMS (i.e. what goes into the EMS). The document need not be long; it is a policy, not a statement of what is to be done in detail or how it is to be done. The final version should have a number of key features; the statements are derived from ISO 14001: 2015 although similar requirements are made for EMAS.

It should be appropriate to the purpose and context of the organisation, including the nature and scale and environmental impacts of the activities, products and services. It should briefly outline the main business and aspects of the organisation and the length and detail should reflect that. A bigger more complex organisation such as an oil prospecting and refining company will have more to explain than a coffee shop. However many are just one page.

It should contain commitments to protect the environment, prevent pollution, to comply with all relevant legislation and similar requirements and to show continual improvement. It is, of course, illegal to cause pollution and not to comply with legal requirements, and penalties are likely to follow if you do. The policy explicitly states that this will not happen and, more importantly, it implies that the organisation will aim to do better than mere compliance by looking for ways to improve performance every year.

The policy provides a framework for setting and reviewing environmental objectives and targets and these form the means by which continual improvement can be achieved. Note that the policy provides the framework – it does not normally contain the actual objectives and targets. They are set in the planning stage as outlined in Section 2.2.3. However, some policies do contain specific pledges. These could come out of something uncovered during the initial review requiring prompt action (such as getting into legal compliance) or something more general but of specific concern to the organisation such as moving towards zero waste to landfill or towards completely renewable sources of energy.

The policy must also contain a statement that the policy will be documented, implemented and maintained; communicated to all persons working for or on behalf of the organisation and be available to the public. Document control comes into play again here as it is expected to be signed and dated by the most senior manager and the details could change from time to time. It should also state that it will be reviewed on a regular basis such as annually.

A sample policy is given in Box 2.2 but many others can be found on the internet sites of major companies.

2.2.3 Planning

The planning of the EMS should start to bring together the actions to address the risks and opportunities that arise during initiation of the process. These have already been mentioned earlier in the chapter. The risks are those that have potential adverse effects on the environment, the potential for legal action, on the way that the business operates or on costs. The opportunities for the organisation may be cost savings, improved performance, enhanced pubic image or new business or may benefit the wider community by enhancing their environment, providing facilities or employment.

Identifying the aspects of the organisation

The starting point is usually to identify the environmental aspects – defined above. Most aspects can be identified under the following headings:

- Raw materials – sources and amounts used;
- Energy types – sources and use;
- Water – sources and use;
- By-products produced from processes;
- Waste produced and destinations;
- Emissions to air;
- Emissions to water;
- Risks from hazardous materials on site (such as toxicity, fire);
- Activities likely to cause land contamination (such as spills and leaks or poor management of materials);
- Life span of products;
- End of life disposal or other use of products;
- Noise, light, visual offence, odours and similar sources of nuisance;
- Packaging;
- Transport (deliveries, staff vehicles, etc.).

All of these can interact with the environment in ways which should be becoming familiar by now and which will be expanded upon in later chapters. The wide range of potential aspects highlights the detail required in preparing to implement an EMS. Not all of these may be relevant to every organisation but just looking down the list brings out the aspects that could easily be overlooked in a superficial examination.

Aspects can be categorised as direct or indirect. A direct aspect is directly attributable to the activity, product or service and is most easily identified with a site on which an activity takes place. Energy used to run a machine to make a product is a direct aspect. Indirect aspects are something over which the organisation has little or no influence. Design could directly influence the aspect of a product (design to minimise use of materials of construction, for example) but how the customer uses it is outside the control of the organisation (using it in a way that is

Box 2.2 An example of a simple draft environmental policy

John Smith Manufacturing is a producer of fabricated aluminium assembled parts for windows and doors for house builders and property refurbishment in the UK. It has a factory for production in Birmingham and distribution depots in Leeds and London. This policy applies to all of our facilities.

We are committed to:

- Protect the environment and minimise our impact on it;
- Comply with all environmental legislation and standards for our business;
- Use energy and materials efficiently to provide our products and services;
- Set objectives and targets to monitor our performance in the use of energy and raw materials and the production of waste;
- Continuous improvement in our environmental performance through regular appraisal of the achievement of objectives and targets and their annual review;
- Involve our employees and suppliers in achieving environmental excellence.

We will report publicly on our performance on an annual basis and consult with interested parties on the results and plans for the future.

Note that the draft policy may be reviewed further following further stages in the implementation of an EMS to bring in new commitments or areas for improvement.

wasteful of energy, for example); in this case the aspect would be indirect. The fact that some customers leave their televisions switched on all day wasting energy is not the fault of the manufacturer (hence an indirect aspect) although they could be designed to switch off automatically after a certain time in the absence of any use of the controls and be designed to be energy efficient – both of which are under the control of the manufacturer.

It is useful at this stage to also identify the impacts – defined above. They can be associated with one or more aspects. Impacts are generally limited to:

- Resource impacts (depletion of non-renewable resources and the effects of gaining and transporting materials – Chapters 1 and 7);
- Atmospheric impacts (air pollution, and the consequential impacts such as acid rain, climate change, ozone depletion – Chapter 1);
- Aquatic impacts (water pollution and consequential impacts such as eutrophication – explained in Chapters 1 and 4);
- Land impacts (contaminated land – Chapter 6);
- Damage to humans, wildlife and ecosystems (see below and Box 2.3);
- Community impacts (such as nuisance from noise, dust or visual impact and loss or gain of amenity).

Environmental impact assessment is the main topic in Chapter 3 and more details about the specific impacts are in the later chapters on the media involved. Note that the definition includes beneficial or harmful impacts. Most people would quickly recognise that harmful impacts such as toxicity or depletion of resources need to be managed. However, some aspects could have beneficial impacts: a new product that is more energy efficient has a lower impact than those that it replaces; replacing lead compounds in petrol with lead-free alternatives to improve the octane rating reduced air pollution; new factories provide jobs and income to the local community. Think of a few more recent activities, products or services that have had a beneficial impact.

The definition also states that the impacts can wholly or partially result from the aspects. The impact on a local stream entirely due to a discharge from a business would be wholly due to that discharge. However, climate change due to emissions of carbon dioxide from the same business's vehicles would only be partially to blame, as explained in Chapter 1.

Remember that impacts are caused by the aspects: the two lists should be capable of being cross-referenced. Box 2.4 elaborates on these definitions with some examples. Because likely impacts include contam-

Box 2.3 Ecosystems and food chains

The word 'ecosystems' is used to describe the inter-relationship between a range of plants and animals in the natural environment. The simplest way to understand it is to consider the example of a forest. There will be a range of species of trees and other plants growing in the soil and birds, animals, insects, bacteria and fungi in and around the plants and soil. Many are dependent on others for their survival. The trees and plants take water and nutrients from the soil where leaves and dead organisms are broken down by bacteria and fungi. Insects feed on other insects or leaves. Birds feed on insects and seeds. Animals feed on birds, plants, insects and other animals. Humans may take wood from trees and take plants, birds and animals for food. The whole inter-related system is called an ecosystem. The dependency on other species for food is known as a food chain. Carnivores and humans are at the top of the food chains.

A different range of species will be found in a different habitat such as a beach or a desert or a river. In each habitat there is a different ecosystem.

Box 2.4 Aspects and impacts

The definition of **aspects** refers to **activities, products and services.** The interpretation of these will depend on the nature of the organisation and each of these could have several elements. Here are some examples that should help to understand the definitions using A, P or S to indicate which type of aspect is listed. These are just examples, not a full list of potential aspects! The associated impacts are also given.

Open cast mining

Removal of cover (A) causes noise, creates dust, spoil for disposal, visual intrusion; extraction of ore (A) causes noise, generates dust, waste material for disposal; transporting material (A) generates noise, emissions from exhaust, heavy goods vehicles.

The impacts are consumption of non-renewable resources (the ore and energy such as oil), noise pollution, air pollution from dust and vehicle emissions, road damage, loss of amenity from wastes and visual aspects.

Supermarket

Receipt of goods (A) causes disruption to neighbours, emissions from exhaust; refrigeration (A) consumes electricity, potential for leak of refrigerant; lighting and heating (A) consumes electricity and gas; home delivery (S) causes disruption to neighbours, emissions from exhaust; removing out of date produce (A) results in food waste which needs to be disposed of; surplus packaging from incoming goods (P) needs disposal.

The impacts are noise and community disruption from vehicles, air pollution from vehicle emissions and refrigerant, various wastes such as food and packaging with impact depending on disposal route, consumption of raw materials for energy use.

Food preparation for market

Washing raw vegetables (A) produces contaminated waste water; peeling vegetables (A) produces solid and liquid waste; cooking (A) consumes fuel; packing finished food (P) consumes energy and packaging material; waste offcuts from packaging; transport of goods (A) causes noise and vehicle emissions.

The impacts include air pollution from vehicles, noise, consumption of water and energy, potentially polluted water and land contamination if waste is landfilled.

Running a golf course

Creating the course (A) involves excavation and shifting soil with heavy vehicles causing noise and vehicle emissions; grass cutting (A) causes noise and vehicle emissions, use of herbicides (A) has potential to cause polluted run-off and to kill other plants.

The impacts include air pollution and noise from vehicles, potential for water pollution and loss of beneficial plants and the animals that depend on them. Whether there is a loss of amenity depends on your attitude to golf.

To avoid too much repetition, here are a couple of simplified examples.

Running a bank branch office

Aspects will revolve around the energy use associated with the use of buildings, computers, etc., paper usage and waste with impacts as above.

Creating a nature reserve

Aspects will include changes to land use and land management practices and possibly limiting access or other activities that are detrimental. This is an example where the impacts are beneficial by promoting and protecting the local flora and fauna and reducing pollution.

ination of the environment, consumption of non-renewable resources, damage to human health and wildlife and community impacts, bringing in the views of external stakeholders such as neighbours, wild life organisations, regulators and business associates can help develop these stages.

Compliance obligations

The other ground work at the planning stage is to identify the key legal and similar requirements such as contracts with customers that may specify, for example, that only timber from renewable sources may be used. Some of this may be already familiar if there are permits in place for emissions. However, such knowledge may be distributed across the organisation in production, purchasing, marketing and other departments. The point of this exercise is to produce a central register that is available to all and can be kept up to date as changes occur. It is fairly common to flush out some new requirements – in the sense that they were previously not recognised when they should have been. Finding these may be difficult but government and regulatory bodies, trade associations and colleagues in similar organisations should be able to help. The finer points of waste regulation and rules concerning packaging or the developing regulations and control systems for managing climate change and energy efficiency are often overlooked.

The main obligations may be any or all of the following:

- Legal requirements such as international treaties, national laws, regulations and standards touched on in Chapter 1 that could apply to aspects or impacts;
- Conditions applied by regulators in the form of permits or licences which regulate the operation of an organisation or its emissions;
- Contractual obligations from customers;
- Agreements with local authorities, regulators, neighbours, NGOs, suppliers, customers or other bodies to limit activities to particular times, to avoid nuisance, to use or avoid particular products, etc.;
- Codes of practice adopted by the business or relevant trade organisations.

These are included in more detail in later chapters.

Sources of information

Some of the sources of this information have already been mentioned. The internal ones include: permits from regulators, compliance data for emissions, information about incidents and complaints, usage data for raw materials and utilities, waste records, results from environmental monitoring, information from risk assessment and site inspections and, of course, the employees may have a lot to contribute. Some of this may not be easily available if there has not been a systematic approach to collecting information in the past. External sources of information include government departments, regulators and local authorities, legislation such as statutes and regulations, standards bodies, trade organisations, professional bodies, relevant codes of practice and guidelines, and manufacturers and suppliers. A lot may be published in trade or specialist journals and reference books or can be found on the internet. It may even be possible to gain information directly from business associates in similar organisations or benchmarking against their publicly available reports.

Interpretation of the results

A risk assessment is also useful at this point as it helps the business case as described earlier. It can bring together the legal register and the aspect and impact assessments so that the potential for causing pollution or breaching regulations is better understood, including from abnormal conditions such as plant breakdown, power failure, severe weather or fire. It is not unknown at this stage to find some breach of a permit or other regulation that has not been dealt with or that the organisation is so close to breaching one of these that it represents a considerable risk. These sorts of issues need to be taken into account so that the risks can be managed and problems avoided and priorities established for the next stages.

The approach to this initial review is one form of audit – a systematic look at all the relevant issues. More on audits can be found in Section 2.2.6 where we will see how regular audits are used to check the implementation and operation of the EMS.

Environmental objectives

At this stage the organisation is starting to get an understanding of where action is going to be needed and which will eventually form part of an action plan. The EMS requires that the organisation prevents pollution or avoids other environmental damage. The list of aspects and impacts may be quantified if possible although they may often be in general terms (e.g. it is recognised that waste production is high but there are insufficient data). If there is clear responsibility for causing water pollution in a local stream then this needs to be identified as requiring action and it should be possible to identify what the extent of the problem is and how it can be remedied. However, when it comes to emissions of carbon dioxide and consequential climate change, for most organisations their contribution is relatively small. The emissions may be quantified but the impact cannot. That does not mean that action is not required but that the priority may be lower or that the driver is less obvious.

Another requirement is to look at the register of legal and other requirements and establish current compliance. Compliance with the law is the absolute minimum so any non-compliance needs to be seen as a priority for action. This may require looking back over several years of data to check whether there is already a problem or that one is developing, as performance deteriorates due to ageing plant or increased production. Many organisations leave investment in new or improved plant or machinery for production or treatment plant to deal with emissions until there is a problem. The EMS helps to pre-empt this.

These two stages help to get into the process of setting objectives and targets for the EMS. There is no point in trying to set objectives and targets for every conceivable topic; it would be too demanding and if there are too many they are unlikely to be achieved due to pressure on the resources required to tackle them. So they need to be prioritised according to their significance. A priority objective may be to get into compliance with any legal or other requirements. Any significant environmental impact may also need to be dealt with urgently. For other objectives there may be pressure from the stakeholders such as neighbours or customers. Other features that affect the significance of the aspect or impact also come into play. The determination of the significance of an impact is dealt with in more detail in Section 3.4 in the next chapter.

2

Box 2.5 SMART objectives

The use of SMART objectives applies across all business activities. The acronym stands for:

Specific: the example in the text refers to the reduction of waste produced.

Measurable: we are setting targets in percentage terms which can be measured by weighing the waste produced.

Agreed: the people responsible for achieving the targets need to be involved in setting them so that they have some ownership and are more likely to commit to their achievement.

Realistic: the targets need to be realistic in both quantity and timescale.

Time bound: there is a date by which targets should be achieved; open-ended targets are unlikely to be met.

Objectives – as included in the policy – are strategic and could be a general statement, for example, to reduce waste sent to landfill or relatively specific such as to reduce waste sent to landfill by 25 per cent by 2025 compared with 2019. Targets are more detailed. In management terms they should be SMART as described in Box 2.5. Developing the waste example, they could include targets to achieve a 5 per cent reduction in 2021, a further 10 per cent reduction in 2022, and further 5 per cent reductions in 2023 and 2024. This progressive reduction in waste production also represents continual improvement. The objectives and targets need to be realistic and relate to the costs and other resources available. These are covered later.

2.2.4 Support

Collecting the information together in order to identify the aspects, impacts, legal and other obligations may seem a daunting task. Up until this stage it is likely that much of it was scattered around multiple documents, not properly documented at all, or just not known. One advantage of introducing any international standard is that this information is collated in one place. This means that it should not be possible to miss something in the future and, as mentioned before, much information may be common to more than one standard.

Resources

By this stage it should be clear that the resources to establish, implement, maintain and continually improve the EMS may be significant, especially in the early stage for a large organisation. As well as collecting and collating information, the process needs to be communicated both internally and externally and people will require training. In later stages there may be need for research to find means to improve and changes to plant or processes and monitoring procedures.

In many situations it is unlikely that this work could be added to the daily routine of an existing employee without jeopardising their normal work or risking slow delivery of the EMS. Staff time and money resources will be needed. The early stages of data collection and training can be time-consuming unless managed well. For example, well-thought-out plans and recording systems able to be computerised will save effort: training about the EMS can be added to existing training or briefing programmes.

Competence and training

Training to improve competence can include two different types: training of those involved in the EMS itself and training of the rest of the employees about the implications for them. Usually the introduction of an EMS requires a lead person to take control and they will need to understand how it works and how to implement it. The task may be done by inviting a consultant who is already skilled in the role. This reduces the internal burden at the cost of failing to gain inside knowledge for the future. An alternative is to pick an existing employee and train them. A common choice is the Health and Safety Manager or Quality Manager who have Environment added to their portfolio. If the organisation already has other ISO accreditations this could be a sensible choice. In the absence or unsuitability of these roles, someone else may need to be trained from scratch. More detailed and specialist training will be needed in this case, potentially involving training courses offered by accredited providers.

Training will also be required for the rest of the workforce who may be involved in the development of the EMS. Different training programmes will also be required for those that have to implement it. For example, changes to work practices, use of energy or materials, waste management, transport, monitoring and reporting may be required. It is often the case that current ways of working have developed over many years and become custom and practice. Changes are likely to be met with scepticism or even hostility. The management of change is a topic in its own right and this is not the place to go into details.

The present level of competence needs to be established for both sets of potential trainees in order to develop the appropriate training programmes to improve the level of education or provide relevant experience. Training should extend to broader education about why an EMS is being introduced with a more detailed understanding of the business and how its aspects relate to the environment – for example, the production, management and disposal of waste materials.

Preparing for the training programme will require some means of identifying the areas to be covered which should largely come from the earlier work in the planning phases but there will need to be some work on identifying any skills gaps and preparing for changes in operational procedures, the introduction of new technology, etc. The use of consultants or checklists may help here. The keys to successful implementation are largely about involving the employees: they need to be motivated, understand what is happening and the reasons behind it and what they are going to have to do or change. They need to have the right competencies for any initial involvement in preparing the EMS but also later for carrying out new or different tasks (such as dealing with incidents or checking the calibration of instruments). Some suggested topics for a programme are shown in Box 2.6. More details about the relevant technical issues are in later chapters.

Box 2.6 Topics for a training programme for the EMS

Why we are introducing an EMS

Outline the reasons based on environmental legal requirements and business opportunities such as cost savings.

The background to some important environmental issues

Covering climate change, water and other resources, energy use, etc. influenced by the nature of the business.

The activities, products and services of the organisation

A brief outline of the main business of the organisation.

Our organisation's impact on the environment

Pick out the significant impacts and put them in context of the important environmental issues.

The legal position in respect of our environmental impact

Highlight any permits, legal cases or similar pressures that influence the EMS.

How we can reduce our organisation's impact on the environment

Focus on key issues such as energy efficiency, water use and waste production as appropriate to the main business activities.

The organisation's environmental policy

Introduce the policy and the main contents and invite comments.

The organisation's environmental objectives and targets

Describe the broader objectives and the initial targets that all are expected to help to achieve.

The roles and responsibilities in achieving the environmental policy

Explain who is taking the lead and others involved in the EMS. Develop this into the responsibilities of everyone in the organisation.

Key actions required in support of the policy and targets

Describe what is going to happen to achieve the targets; e.g. new utility meters, waste recycling, managing water and energy use, improved controls on effluent treatment plant, changes in working practices, etc.

How all employees can help to achieve the policy and targets

Stress the importance of everyone's contribution.

Changes required in ways of working

Introduce any new procedures such as storage of chemicals or waste segregation.

Managing the transition

Describe what will happen and when and how employees will get involved.

Dealing with emergencies or other atypical events

Stress the importance of prompt action and set out expectations and actions to take in order to prevent pollution.

Communicating with and involving the employees

How further information will be provided such as new changes or progress reports.

Training requirements

Outline any additional training programmes to help employees adapt to the new EMS.

Awareness

Raising awareness is a crucial first step. All the stakeholders need to be informed about the planned introduction. The internal stakeholders, essentially the employees, will respond favourably if kept in the picture right from the start. Indeed, some may volunteer to be involved or provide information or ideas to help the process. Some external stakeholders such as regulators, suppliers and customers should also be made aware. They are likely to be affected in due course and may have something to contribute from their own knowledge or experience that could be useful. The awareness may well cover some of the early topics in Box 2.6 which should be communicated to all employees. Awareness should include the policy, the significant aspects and impacts relevant to the organisation, how they can contribute to the effectiveness of the EMS and the implications (legal, financial, PR) of not conforming.

Communications

In support of this there will need to be a communication plan to brief all employees and some other external stakeholders. Some will just need an overview and be briefed on what they may be required to do to help with the implementation such as providing information or changing the way that business may be conducted in the future. Others that are more directly involved may need more information. Later communications will inform on progress on implementation and also issues that have arisen and how they will be dealt with. Finally regular communication should show compliance with or progress towards the expected outcomes of the EMS. The communication plan should identify what to include, when to communicate (such as at key stages), who should be informed and how it will be done.

Various outlets are available and the choices will depend on the size of the organisation and its current methods of communication. However, the introduction of an EMS may be the opportunity to review these. Common methods for reaching the wider public are:

- Corporate reports such as the annual report or an environmental report;
- Brochures and leaflets;
- Press releases and use of media such as newspapers and radio;
- Newsletters to employees or customers;
- Attendance at conferences, seminars and exhibitions;
- Contracts and tender documents;
- Open days;
- Addresses to the Annual General Meeting or other lectures
- Use of the internet and social media sites.

Employees also need to be engaged and this can be achieved through:

- Notice boards;
- Employee handbook;
- Training and induction days;
- Team briefs (sometimes called toolbox briefs);
- Trade unions and works councils;
- The company intranet.

Involving employees and external stakeholders in the process also aids communication as well as gaining their confidence and support.

Document control

Although at this stage we are mainly in the planning and early implementation phases, what has been achieved already will form the basis for more detailed work in the operational phase. Therefore it is appropriate to get into the habit of documenting the work and implementing a document control system from the start. Document control is a requirement of ISO 14001 and EMAS and the standards specify the minimum requirements. However, even if there is no immediate intent to certify or validate the EMS, the discipline is a good one. So far we should have documents at least recording the decision to proceed, the roles, the budget, the policy – including objectives, the aspects and impacts, the register of legal and other obligations, the review of competences, any training programmes in place and who participated, who has been informed in awareness programmes, the sources of information used, progress reports and probably more. The documentation need not be too complicated but it helps if they are to a common format and style and include as a minimum a version number, a date and an author reference. The documents need to be readily available but not subject to random changes by anyone not authorised. It is very easy to get lost in changing paperwork and not know which is the most up to date if there is no system of control. The use of computers to create and store the information makes the task easier provided it is well managed and discipline is applied in the process such that confidentiality is not compromised and improper use is not allowed. Secure back up of either paper or computer records is obviously essential. Documents will require periodic review and updating as people and procedures or other features change and ISO 14001 and EMAS require these to be available and current. The person (or post) responsible for the review and updating of the EMS should be identified on any documents and available to answer any queries.

By the end of the process the EMS document should contain the basics of:

- The final environmental policy;
- A statement of the scope of the EMS;
- Objectives and targets;
- Version number, author and date.

Other parts, some of which are described below, may be in the same document or referenced as to where they may be found (as some may fulfil other organisational purposes). They include:

- Management structure and responsibilities;
- Register of legal and other obligations;
- Relevant standards, codes of practice, etc.;
- Registers of aspects and impacts;
- Plan to implement the EMS;
- Operational procedures;
- Control procedures;
- Monitoring, inspection and auditing arrangements and where to find the results and reports;
- Emergency plans;
- Training programmes and records of employee competencies;
- Management reviews and reports on performance against the EMS.

Procedures for review and replacement of the various parts need to be included and previous versions kept at least for an agreed period before destruction.

2.2.5 Operation

So far we have looked mainly at the administrative parts of introducing an EMS. The intended outcome is defined in the policy as such things as reduced pollution and improved legal compliance. These do not come from documentary systems but from actual activities on the ground – the 'Do' phase in the PDCA cycle. It is time to move on from the 'Plan' phase: the things that we need to do were identified in the plan produced in the last section.

Operational planning and control

There are two angles on the operations of an EMS: implementation of the EMS itself and the way that it influences the operations of the organisation in order to comply. The operation of the EMS is about the documentation, monitoring, auditing, etc. The operation of the organisation is about delivering the activities, products and services. These should become one: the EMS influences the organisation to the extent that it effectively becomes the way it is run; the requirements become the routine. However, in the initial phase, changes will be implemented that may only affect some parts of the organisation. Many organisations find that it eventually affects the whole business. It may start in production and end up including all of the support services.

The complexity of this plan will be affected by several factors the most obvious of which is the complexity of the organisation and the range and difficulty of any problems that have been identified. Plans should not just be a list of what needs to be done but include who is going to do it, by when and including the resources (people, money, etc.) that are required. The plan for an oil refinery will be more complex than that for a local shop as it will involve more people and many types of resources. The production of the resources elements of the plan could test the determination of the senior management to implement an EMS so there is no benefit in making it so ambitious in timescale and resources that it gets bogged down in arguments about affordability or practicality unless the circumstances warrant it. ISO 14001 describes the operational phase as establishing, implementing, controlling, and maintaining the processes required to meet the EMS and to implement any actions previously identified. It goes on to describe these as establishing operating criteria for processes, and implementing control of the processes in accordance with operating criteria. In more practical terms the actions required in the plan may cover a very wide range of types of activities.

Examples could be:

- Refining the detail of performance objectives and targets where these can already be determined;
- More data collection on e.g. emissions performance or environmental impact;
- Research into better ways of dealing with a problem;
- Changes to activities, products or services;
- Recruitment of specialist knowledge directly or through advice;
- Investment in new plant or machinery;
- Changes to processes in production or other aspects;
- Training of staff, contractors and suppliers;
- Renegotiation of contracts with suppliers;

This will require resources which could include:

- Time commitment of existing staff or additional staff;
- Consultancy costs or working with a local university;
- Sampling and analysis costs – initial or ongoing;
- New, improved or extended monitoring equipment such as electricity sub-meters;
- Sub-contracting of research and development;
- Investment in new products and services;
- Cost of new production plant, transport, emissions treatment;
- Improved inspection and maintenance regimes.

This list is not exhaustive but gives a flavour of what may be involved and brings into focus the business case and how these costs should lead in the longer term to savings in other costs or improved benefits. Specific details for various types of issues are included in the later chapters.

Any changes need to be monitored, controlled and documented and any unintended consequences dealt with. Changes may need to be applied to external suppliers of goods and services and built into procurement processes and contracts; hence the importance of including external stakeholders in the communication programme. It is also important to include external factors such as the use of products, transport or end of life implications (such as reuse or disposal as waste – included under life cycle analysis in Chapter 3) in this stage as this will also certainly involve external stakeholders. Some of this may be communicated on labels for end consumers – energy efficiency or recycling options being the most common.

New ways of delivering products and services may need to be researched and then implemented. Changes in sourcing of materials or handling waste, changes to monitoring or maintenance schedules and a higher public profile are some examples that should eventually become the norm. The detail will vary from organisation to organisation but in most cases it will involve changes to the way that people do their jobs. The changes may be brought about by others but should involve those directly affected. Many existing practices probably developed over time and so became embedded in an organisation. However, the people who undertake the operations often have good ideas about how they can be improved; it is just that no one has ever asked them. This is where the importance of raising awareness, training and employee involvement will make the introduction of the EMS easier and could lead to enthusiasm and an extension of the principles into their own lives and activities.

The operational changes need to be documented and kept up to date as with the rest of the EMS. As far as running the business is concerned, this will be in operational procedures or schedules of inspections, maintenance, sampling and other critical activities. They will also need to cover how to deal with non-compliances, incidents and complaints or emergency situations and many of these documents should be relevant to other factors needing control for reasons of quality or health and safety. It is not necessary to document every activity or procedure in fine detail. A risk-based approach should be adopted taking account of the potential for environmental impact and the relative risk (similar to safety hazard and risk analysis) as well as the need for conformity. Handling toxic or flammable chemicals or controlling an effluent treatment plant would be examples requiring such a detailed approach. For some operations procedures may need to give special emphasis to particularly risky stages in an operation such as start up or shut down.

Responsibilities

The example in Box 2.7 shows a simple plan of work to reduce waste going to landfill with a number of tasks allocated to individuals along with costs in staff time and money. Implementation of the plan at one level is just carrying out those tasks but, of course, there is more to it than that. The tasks are at an individual level – one or more persons are tasked with collecting data or similar activities. Yet this has to take place within a framework of an organisation that has other priorities and calls on its resources. This means that there has to be a matching framework of responsibilities set out within the wider organisational structure.

To be successful the EMS has to have senior management ownership so the most senior manager should have overall responsibility for its delivery. They have the similar level of responsibility for ensuring the achievement of other targets such as production and quality and for raising and balancing the resources required to meet all of them. The detail is likely to be delegated to a subordinate such as the environmental manager or someone else in the hierarchy. They, in turn, may well have to allocate some responsibilities to others in the

Box 2.7 A simple plan for waste reduction

Objective: to reduce waste to landfill by 25 per cent in 2025 compared with 2019

Targets	2021	2022	2023	2024
Percentage reduction waste to landfill by:	5	10	5	5
Additional targets for year 1:		Who	When	Cost
Segregate waste by type (metal, paper, card, etc.)		AB	Immediate	X
Train staff in importance of waste management		AB	March 2020	X
Investigate current waste types and quantities		AB	March 2020	X
Seek staff suggestions		AB	April 2020	X
Explore routes to recycle any waste in house		CD	June 2020	X
Implement incentive scheme		AB	July 2020	X
Negotiate recycling opportunities with others		CD	Aug. 2020	X
Review production methods to minimise waste		EF	Aug. 2020	X
Report on production changes required		EF	Sept. 2020	X
Board approval for investment		EF	Oct. 2020	X
Produce a detailed plan for 2017		AB	Oct. 2020	X

This is not intended to be complete but it gives a flavour of what is required for the first year. It may also include an outline plan in Year 1 for subsequent years, especially if major investments requiring capital are likely to be required. It should also include an estimate of the time required of AB, CD and EF under cost in hours or actual cost and any other costs such as consultancy fees or for purchasing recycling containers.

organisation such as production, purchasing, quality control, etc. So you end up with a management structure for implementing an EMS which is a subset of the whole management structure. The people involved are likely to have other tasks to do as well which means that progress monitoring and reporting across the whole implementation is also an essential feature. This should present no problem to a well-run organisation. The documentation in support of the EMS should include this structure and a definition of the responsibilities of everyone involved. These responsibilities may start with the implementation of the plan but will eventually be mainly concerned with the continuing maintenance of the EMS, covering such tasks as are set out later in this chapter including operations, monitoring, auditing and emergency planning.

Emergency preparedness and response

So far we have assumed that the EMS is dealing with the normal operation of the organisation. This may not always be the case as there could be unusual incidents that cause some damage to the environment. An emission to air, land or water that causes pollution is the most common. Noise or similar nuisances, a fire on site, litter from blown waste material, loss of water from a burst water main or a report of spillage of material in transport could be other examples. Such an event is in breach of two principles of the EMS – to prevent pollution and to comply with legislation – and may also be in conflict with policies associated with the effective management of resources or the organisation's CSR policy. Causing pollution is usually a criminal offence and so could lead to a prosecution and there is also the risk of a civil action for damages. ISO 14001 requires plans to prevent or mitigate such emergency situations, activate a response, take action to mitigate any consequences and associated details. These are covered in detail in Chapter 9.

The investigative response to an incident should be similar to that for a breach of permit although there is also the imperative of dealing with the incident itself and clearing up, as discussed further in Chapter 9. The results of the investigation should be reported internally but also to the person or organisation that reported it and to any regulators with an interest.

The term 'near misses' is used to describe an occurrence that has the potential to cause an incident but somehow that was avoided so that there was no environmental impact. So a failure in the plant to control emissions that was spotted in time to avoid an actual discharge, a spill that was contained on site or damage to an installation by vehicles or fire that does not result in loss of material or pollution would be examples. These should be investigated in a similar way to actual incidents to learn any lessons, as if there is a recurrence the next event may lead to a real incident.

The actions taken in response to incidents and near misses to avoid any recurrence may involve changes in design or operation of plant, additional safety features to protect installations from damage or vandalism, the installation of automatic monitoring or control systems, more frequent inspections or maintenance, additional training for employees or anything else relevant. Again, all of the results of the investigation and the actions taken need to be fully documented and reported.

Complaints

Many complaints are associated with emergencies or minor incidents – smells, noise, dust and smoke affecting employees or neighbours as well as complaints of pollution to the wider environment. However, they may actually be the response to what is considered normal operation by the organisation causing them. From the complainant's point of view this is a fine distinction. Examples could be the smell from a sewage effluent treatment plant when desludging takes place or noise and dust from a construction site or waste blown from a site. No breaches of permit or other regulation may have occurred and the operator may not have been aware of the impact.

However, once the operator is aware, the response should be similar to that of an incident – an investigation and then measures to mitigate the cause. These could involve changes to plant or operating times and procedures, the installation of new equipment or simple measures such as damping down dust or cleaning the site more frequently. Some of these are covered in later chapters. As well as documenting what has been done, it pays to feedback the actions taken to the complainants and build relationships, so that similar events do not result in conflict if this can be avoided (this may not always work but failure to consult in this way is more likely to result in conflict). The organisation's environmental policy may not mention complaints specifically (although it could if there were activities that were likely to cause them – such as for a demolition company) but the commitment to reporting to the public will eventually bring it out into the open. Regular or frequent causes of complaint may result in civil action as discussed in Chapter 1 but in any event their avoidance ought to be part of good governance and a feature of CSR.

2.2.6 Performance evaluation

The next step in the PDCA cycle is 'Check' – ensuring that the actions in the 'Do' step are taking place and delivering the intended outcomes. The EMS itself is just a document and does not guarantee success! This comes from a rigorous approach to identifying the main drivers behind the EMS, putting in place the right procedures and controls and actually doing what is required well and consistently. Before looking at the requirements of ISO 14001 in detail there are some general comments that apply across most organisations.

Self-checking

A large part of this is down to people again; if they do not implement the changes or procedures as required then the outcomes will not be as intended however good the paperwork may be. So checking starts at the bottom:

- Do I know what I am meant to be doing?
- Can I do it?
- Have I got all that I need at my disposal?
- Is what I am doing delivering the expected results?
- Are there improvements I need to make?

Self-checking can only deliver the desired outcomes if the operator knows what they are and how to see or measure them. In answering the questions above, the operator needs to have confidence in the process overall and that they have had the right training and equipment to do the job. If there are shortcomings then they need to expect a supportive approach in helping them to achieve success.

As an additional part of the self-checking process, operators should be regularly checking their own work space and equipment. Examples from an environmental management point of view would be:

- Are there leaks or spills that require mopping up?
- Is all waste tidy and in the correct containers?
- Is the plant working properly or are there symptoms of imminent failure?
- Is maintenance required due to poor performance or is it scheduled?

These would supplement other spot checks for quality or health and safety that probably need to be in place as well.

Supervision

Operator checks do not negate the need for effective supervision. Supervision is rarely present all the time and often poor performance does not become apparent until some time has elapsed. Supervisors and managers should be asking similar questions about themselves and the operations that they manage. Good habits slip into bad; new procedures lose some of their initial sparkle; new employees do not perform as well; the results back from monitoring are not as expected. Regular reviews of the outcomes with the operators whilst checking their performance will ensure few surprises.

Measurement and monitoring

Part of the checking process may involve taking measurements and monitoring processes. There is more on this later in the chapter but for now the emphasis is on the importance of having a process in place for the people involved to ensure that it is carried out and properly recorded. Some measurements (e.g. for process control or monitoring energy use) may be recorded continuously but the operator needs to check that the instruments and recorders are working properly and carry out any calibration on a scheduled basis. Other measurements may need to be taken from a meter, by testing samples or recording materials used or waste produced. Again these should be to a schedule and the methods of measurement written into the documented procedures.

These principles apply to all aspects covered by the EMS. It is not just relevant to operators of process plant: the facilities manager of an office or a hospital should be collecting the relevant information for example on energy and water use or waste collected.

Record keeping

The results of the measurements described in the previous section and records of other scheduled activities such as site inspections or maintenance need to be properly recorded. This may be done on daily record sheets or entered into a computer (or often both – an opportunity to streamline data collection using modern technology). The records need to include date and time and the name of the person collecting the information as well as the results of any measurements or operation and any action taken as a result.

The record sheets need to be organised according to process or some other logical indicator and kept secure. They may be needed at a future date if there is an audit or for an investigation of an incident or plant or process failure.

Monitoring has been mentioned several times in previous sections and it is important to recognise two types: active and reactive monitoring.

Active monitoring

Monitoring in this context is the checking of an organisation's performance against the standards that apply. It is carried out at the time the relevant processes are taking place. The standards could take several forms:

- EMS objectives and targets (e.g. waste reduction or recycling);
- Conditions written into a permit such as emission standards;
- The terms of a licence or permit such as for waste disposal;
- Terms of contracts (e.g. requiring timber to be from a sustainable source);
- Local regulations such as limits on noise;
- Legal notices such as an injunction;
- The terms of a local agreement such as restricting hours of work;
- Operating procedures such as storage of chemicals
- Internal standards such as maintenance schedules and calibrations.

These are the main examples and they illustrate the point that the monitoring is of ongoing activities and is likely to be required on a regular and frequent basis. Note that the standards against which performance is measured may be numerical or descriptive. Some permits and licences may specify the required frequency or they may be agreed within the terms of a contract or an agreement. Otherwise frequency may be established by best practice or custom or, more scientifically, by consideration of the risks involved of non-compliance or to the environment and the variability of the parameter being measured.

It must be stressed that monitoring is to compare actual performance with the standards that apply and any non-compliances should be promptly reported. The onus is on the manager receiving this advice to respond quickly and effectively. Failure to do so is not just in breach of the EMS, it discourages future reporting as just a waste of time.

Inspections

A key role for the manager or a third party such as the environmental team is inspection of premises or plant. This may be part of a wider remit looking at the quality of product or service, health and safety requirements and behaviour. It entails checking on the aspects of the activity, product or service with the knowledge to identify and remedy any shortcomings. It could cover such things as:

- Storage of chemicals, waste or other substances that represent a risk to the environment;
- Site tidiness;
- Operators following procedures as laid down;
- Supervision is adequate and responding to issues as they arise;
- Plant and equipment adequately maintained and performing properly;
- Monitoring equipment or procedures in place and working;
- Emission control equipment working;
- Prompt and effective responses to alarms or malfunctions;
- Results of performance monitoring available and displayed,
- That the area adjacent to the site is not being impacted by operations.

The detail of what is inspected, the frequency of inspection and the depth of inspection should reflect the risk. For example, storage of some chemicals will be of greater significance than others based on their properties and the volumes on site (compare heating fuel oil and printer toner). Formal site inspections are usually carried out with a systematic look at a site or a process or part of these on a regular basis with a record kept of the findings and actions required. Frequency could be as high as daily or as low as annually and may vary across activities. The inspection should always look at the output from previous inspections to check that any actions have been implemented and the problems previously identified are now under control.

The formal inspections may require a special action such as lifting a cover or looking at records but much of the routine such as waste management should be continuously under inspection as part of the daily supervisory role. Lapses must be promptly remedied and noted down with dates and actions taken. The records fulfil several purposes ranging from performance management of employees to looking for opportunities for improvement but in the context of an EMS the records need to be available for management purposes such as checking the robustness of the plant or the procedures and for audit.

Again it should be mentioned that inspections apply beyond any process plant or site boundary: for example, purchasing officers should be checking that suppliers are meeting the terms of their contracts. This could involve site visits or inspection of documents given to or provided by the supplier (e.g. certificates of origin and compliance with local legislation).

Reactive monitoring

Active monitoring is pre-programmed on a routine regular basis. In contrast, reactive monitoring is the response to events. At the lowest level it is recording the events themselves but that could trigger further actions such as special site inspections or sampling of processes and emissions.

The events that could be the trigger are those that are outside the normal routine such as:

- Non-compliance with a permit found as a result of the active monitoring;
- Enforcement action by a regulator;
- An incident that has an environmental impact;
- A 'near miss' that could have had an environmental impact but for some circumstance that was avoided;
- A complaint from the public, an employee or a neighbour;
- A sudden change in a measure such as water or energy use that could imply a potential problem.

The event itself needs to be recorded: what happened, where, when, who reported it or how it was noticed, the impact (especially any environmental impact) and actions taken. The actions would include any further inspections or monitoring as well as dealing with the cause and the effects. The monitoring could well extend beyond the normal boundaries of active monitoring to include the wider environment, such as a river or the flora and fauna that may be at risk in the vicinity or the damage to a neighbour's property.

Reactive monitoring is usually aimed at establishing the cause of any problem and should be followed up with some actions. Depending on the causes found, these could include:

- A more detailed investigation of the likely causes of the problems;
- Additional training of employees in processes and procedures;
- Practice drills in emergency procedures;
- Changes to procedures;
- Improvements to plant and equipment;
- The introduction of new technology;
- Improvements in monitoring and control systems;
- A review of the EMS.

Review of environmental performance

The purpose of monitoring, whether by collecting samples, reading instruments or site inspections, is to evaluate performance against the EMS and any other standards that may apply. The reliability of the results depend on the quality of the information collected so it is important that these activities are controlled and documented in the same way as all the other activities. The information needs to be in a standardised form that lends itself to easy management and interpretation.

Sampling and analysis

One aspect of monitoring may depend on sampling and analysis of raw materials, emissions, environmental media (air, water or land), noise or controls such as pH or temperature within the plant. Sampling procedures and the way the samples are treated and stored can have an effect on the results so training and inspection and supervision of the sampling process are also required. The analysis may take place in a laboratory which needs to be suitable for the range of analyses required and the sensitivity of its methods should ensure compliance with low environmental standards.

Some analysis depends on instrumentation that is sensing continuously and measuring and recording the output from a sensor. Common applications include pH, suspended solids and temperature in liquids or temperature and the presence of some gases such as sulphur dioxide in an air flow. Here it is important to ensure that the sensors are kept clean and by their position truly reflect the quality of the flow that they are meant to be measuring. Calibration is also required, often on a frequent basis which may be more than daily. Inspection regimes need to ensure that the calibration is taking place and that the results are again recorded. The need for frequent recalibration may be an indication of imminent failure of the instrument or of interference by the process being monitored.

The results of sampling and analysis need to be considered in the same way as those for the use of resources or the production of waste. The individual results need to be consolidated into a system that allows statistical analysis, the monitoring of trends and the production of reports on performance. The performance should be benchmarked against previous trends, public information from other companies or the expected performance based on the supplier's published data.

Monitoring of resources and waste production

Part of most organisations' objectives and targets include measures of use of resources such as water and energy and the reduction of waste production or the routes by which it may be disposed. The details of these are in later chapters but for the purposes of ensuring compliance with the EMS they need to be included within the monitoring processes. Water, gas and electricity are monitored by the suppliers but usually at the point of entry to a site. For a more detailed assessment of use there may be a requirement for sub-metering so that usage can be allocated to a process or an activity. Some meters may have data loggers or charts attached; others require regular reading and recording of the results manually.

Waste production and disposal are often poorly recorded. Some waste may go into bins or skips with no record as to type or quantities. Some may be weighed or measured by volume but again the information may be limited, especially if the waste is mixed. Detailed records of waste may be a requirement of local legislation such as in the UK. Improving the management of waste is covered in Chapter 6; at this stage we just need to register that the information is likely to be in the form of paper or digital records.

Bearing in mind that these measurements are affected by the amount of production or other factors such as area heated, these may need to be measured and recorded as well. The collected data, or derivatives of them such as water used per unit of production, are usually referred to as key performance indicators (KPIs). These may well be collected already for other purposes but the quality of the data or the frequency of collection may need to be changed. If electricity use is of concern then monthly meter readings taken from one meter for billing purposes are unlikely to be sufficient to get use under control and achieve a target reduction. Some examples are shown in Box 2.8.

Don't forget also that the implementation plan and routine operation need to be monitored for progress: achievement of the aims and targets, resource utilisation, timescale, cost, etc. This would be achieved by additional monitoring and building progress reports into the whole programme – reporting progress on performance targets for key outputs or steps in the plan. There will be dependencies within the plan that need to be taken account of. The investment in new plant may depend on research and development which in turn may depend on the collection of better data in the first place so a plan may extend to several years and progress monitoring is required to ensure that it stays on track or has to be adjusted.

The monitoring of resources used and waste production requires that the measures be properly recorded and the results kept in a format that can be easily accessed and interpreted. A stack of papers or charts on its own or a data base just of numbers does not meet that need. Somehow the data need to be input to a database or spread sheet that allows analysis so that the compliance with any standards can be demonstrated and trends established. This is most easily demonstrated with graphs showing performance against time and the change in slope of the graph gives a visual impression that is easy to follow. An upward drift in energy consumption, for example, is immediately apparent.

Box 2.8 Suggested performance indicators to monitor progress with the EMS

Inputs	Outputs
Gas used in KWh	Space/area heated or production of goods
Electricity used in KWh	Space/area illuminated; other uses
Petrol, diesel etc. used	Transport/vehicle km travelled
Water used m^3	Liquid effluent volumes treated/sent to sewer m^3
	Load to sewer (if charged) BOD/SS
Quantities of raw materials by type	Number/volume/weight of products by type
	Emissions to air
Packaging materials by weight/volume	Total waste produced kg/tonnes
	Waste reused kg/tonnes
	Waste recycled kg/tonnes

As a cross check it is sensible to conduct a balance calculation on some measures. If oil is stored on site, the volume delivered should match the sum of that used and that remaining in storage. If it does not it may be going astray by various means but for our purposes it is important to check that it is not leaking away unnoticed and causing pollution. (Escapes such as this are called fugitive emissions.) Similar principles apply to any chemicals used or wastes produced.

Non-compliances and enforcement action

These must be taken seriously. The non-compliance may be found as a result of the organisation's own active monitoring or reported by a regulator. The regulator may have served some form of enforcement notice (see Section 1.4) as a result or may just have drawn attention to the issue. Either way there is a risk of some form of further legal action and this event is in conflict with one of the principles of the EMS – to comply with legal requirements. The event may be a breach of a numerical standard applied to an emission or some other breach of a permit condition such as damage to a sampling point or failure to submit a report. An enforcement notice may also be as a result of an inspection by the regulator who has found something that is cause for concern. This could be storage of chemicals or waste, a leak or spill or some malpractice.

Any of these merits immediate investigation to determine the cause and to put matters right. Additional inspections and reactive monitoring of the installation and the local environment may be required, possibly extending over several days until there is confidence that the situation has returned to normal. All actions should be recorded as part of the EMS and to help avoid a similar recurrence in the future. There is also a need to keep the records to demonstrate responsible action to the regulators and to use in evidence if any case ends up in court.

Sudden changes in use of resources

The management of the utilisation of resources is an important feature of the EMS and should be regularly monitored. Utilities such as gas, water and electricity are usually metered but others depend on regular recording of weight or volume. The amount of waste for disposal should also be recorded. Sudden changes in any of these measures may be the result of a known cause such as a change in production processes or rates of production. However if there is no ready explanation an investigation is required as it may reveal a cause that requires action. Examples could include an underground leak in a pipe or tank, drift in practices by the workforce or breakdown in control instrumentation. There will be costs associated with loss of material as well as being in breach of the EMS and there is also the possibility of causing an incident or breach of permit.

Conducting investigations

The responsibility for the investigations mentioned above needs to be allocated to someone with the relevant knowledge and skills. In many cases the investigation could be quite simple and brief and fairly straightforward. In some cases involving significant events in scale or impact the investigation could be quite demanding. It could cover such things as the legislative requirements, the process being investigated, interpretation of analytical information, the availability and performance of various technologies and the environmental sciences. In addition skills are required in investigation, reporting and presentation and the interpersonal skills to relate to others involved. The investigation often falls to the supervisors or managers or the environmental manager or team if they exist within the organisation. Training will be required to support them or there may be a need to involve external investigators to bring the knowledge and skills and some independence to the process, especially in those instances with high public interest.

The conduct of an investigation requires a systematic approach and in many ways is similar to conducting an audit. Initial planning will help to make this efficient but plans need to be flexible to accommodate changes that may become apparent during the course of the investigation. It is important that investigations start as soon as possible so that evidence of what happened is still available and memories of those involved or affected are fresh. There may also be the need to identify the causes of problems and put them right to avoid any repetitions. The use of questionnaires to structure interviews and ensure that nothing is missed and the rigorous recording of information are required in order to clarify and justify any conclusions.

2.2.7 Environmental auditing

A more formal way of checking is an audit. There are two types of audit that may be required.

Auditing the environmental performance

The first type is an environmental audit of the organisation focusing on performance against standards and targets. This may have been carried out at the start of the initial review to establish a base line and find out what the main areas for improvement are. This can be repeated at any time and may be on a regular basis. The purpose is to check that the organisation is complying with standards or is on course to meet its own objectives and targets. This audit may extend the monitoring of a process to the wider environment, for example, to check the water quality in a local watercourse or that noise or dust is not causing a local nuisance. These audits would normally be by an independent person although they could be from within the organisation.

This audit may just focus on one aspect or part of an organisation or could cover a range. It could include:

- Use of resources such as raw materials and utilities;
- Waste produced;
- Emissions to air and water;
- Energy usage and efficiency;
- Transport energy use and efficiency;
- Nuisance factors such as odour and noise;
- Costs associated with any of the above.

Auditing the EMS

The audit of the EMS focuses more on the system itself and the controls in place to monitor and ensure compliance. In ISO 14001 the audit is of the system rather than environmental compliance (i.e. the EMS that is the subject of this chapter rather than whether the organisation complies with its permits or other controls) although, if relevant, both may be done at the same time. This focus is on the policy and management systems. It may be done internally or contracted to an outside body. Internal audits need to be carried out by people independent of those directly responsible for the EMS or the organisation's operations. No one should audit themselves as there is a risk of missing again those things previously missed or of not investigating as rigorously as someone independent.

There are many organisations offering a service and the independence of an external audit (at a price) should ensure that it carries weight with the senior management and stakeholders. An external auditor can also bring expertise that may be missing in the organisation such as changes to legislation or other requirements or best practice from similar organisations. The choice between internal or external audits may be influenced by several factors and Table 2.1 summarises some of the advantages and disadvantages of the two approaches. It may be appropriate to use a mix of internal and external audits. Internal audits could be used with a limited scope to just a part of the EMS or the business or carried out annually with a less frequent external audit as an independent check.

Table 2.1 Internal or external audit?

Internal audit **Advantages**	**Disadvantages**
Cheaper, although hidden cost for staff time	Too close to the business and lack independence
Internal knowledge of the organisation and the personalities involved	May inhibit approach if know people well and do not want to offend them
More likely to be aware of the internal risks	Could be unaware of some risks not previously identified
Detailed knowledge of business	May lack external knowledge of best practice
Can be flexible in timescale	Staff taken from other roles; timescale could drift due to other pressures
Opportunity to learn or share knowledge and experience	May need training in audit process and procedures
	Could be subject to internal management pressure
External audit **Advantages**	**Disadvantages**
Independent of management and internal politics	May be ignored if recommendations not liked
Detailed knowledge of audit process	May be unfamiliar with this business
Brings external knowledge and expertise from other businesses	Need to ensure get appropriate staff and not juniors or trainees
Access to a wider team of expertise	The more people the higher the cost
Should be able to deliver against an agreed timescale	Will require access to employees to fit their schedule
Carry more weight with external stakeholders	May meet opposition from employees
	Their recommendations may not constrained by cost or practicality

Whichever route is taken there is a commitment of resources in undertaking an audit. The time of employees will be required to support the audit team wherever they come from as the audit process involves interviewing various people responsible for producing the EMS and implementing it. The team will require space to interview and work on documents and may require personal protection equipment if going in site. If several sites are involved then travel expenses could be significant. The costs can be minimised by good preparation, avoiding wasted time searching for documents or waiting for or finding interviewees.

The general purpose of this audit is to check the compliance with the organisation's EMS:

- Are all the relevant parts present (policy, legal register, etc.)?
- Are these complete and up to date?
- Are the management structure, roles and responsibilities identified?
- Have employees been briefed and trained?
- Are relevant operational procedures in place, documented and followed?
- Is there evidence of compliance against standards and targets?
- Are there proper records of monitoring of environmental information?
- Are non-compliances with standards identified and acted upon?

- Are site inspections recorded and actions followed up?
- Is maintenance properly scheduled and recorded?
- Have all likely potential risks to the environment been identified?
- Are there emergency procedures in place to respond to all the possible risks?
- Are the responses to incidents and complaints adequate and followed through?
- Is document control in place and properly managed?
- Have previous audit or inspection findings been adequately dealt with?
- What are the results from management reviews and have they been followed up?

This is a long list and for a large organisation quite an undertaking. It may not be necessary or appropriate to apply all of these to the whole organisation every time. A risk-based approach may make more sense applying audit to one part of an organisation or activities of particular concern.

Scope and purpose of the audit

The aims and objectives of the audit need to be defined in advance and the scope is the first point to consider and agreed with the manager commissioning it; after all they will be committing the organisation's resources to support it. It could apply to all the activities, products and services of a whole organisation; it could be restricted to one factory or site; or one activity, product or service; or even one unit within one of those. The risks from a large-scale chemical manufacturing process may be much higher than that from the packaging end of the plant and so merit more frequent attention.

The purpose also needs to be agreed. It could be just to ensure that particular plant is following some procedures, maybe following a series of problems. It could be the full EMS to ensure that it is fit for purpose and being fully implemented.

Preparation for audit

Once the scope and purpose have been agreed a plan for the audit can be put together which identifies the tasks to be undertaken and the timetable and resources to carry them out. Auditors often work from standard check lists which they can adapt for the audit in hand or they may use proprietary software. These help them remember which documents to see and which questions to ask and may even provide a format for recording the results. An early step is to identify the documents to be reviewed and these may need to be gathered together from more than one place. Preparations need to be made for the people involved to be interviewed. Planning interviews may need to take account of people in different locations working different shifts and liable to be off sick or on holiday at critical times.

Various other issues need to be considered at the pre-audit stage:

- Accommodation for auditors and interviews;
- Administrative support such as copying and telephone and computer access;
- Communications with those directly involved and the wider staff on the purpose of the audit and what they may be required to do;
- Scheduled meetings with the management to report progress and findings
- Any training requirements in the auditing process or the processes, etc. being audited.

The audit can then begin with a systematic examination into the adequacy of the organisation's EMS and its performance against the criteria set out in it. During the audit there may be interim reports or informal discussions about progress and any problems that are found. The final output of the audit is a report which records what was carried out and identifies what is satisfactory and where improvements or changes are required. It should conclude with recommendations for any actions to bring the EMS into compliance. This is sent to the person who commissioned the audit, normally a senior manager who would have the responsibility of implementing any recommendations or coming up with reasons for not doing so.

It is also essential to ensure that the findings, or at least a summary, are communicated to other interested parties. This could include the employees performing the processes or at the sites being audited as well as those responsible for their supervision and management. They need to be complimented on good performance and any critical findings explained along with the remedial actions proposed.

2.2.8 Review of environmental performance

The accumulation of all these active and reactive monitoring data is not the end of the process. To be of value it needs to be analysed, compared with standards and reported on to managers. This represents the start of the Act step in the PDCA cycle. The activities in the 'Check' step should have led to prompt action where it was required: if not it is up to the management to take action to put this omission right.

The review reports can take several forms:

- Performance against the EMS;
- Compliance with standards;
- Achievement of objectives and targets;
- Enforcement actions;
- Incidents and complaints;
- Use of resources;
- Results of audits, inspections and emergency exercises.

To get meaningful information requires more than summary statistics. Two important measures are exception reports and trends over time.

Reporting to management

An exception report will cover anything out of the normal or expected range. So incidents, complaints, breaches of permits and enforcement actions should all be summarised and any patterns or changes in frequency highlighted. Failure to achieve standards or other objectives and targets or any other aspect of the EMS should be treated similarly. These failures should be treated seriously and managers should expect to see that actions are being taken and the situation being brought under control.

Analysing trends is just as important. Daily, weekly, monthly or annual statistics should be compared over as long a period as possible. There may be compliance with permit conditions now but if the concentration of a substance that is controlled by the permit is increasing, then measures may be required to avoid a future failure. Similarly, changes in consumption of utilities or raw materials or waste recycling patterns need investigation and comment. Changes ought to be related to changes in activity such as production; if they cannot be easily explained then they may indicate a failure that requires further investigation and action.

Analytical data from instruments or laboratories need to be subject to additional checks. There should be some information available about the quality assurance behind the results; the results of calibrations and tests or quality control procedures in the laboratory. Any deviances or other problems need highlighting and explanation along with actions taken.

There should also be a programme of audits, inspections and emergency exercises (the last have not been discussed so far; they feature in Chapter 9). The review should report on the achievement of these programmes and the outcomes, good and bad. The good outcomes should be recorded and reported not just to managers but to those who have contributed to the success. Any bad outcomes need comment on how they are being addressed and whether the situation has now been put right or if further action is required.

Management action in response to reports

As the information progresses up the management chain it tends to get consolidated and refined. Supervisors of an activity or service will be interested in their particular area of responsibility – to correct any problems and to head off any potential problems. The next tier, including any environmental management staff in the organisation, will be looking across the various departments assessing performance and compliance with the EMS and any standards that apply. The most senior managers will also be looking at the implications in terms of planning for investment: the impact of new business on performance and the need for investment in environmental management now or in the future. The chief executive or equivalent has put their name to the EMS and so will be taking a view on the performance, the public relations implications of the information and planning the future for the EMS itself. The board may have to approve significant expenditure as well as be prepared for high-profile PR issues and dealing with shareholders or other stakeholders.

The senior management and board will also be looking at the EMS in the context of the performance of the rest of the organisation. The organisation exists primarily to carry out activities or provide products and services. The EMS needs to be considered alongside performance and investment in these and the requirements of other support activities such as quality, health and safety, marketing, sales, human resources, etc. The balancing of the competing requirements and available resources using the relative priorities may be difficult, especially in times of recession or other external pressures.

The reviews of environmental performance are a regular feature of an EMS, with monthly progress reports and annual summaries being a common way of working. Everyone in the organisation has a role and this is only to be expected given the high level commitment required at the introduction of the EMS. Remember that the EMS also has a commitment to continual improvement and so any problems identified should feature as requiring improvement and be included in the management review of the EMS.

2.2.9 Management review

The management review is the name given to a requirement of an EMS and incorporates the formal review of environmental performance. It may be limited to that, but periodically the EMS itself needs a review. Some of this may be driven by the results of the reviews of environmental performance if significant issues are being found as already described. However, a review may be prompted by other factors that could affect the business. Examples are:

- Evidence of environmental damage caused by the organisation;
- The introduction of new activities, products or services;
- Takeover of a business or change of ownership or management;
- The acquisition of new premises or changes in location and layout;
- Changes in the local environment such as declaration of a protected status or a new development of houses nearby;
- Changes in legislation or permit conditions;
- The availability of new technology;
- The availability of new sources of materials or energy;
- The availability of new recycling opportunities;
- Business pressures such as competition, claims by competitors or supply chain pressure;

- The availability of new markets or customers;
- Legal action, incidents or complaints;
- As the result of an audit;
- Pressure from stakeholders such as shareholders or environmental activists;
- Publicity resulting from external events involving the organisation or its peers.

Hopefully these are not all going to happen at once or too frequently and the parts of the EMS that need review should be limited. Even without these pressures there is still a requirement for a regular review on an annual basis, or maybe less frequently for a small organisation or one in which little else changes.

This review looks back at first principles and revisits all of the stages in the initial preparation. This need not be in such detail but it needs to consider the information collected as part of the initial review as described in Section 2.2.1, reminding us again of the importance of document control. It may be quite superficial if there is confidence that little has changed. In particular a check is needed for any changes in the aspects and impacts and the register of legal and other requirements. These are the fundamentals that support the EMS and could be vulnerable to small changes that have not been noticed during the year.

The management review also needs to look at performance and particularly against the objectives and targets. The first point is have they been achieved? The objectives may be longer term but the targets ought to have something SMART to be achieved by the review period. If they have not been achieved, why not? Were they too ambitious or was there insufficient effort or did some unexpected factors get in the way? Independent critical thinking is needed here – it is too easy to opt for the 'too ambitious' approach or to find an excuse. In reviewing the objectives and targets, they may need to be considered taking account of new information such as:

- New standards arising from legislation or other requirements;
- The availability of new technology or other opportunities to improve;
- Benchmarking against best practice or competitors;
- Information gleaned from audits or other external sources.

Overall, even if none of these factors require a change in the objectives and targets, the requirement of the EMS for continual improvement will.

The review also needs to look at the policy to check that it is still relevant. The policy ought not to require frequent change but there may be new information about the company, a change in signatory or issues that have arisen in the management review that need to be considered. External pressures on topics such as waste recycling, energy use or public concerns must be taken into account as well.

The output from the management review should be documented as part of the EMS and consideration of previous reviews should be part of the next one. Were any changes last time appropriate or have they subsequently caused problems? Have all the changes been implemented? If not, why not? Are there lessons from the review process that need to be learnt?

2.2.10 Continual improvement

Continual improvement is a commitment within the EMS and requirement of ISO 14001. The principles should be applied across the business in such factors as quality and health and safety so the concept should be familiar. As far as the EMS is concerned, it is about:

- Dealing with non-conformances already identified as a priority
- Looking for opportunities to improve the environment or reduce the damage caused;
- Setting out to overachieve against legal or other requirements (possibly to create more headroom) or pre-empting known future changes;
- Changing the scope to include a wider range of sites or business activities;
- Setting more challenging objectives and targets;

A progressive organisation sets challenging objectives and targets at every review but they need to be made more challenging as time passes. The initial improvements are often relatively easy to achieve as there are usually some simple measures that can deliver impressive results. Reducing water and energy use or the amount of waste to landfill are covered in later chapters but many organisations find that up to 25 per cent savings are possible with basic housekeeping measures. Reducing the numbers of incidents or complaints or critical reports from regulators or inspections and audits are other areas for improvement and can be quantified in a target on a performance indicator. Benchmarking against other similar organisations can show which are the ones to emulate in order to be seen to be amongst the best. Better still, why not set out to be the best in the long term?

Continual improvement results in better environmental performance but it should also deliver some of the other benefits (such as financial and reputational) that were mentioned at the start of this process.

2.3 Benefits and limitations of introducing a formal EMS into the workplace

Much of this has been covered in Section 2.1 where the reasons for implementing an EMS were discussed with some further discussion in later sections. In summary the

benefits of introducing ISO 14001, BS 8555 or EMAS into an organisation are:

- **Increased compliance with legislative and other requirements;** less chance of breaking the law or not complying with contracts or other agreements.
- **It may be a condition of a permit;** so an effective EMS is essential.
- **Competitive edge over non-certified businesses;** leading to more business or retaining existing business.
- **Improved management of environmental risk;** meaning less likely to cause pollution or other environmental damage.
- **Increased credibility that comes from independent assessment;** ISO 14001 is accepted world-wide, EMAS is Europe-wide.
- **Savings from reduced non-compliance with environmental regulations;** mainly legal and court costs and fines with also potential clean-up costs or damages as well.
- **Heightened employee, shareholder and supply chain satisfaction and morale:** with an improved public image and lower costs for insurance.
- **Improved relationships with regulators;** resulting in lower inspection frequencies.
- **Meeting modern environmental ethics;** the expectations of the wider society and the requirements of CSR.
- **Increased resource productivity;** saving the unnecessary consumption of the earth's resources such as energy and materials and reduced cost of waste management. Overall there should be cost savings for the organisation.
- **Provides a framework for continual improvement of environmental performance;** which should compound the benefits above.
- **It can be integrated with other management systems;** such as quality and health and safety leading to improved business efficiency.

The **limitations** are:

- **Prescriptive environmental performance levels are not included within the standard;** although they will come from permits and other legal instruments.
- **Improvements in environmental performance can be negligible;** particularly for an organisation with few significant aspects and potential impacts or if it is already well managed and operated.
- **Inconsistency of external auditors;** this can be due to differing levels of knowledge and experience of the various activities or industries: they need to be selected with care.
- **Implementing an EMS may have costs that are too high for smaller enterprises;** although this can be partially addressed by managing the scope and adopting the BS 8555/ACORN approach referred to in Section 2.1.

2.4 Key members of the ISO 14000 family of standards and their purpose

This chapter has been mainly about ISO 14001 as the model for the introduction of an EMS. It specifies the requirements with guidance for use. This standard is part of a family of standards within the ISO 14000 family. Those that are directly relevant to implementing an EMS are:

- ISO 14001: 2015 *Environmental management systems – Requirements with guidance for use*.
- ISO 14004: 2016 *Environmental management systems – General guidelines on implementation*.
- ISO 14005: 2019 *Environmental management systems – Guidelines for the phased implementation of an environmental management system.* Described over six phases.
- ISO 14006: 2011 *Environmental management systems – Guidelines for incorporating eco-design*. This is under review.
- ISO 19011: 2011 *Guidelines for auditing management systems.* This applies to quality management systems as well as an EMS.
- ISO 14031: 2000 *Environmental performance evaluation. Guidelines on the selection and use of environmental performance indicators*.

There are other standards which complement those concerned with an EMS and are relevant to other topics covered in this book. They are aimed at professional environmental staff but it is useful to know the types of standards available if you need to get involved in a topic in depth. The main ones are:

- ISO 14020 series on environmental labels and declarations. This is concerned with the eco labels commonly found on items to promote their energy efficiency or other environmental claims.
- ISO 14040 series on life cycle assessment (LCA). LCA will feature in Chapter 3.
- ISO 14063: 2006 on environmental communication.
- ISO 14064: 2006 on greenhouse gas accounting and verification and ISO 14065: 2007 on accrediting requirements for greenhouse gas accounting.
- ISO 14067 *Greenhouse gases – carbon footprint of products – requirements and guidelines for quantification and communication*.
- ISO 50001: 2011 Energy management systems takes a similar approach to ISO 14001 but applies it just to energy management. (This replaces BS 16001). Energy use is covered in Chapter 7.

The full list and status of standards can be found on the ISO website at www.ISO.org. Be aware that the prices of the standards are relatively high, typically 138 Swiss Francs each (£115, prices in August 2019).

CHAPTER 3

Environmental impact assessments

After this chapter you should be able to:

1. Explain the reasons for carrying out environmental impact assessments
2. Describe the types of environmental impact
3. Identify the nature and key sources of environmental information
4. Explain the principles and practice of impact assessment.

Chapter Contents

Introduction **48**
3.1 Reasons for carrying out environmental impact assessments **48**
3.2 Types of environmental impact **53**
3.3 Nature and key sources of environmental information **56**
3.4 Principle and practice of environmental assessments **60**

INTRODUCTION

The term 'environmental impact' was mentioned in the two previous chapters. In Chapter 1 there was an introduction to some of the main environmental issues such as climate change and sourcing raw materials. There the discussion was of impact on the environment in broad general terms. In Chapter 2 the term was used in the context of developing an environmental management system (EMS) and we saw how Environmental Impact Assessment (EIA) was used in ISO 14001. The reason for introducing an EMS into an organisation was to manage and reduce its environmental impact. This chapter expands on the use of EIA.

The International Association for Impact Assessment (IAIA) states that an Impact Assessment is simply defined 'as the process of identifying the future consequences of a current or proposed action'. The 'impact' is the 'difference between what would happen with the action and what would happen without it' (IAIA 2009). This definition applies to any type of impact assessment and covers more than the environment: it could be social or financial impacts for example (such as the impact of changing the tax regime). The Association goes on to describe impact assessment as 'a technical tool for the analysis of the consequences of a planned intervention (policy, plan, program, project)' or for unplanned events (such as an accident) and as 'a legal and institutional procedure linked to the decision-making process of a planned intervention'. It can also contribute to public understanding and participation and identify monitoring or mitigation measures that may be required. Because it can cover the economic and social impacts as well as the environmental, it is seen as a necessary step to promote sustainable development – as described in Chapter 1. As applied to the environment, the definition of EIA is 'the process of identifying, predicting, evaluating and mitigating the biophysical, social, and other relevant effects of development proposals prior to major decisions being taken and commitments made' (International Association for Impact Assessment 2009). Different jurisdictions have different requirements as to the way any type of impact assessment is applied and this needs to be known for any particular application. In the rest of this chapter we will look at the use of EIA in the broadest sense.

An EIA, therefore, is a systematic assessment of the possible impact of a proposal on the environment and the purpose is to decide whether to proceed at all or which of possible alternatives to choose. The impact can be positive (such as a new wildlife reserve) or negative (such as a heavy industrial development). Where impact assessment is applied to policies, legislation and strategic plans and programmes, it is known as Strategic Environmental Assessment (SEA) and is applied to international activities by the World Bank, for example, as much as regional or national strategies.

3.1 Reasons for carrying out environmental impact assessments

The use of EIA in preparing an EMS was introduced in Section 2.2 along with the meanings of 'aspects' and 'impacts' as defined in ISO 14001. There are other reasons apart from developing an EMS for carrying out an EIA. There may be a legal requirement as part of an application for a permit to make a discharge or for permission to build a development that could cause a change to the environment. It will not be required in every case; the need will depend on the significance of the impact (this is explained in Section 3.4). It could also be used by an organisation in assessing the impact of its aspects with the aim of exploring alternative ways of producing its activities, products and services in order to select the one with the lowest potential environmental impact. In this case the purpose of EIA is to look at how different ways of ways of working or development affect the impact with the intention of adopting the most beneficial, although at some stage costs will also have to be taken into account. The most beneficial path may be extortionately expensive. This is developed further in Section 3.1.3.

3.1.1 Legal requirements

Many countries have a legal requirement for EIA. Within the EU there is a legal requirement in accordance with the EIA Directive. This first directive came into force in 1985 and there have been amendments since (European Commission 2008b). The latest consolidated version is Directive 2014/52/EU on the assessment of the effects of certain public and private projects on the environment (European Commission 2014a) and it lists the projects for which an EIA is required in an Annexe. The list is long and the requirement for an EIA is dependent on size or capacity reflecting the likely significance. A simplified list is given in Box 3.1 in Appendix 4 but the scope includes most major infra-structure developments and potentially polluting industries. The EIA Directive also gives member states flexibility in requiring an EIA for other projects ranging from agriculture to tourism on a case by case basis or according to local legislation and potential impact and these are in a separate Annexe. It is notable that the Directive itself was subject to an impact assessment.

The principles behind EIA are also applied under other EU directives, for example, the IPPC Directive (European Commission 2010) which is concerned with reducing emissions and the Seveso III Directive (European Commission 2012) which is concerned with managing the risks from accidents involving dangerous substances. These will be referred to in later chapters. A broader European Convention on EIA in a trans-boundary context (United Nations Economic Commission for Europe 1992) and the Rio Declaration on Environment and Development (United Nations 1992) are other examples of international agreements with application outside of the EU that refer to or require EIA when considering significant economic developments. Within the EU the Strategic Environmental Assessment is covered by a Directive known as the SEA Directive although the title is 'on the assessment of the effects of certain plans and programmes on the environment' (European Commission 2001a).

The requirements given above are mostly from the EU but other states have similar legislation based on the Rio Convention or national requirements. The UK has implemented the directives in local legislation in the examples given in Box 3.2 in Appendix 4.

Voluntary assessments

In some circumstances voluntary assessments may be undertaken even if there is no legal requirement. This may be for internal purposes such as preparation for the introduction of an EMS but it can also be done for public purposes. In effect it is applying EIA to justify a proposal or the means of implementing it to a wider audience but on a voluntary basis. This could be to gain initial support for a proposal or to seek views or ideas on alternative ways of achieving something. A simple example would be justifying the need for a new tennis court in a public space and the choice of alternative playing surfaces where the appraisal may also bring in issues other than the impact on the environment. There is more on the voluntary assessments in the next section.

3.1.2 Aims and objectives of impact assessment

In the introduction to this section the IAIA was quoted as defining EIA as 'identifying the consequences of current or proposed action'. That is not the main purpose of conducting an EIA; the results should be used to influence a decision. The main thrust of this chapter so far has been about seeking permission to undertake investment in a new development (planning permission in the UK) or to gain a permit for a plant to operate or make emissions. In either case the results of the EIA might cause the regulatory body to refuse the application or to put conditions on to mitigate or minimise the impact. The EIA also enables a comparison of alternative routes to the same development and could be used in deciding the outcome.

The other main purpose is to enable an organisation to make a rational choice among several ways of achieving the same outcome or for customers to make a choice about which product or service to buy. In this case it is more usual to use life cycle analysis (LCA) which is covered in the next section. Eco-labelling is the use of labels on consumer products with claims about their environmental credentials such as energy efficiency. This also relies on EIA or LCA to justify the environmental claims for products or their suitability to be included in a labelling scheme.

EIA is used in establishing an EMS and then in its continuing operation. Identifying the impacts enables the initial objectives and targets to be set for the aspects so that both can be brought under better control and the principle of continual improvement applied to reduce the impacts over time.

A feature of EIA (required if it is part of a statutory procedure) is to invite participation by other interested parties, including the public, so that they can participate in the decision making. An Environmental Impact Statement is produced by the applicant and is made available for inspection and comment. In England, for example, there are statutory consultees such as local authorities, the Environment Agency, Natural England and other similar public bodies which would see all applications. Other representative bodies may be consulted depending on the nature of the project. These could include wildlife trusts, organisations representing specialist interests such as fishing or local community associations. In addition the availability of the statement would be publicised in the local media with copies to be inspected or taken away by anyone interested. Anyone can generally make a representation to the decision-making body, drawing attention to anything that they think is wrong or has not been included in the assessment when it ought to have been. This can result in many comments for contentious projects.

In summary, the aims of EIA can be about making choices. If the assessment is confined to the environmental impacts, it is to choose the route with the best (or least damaging) environmental outcome. Most projects or plans requiring an EIA also have some social or economic purpose, or both, and extending the assessment to include economic and social impacts will facilitate a route to sustainable development, although this could be a very complex process.

3.1.3 Life cycle analysis

Life cycle analysis (LCA) is also known as life cycle assessment and cradle-to-grave analysis (see also reference to the circular economy in Chapter 1). It has

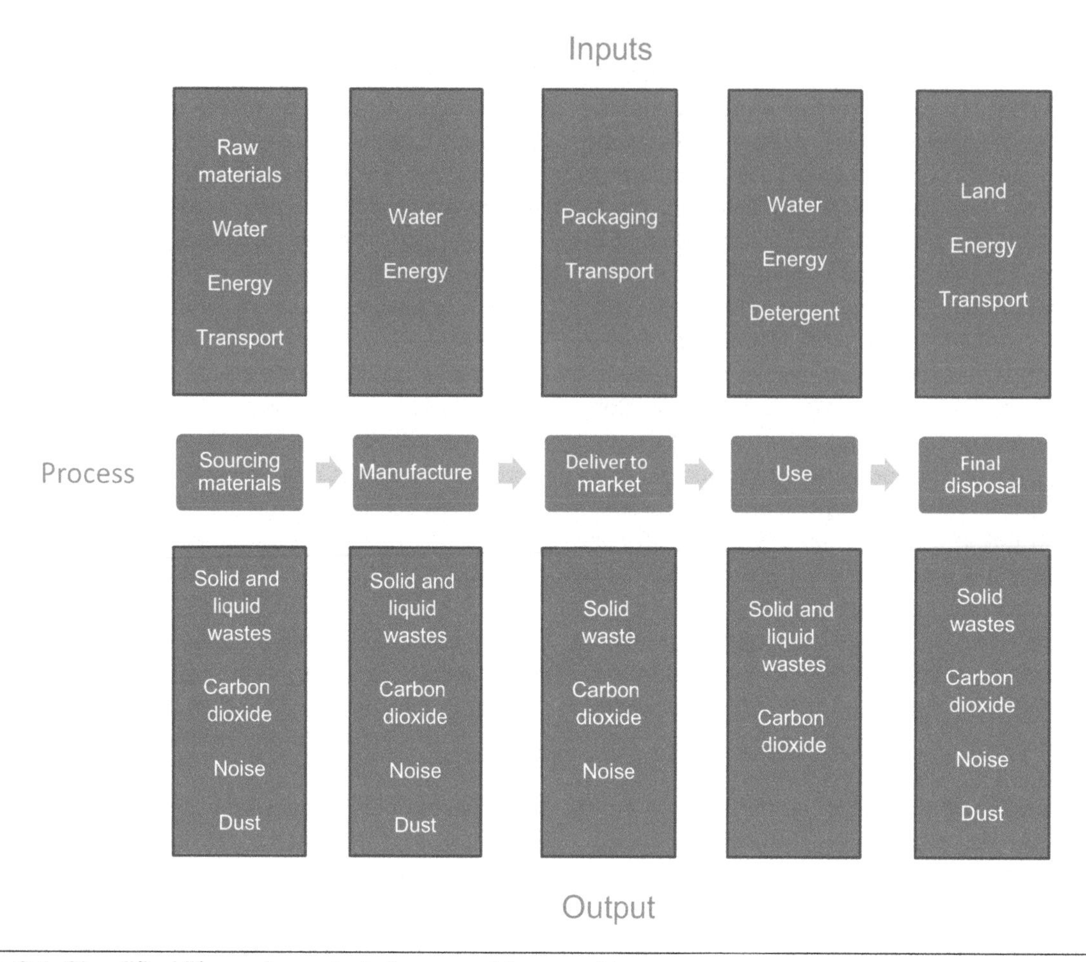

Figure 3.1 Simplified life cycle analysis for a washing machine

already been briefly mentioned in the previous section as a tool for the comparison of different routes to the same end product and in support of claims made by manufacturers.

The technique analyses steps in the life cycle of a product or service from the sourcing of the inputs to the final disposal at the end of life. The steps can include all the inputs to and outputs from the complete life cycle or could focus on just a part of the cycle; for example, the manufacture of a product or the alternative waste disposal routes at end of life. LCA tries to get an overall measure of the environmental impact of the cycle or part of the cycle analysed. The application of LCA to the whole life leads to the cradle-to-grave concept.

Figure 3.1 gives a simple overview of the approach for the life cycle of a washing machine. It starts with extracting the raw materials to produce a manufactured product, its use and then disposal at the end of its life. In reality there are many other steps involved such as conversion of extracted materials into intermediates for manufacture (sheet metal or moulded plastic shapes) which themselves could involve several additional steps. It also raises one of the problems of LCA: where to draw the boundaries or the scope of the analysis. In this example the approach is to start with the extraction of raw materials, but should we go further back? What about the inputs and outputs to produce the plant and machinery used in the quarrying process or the inputs and outputs to create the energy sources used throughout the cycle? Some of the inputs and outputs can be classed as direct – the raw materials and energy inputs and the CO_2 emissions to air – or as indirect inputs and outputs – the coal extracted to provide the energy or the climate change resulting from its combustion.

Inventory analysis

The further the analytical cycle is extended the more complex the data collection and analysis becomes such that it could become very difficult to manage. Inputs such as energy from electricity are common to several stages in a cycle and there may be common outputs as well such as waste or emissions to air. However, overall the full inventory of inputs and outputs could be enormous for a large or complex product and going through each one to assess its potential environmental impact will be a long process involving suppliers and other sources of information. To be of any value, the data collected need to be quality assured for accuracy and relevance to the LCA in hand.

As another example, if we want to compare the alternatives of using ores of differing quality in terms of their metal content, we will need to be able to compare the relative volumes to be extracted, the relative volumes of waste produced, the relative areas of land taken and the relative volumes of energy used to extract the ores and convert them to useful metal. In theory if the ore has half the content of metal, the other inputs and outputs should be doubled, but in practice we may have to go deeper into the ground or further afield to find the better quality ore. This will possibly require more land and energy and produce different waste and spoil ratios to the theoretical. This example is shown in Table 3.1 with a simple inventory of inputs and how LCA can be used to compare two routes to the same end product containing equivalent amounts of metal. In this case the total energy used for the two routes can be compared and the decision is likely to be influenced by the trade-off between the energy used to transport the higher content ore against the energy used to process the lower content ore.

The use of equivalences

The inputs and outputs will include measures of weight and volume and, in some cases, a measure of quality. Just considering the simple inputs, we may have weights in tonnes of ore and energy as litres of oil, m^3 of gas, kWh of electricity or tonnes of coal. Each of the energy sources could be converted to a common measure of energy such as kWh but each will have different inputs and outputs associated with its extraction and refinement. Wherever possible it helps to get inputs and outputs into equivalent units.

Table 3.1 Example of extraction of ore to produce one tonne of metal

	Using 1 per cent ore	Using 2 per cent ore
Weight of ore to be extracted	100 Tonnes	50 Tonnes
Energy used to extract ore[a]	X Joules	X/2 Joules
Energy to transport ore to refinery[b]	Y Joules	10Y Joules
Waste spoil produced[c]	99 Tonnes	49 Tonnes
Energy used to transport waste[c]	99W Joules	49W Joules
Energy used to refine ore[d]	Z Joules	Z/2 Joules

The land area taken to extract 1 per cent ore is likely to be about twice that for 2 per cent ore.

[a] The actual energy used will depend on the types of process and fuel used to extract the ore from the ground.
[b] This assumes that it is ten times further to transport the higher quality ore.
[c] This is a simplification that assumes that the weight of waste is the weight of ore less the weight of metal extracted. In practice the chemical composition and water content are likely to be different and so the true weights will be different but the relative amounts will be about as shown as will the energy required to process and transport the waste.
[d] The actual energy used will depend on and the processes to concentrate and refine the metal.

Carbon footprint and embodied carbon, energy and water

LCA has spawned two concepts that are becoming more widely used. The carbon footprint is a measure of the carbon dioxide that is emitted by a process or service, mainly arising from the use of energy derived from fossil fuels. Other emissions of greenhouse gasses may be converted to the CO_2 equivalent by applying factors that allow for their different greenhouse potential. The carbon footprint may be limited to the part of a process that uses energy or could extend, as with LCA, to the whole life cycle bringing in the carbon footprints of all the inputs and outputs.

A similar measure is the embodied carbon or the CO_2 equivalent of the energy used to make a product. If the measure is left in units of energy (Joules or kWh) then it is known as embodied energy and can include all the energy used from the extraction of the raw materials through to the finished product. A similar term 'embodied water' or 'virtual water' referred to in Chapter 1 is also used for the total input of water into a product or service.

Further uses of LCA

Comparing different ways to get to the same end product or service can feature as a method of achieving continuous improvement as part of an EMS. Progressive changes to inputs and processes should help to reduce the undesirable outputs such as emissions and waste; LCA can help to find the optimum solution.

Another use of LCA is to compare the impacts from two different products that have the same purpose. This is becoming increasingly of value as consumers compare different products with the intention of choosing the most environmentally friendly. Commonly quoted comparisons for consumer products include:

- Soft drinks in metal cans, plastic or glass bottles.
- Water or solvent based paints.
- Washable or disposable nappies for babies.
- Hot or cold water fill for a washing machine.
- Plastic, wood or aluminium window frames.
- Natural or synthetic fibres for clothes.
- Diesel or petrol for fuelling a vehicle.

Businesses have to make similar choices:

- Which energy source to use from a variety of fossil fuel and renewable energy sources.
- Choice of fabrication materials such as plastic or aluminium.
- Materials of construction for a vehicle.
- Washable or disposable cups for drinks in an office.
- Copy paper from new pulp or recycled material.

Some of these comparisons will be revisited in later chapters.

Most of these choices are not as simple as an LCA of the inputs and outputs. Other factors such as cost, acceptability to the user and availability may dominate the decision. Box 3.3 looks in more detail at a couple of the above examples.

Box 3.3 Using life cycle analysis to compare products

Containers for soft drinks

There are many reports of the various materials used for packaging drinks. These can be found by a search on the internet. A full life cycle analysis includes the caps, labels and printing as well as the containers. The reports are often from manufacturers or trade associations which have a vested interest and the results are dependent on the assumptions made about reuse or recycling rates. The summary below picks out the main points that affect the analysis.

Aluminium cans

- High energy requirement to manufacture from bauxite ore.
- Used cans easily recycled.
- Moderate in weight so transport energy requirements are low.

Iron cans

- High energy requirement to manufacture from ore.
- Needs corrosion protection with tin and lacquer coatings which use more materials and energy.
- Used cans easily recycled.
- Slightly heavier weight so use more energy to transport than aluminium.

Glass bottles

- High energy requirement to manufacture from raw materials.
- Containers can be reused (not often) or glass recycled.
- Heavier than aluminium or iron so uses most energy for transport.

Plastic bottles

- Made from fossil fuel (oil).
- Energy requirement lowest of the four.
- Plastics can be recycled (although often are not in practice).
- Lightest weight so energy for transport is lowest.

Putting numbers to the materials and energy requirements will enable a detailed comparison to be made. If energy is the main concern, plastic bottles are likely to be the most favoured. However, other factors such as robustness of the container, costs and availability will affect commercial decisions.

Washable or disposable cups

Offices and coffee shops may use china cups or disposable plastic ones (usually made from polystyrene although paper is also used) for drinks. The main points are summarised below.

Washable cups

- High energy use to manufacture cup.
- Hot water and detergent to wash on frequent basis.
- Emissions to waste water plant from washing.
- Detergents may be made from fossil fuel (oil).
- Cup has long life so infrequent replacement.
- End of life material is not recyclable; as it is inert it can go to landfill.

Plastic cups

- Made from fossil fuel (oil).
- Lower energy use per cup than china to manufacture.
- Cup not reused and polystyrene not recycled so goes to landfill.
- Decay in land fill produces methane – a greenhouse gas.

Paper cups

- Paper can be made from renewable resources.
- Lower energy use per cup than china to manufacture.
- Paper cup not reused but paper can be recycled.
- Energy use for recycling.

Most studies show that paper or plastic cups have lower impact overall but it depends on the assumptions made, especially for the life of the china cups. Convenience probably determines most decisions!

Just sketching out the main aspects associated with simple products or services illustrates the complexity of LCA unless it is limited in scope. A car has several hundred components; an aircraft has several thousands. These are made from many materials including composites made from carbon or glass fibre reinforced resins or plastics and metals such as titanium which are produced by energy-intensive processes. Analysing the raw material and energy inputs and the waste emissions is demanding.

ISO has produced standards that apply to the application of LCA in the 14040 series. The main ones are concerned with the principles and framework (ISO 2006a) and requirements and guidelines (ISO 2006b). There are amendments to the original standards and other standards available related to specific products that may help those who get involved in LCA in more detail.

3.2 Types of environmental impact

The types of environmental impact were mentioned in Chapter 2 as part of the assessment required for implementing ISO 14001. This section adds more detail to the variety of impacts that can arise.

It is important when starting to make a list of impacts to consider their extent. Impacts can be very local, effectively confined to the site; low level noise would be an example of this. However, releases to the atmosphere can spread to the neighbourhood (e.g. smells) or have a global impact (e.g. CO_2 causing global warming). The need to consider the extent will depend to some extent on their significance which is dealt with later in this chapter.

The other factor is the consideration of which stages of a product or project to include. For a product, this can start at the design and continue through manufacture, use and disposal. For a building project it could start with design again but continue through construction, operation and decommissioning. The extent to which all of these are taken into account will influence the extent of the impacts to be included.

3.2.1 Direct and indirect impacts

In Chapter 2, in preparing the list of aspects for the EMS they were categorised as direct or indirect: direct aspects are those attributed to the organisation's activities, products and services. Indirect aspects are those not under the control of the organisation, for example, those of a supplier. Similarly, direct impacts are those that can be easily attributed to any aspect whereas indirect impacts are one or more steps removed. An example of a release of a toxic chemical onto land better illustrates the distinction. The direct impacts are contamination of the land and could include the death of flora and fauna in the immediate vicinity. The indirect impacts may take longer to become apparent but could include percolation into groundwater causing pollution of water supplies or into surface water where further toxic effects kill fish and other aquatic organisms. There could also be spread onto adjacent land rendering it unsuitable for growing crops.

Remember also that impacts can be beneficial as well as adverse. People tend to focus on the adverse as they take these more seriously and the impacts may be more apparent. In the following sections we will look at some of the impacts in more detail and search for beneficial ones. The lists presented are not comprehensive; previous sections have highlighted the difficulties of defining the scope and identifying all of the impacts.

3.2.2 Impacts from raw material extraction

This has been mentioned already in previous sections and chapters. It is concerned with the impact from the extraction of raw materials such as ores and timber and including energy sources such as coal and oil. The impacts arise from the disturbance of the ground to extract and the accumulation of spoil and other waste products. There are also emissions from vehicles and machinery, much of it very large. Many of the impacts overlap with those under other headings below. In summary the direct impacts could include:

- Damage to land from excavation.
- Damage to land or water by spills of materials and leaks of oil.
- Damage to habitat by heavy machinery.
- Damage to other adjacent sites such as archaeological sites.
- Loss of amenity for local population.
- Pollution of the air by emissions.
- Visual impact and noise.

- Damage to fragile ecosystems such as the arctic from exploration for oil.
- Creation of local jobs.

The indirect impacts could include:

- There may be an opportunity to reclaim previously contaminated land.
- Pollution of water sources and land by leaching from spoil heaps.
- Damage to wildlife by loss of habitat or by the effects of spreading pollution.
- Harm to humans, animals and plants from air pollution.
- Loss of ecosystems or reductions in biodiversity due to loss of species.
- Pollution of the environment by spills during transport.
- The local economy gets a boost from supplying labour and supplies.

3.2.3 Atmospheric impacts

Several references have already been made to atmospheric impacts. These commonly arise from emissions from processes either through a discharge point such as a chimney or through leaks from pipework and plant. There may also be some general emissions that come from open containers or spills; these are known as fugitive emissions. Emissions also occur from underground sources such as methane from landfill sites and coal mines. Landfill sites can also emit other gases such as ammonia and hydrogen sulphide. Volatile organic compounds (VOCs) can be released from the use of solvents and chlorofluorohydrocarbons (CFCs) from the use of refrigerants. Contamination of the atmosphere can also be caused by releases from volcanoes and other natural sources. Chapter 4 contains more about atmospheric pollutants and their control.

The impacts can take several forms, the commonest direct ones being:

- Human toxicity through inhalation of vapours and dusts or fibres causing acute symptoms such as sickness or death or chronic symptoms such as asthma.
- Toxicity to fauna by inhalation as for humans.
- Toxicity to flora by entering plant tissues during plant respiration or by dust fallout.
- Smoke containing soot, ash and other small particles from the combustion of fossil fuels, mainly coal.
- Odours or smells causing nuisance.
- Explosion risk from methane or other flammable gases.

Indirect impacts would include:

- Erosion of buildings by acid rain caused by emissions of sulphur dioxide or other acidic gases.
- Damage or staining to property caused by fallout of solid particles.
- Smog from photochemical action on a combination of small particles, ozone, nitrogen oxides, and hydrocarbons emitted from the combustion of fossil fuels.
- Global warming caused by greenhouse gases.
- Damage to the ozone layer by CFCs.

Beneficial impacts are unlikely in the atmosphere.

3.2.4 Aquatic impacts

Discharges of effluent from processes and sewage installations and spills and run off from industrial sites and roadways are common sources of water pollution. There is more on the control of contamination of water sources in Chapter 5.

The main types of direct impacts, which would apply to surface waters, are:

- Loss of dissolved oxygen due to biodegradation of organic matter.
- High temperature due to discharges of cooling water.
- Oil film from spills and leaks causing visual offence and fouling of river banks and wildlife.
- Toxicity to fish and other aquatic organisms from ammonia, metals, and other toxic substances.

Indirect impacts, which could apply to surface water (also groundwater in the first two), include:

- Contamination of water sources rendering them unsuitable for public water supply.
- Contamination of water sources rendering them unsuitable for abstraction for agriculture or industry.
- Odours due to anaerobic (low oxygen) conditions.
- Fish and other aquatic organisms are impacted by low oxygen concentrations; if they are too low they will die. High temperatures also affect fish if they rise more than a few degrees above the normal range to which they are adapted. They may not breed or can also die.

Beneficial impacts are unlikely in water unless a project will improve water quality from an existing poor state.

3.2.5 Land impacts

Land impacts can arise from construction projects such as a reservoir, new houses, a road, airport or railway or an industrial development. The main causes of contamination of land are as a result of residues from mining and quarrying, tipping of solid and liquid wastes onto land and from the fallout and disruption from industrial processes. There is more on waste and land use in Chapter 6.

The main types of direct impact are:

- Loss of agricultural land, forest, etc.
- Loss of visual amenity or historic sites and buildings.

- Toxicity to humans and other animals by skin contact with the contamination causing death, injury or sensitisation.
- Toxicity to humans and other animals by transferring solid material to the mouth.
- Toxicity to humans and other animals by inhalation of fine particles or released gases.
- Toxicity to plants by transfer of contaminants into the plant tissue through the root system.
- Loss of amenity or land value for further development.
- Radiation risk from deposits of radioactive material.
- Risk of fire or explosion from volatile or flammable material.

Indirect impacts would include:

- Leaching of soluble contaminants into groundwater or surface water causing further impacts.
- Land contamination further afield due leaching or to windblown dust or waste.
- Damage to property caused by fallout of solid particles.
- Toxicity to humans and other animals by ingestion of contaminated plants or crops.
- Land may be eventually brought back into beneficial use especially if it was cleaned up after being previously contaminated.

3.2.6 Community effects

Some reference has already been made to effects on the community in the previous four sections. Some are positive whilst others are negative and the distinction between direct and indirect impacts may depend on the circumstances. Attention tends to focus on the adverse direct impacts which could include:

- Visual impact from derelict land and property, waste deposits or new installations (such as wind turbines in some areas).
- Loss of access and amenity due to controlled access or contaminated land.
- Loss of access or amenity due to land taken for new development.
- Noise, odours and other nuisances.
- Loss of land rights due to deforestation or development of virgin land for agriculture, mining or industry.

Indirect adverse impacts could include;

- Loss of sources of food or timber due to deforestation or land taken for other uses.
- Damage to property from emissions to air.
- Structural damage to local roads and congestion due to transport;
- Loss of land or property values as an indirect impact of some of the above.

There may be beneficial impacts though these tend to get overlooked or may just benefit a section of the community. Examples of direct benefits are:

- New employment or higher skilled employment leading to economic benefits to the wider community.
- Economic benefits from local sourcing of materials and services.
- New amenities for a few (such as a golf course) or many (such as a public park or a new forest) from the actual development.

Indirect benefits could include:

- New amenities (such as a school, hospital or community centre) provided by a developer as an addition to a larger productive development.
- Other benefits such as new access roads and new or improved sewage treatment built by a developer.

Some developments can have both positive and negative impacts. A new reservoir will result in loss of land used for agriculture or some other purpose but a new amenity may be created which can offer sailing or fishing as well as supplying water for irrigation or consumption.

3.2.7 Effects on the ecosystem

The ecosystem (explained in Box 2.2) is affected in similar ways to communities and is vulnerable to both direct and indirect impacts. Many are indirect because they are the result of another direct impact such as air or water pollution. There are potential positive and negative impacts and some have previously been mentioned above as indirect impacts. Taking the negative ones first, some examples are:

- Loss of habitat due to land development, deforestation or intensive farming.
- Loss of biodiversity (see Box 3.4).
- Toxic effects killing flora and fauna resulting in loss of food sources for other fauna.
- Loss of pollinating insects due to use of insecticides or presence of toxins.
- Toxic compounds building up in the food chain (bioaccumulation – see Box 3.4).
- Disturbance, especially of breeding birds, due to light, noise or activity.
- Potential for injury, especially to birds due to high structures or wind turbines.
- Loss of habitat due to melting glaciers and ice caps caused by global warming due to greenhouse gases (clearly an indirect impact).

Positive impacts can arise although they tend to be forgotten or are less obvious. Typical examples are:

- New habitat deliberately created such as a new nature reserve.
- New habitat as an additional feature of a development such as planting hedges rather than building a fence.

Box 3.4 Biodiversity, bioaccumulation and food chains

Biodiversity is the term used to describe an ecosystem (see Box 2.2) in which there is a wide range of species present. In natural ecosystems there are many bacteria, fungi, plants, insects and animals that often depend on each other for survival. Plants provide cover and food for insects, birds and animals; insects provide pollination of plants and food for birds; birds provide food for other birds and animals; the bacteria and fungi break down organic matter such as dead plants and animals in the soil. Different habitats (e.g. forest and beach) will exhibit different characteristics which result in the range of species present varying but they can be described as having good biodiversity if a good selection from those expected to be present can be found. In contrast, large-scale agriculture or commercial forestry can result in only one species of plant, the crop, which does not support many other species around it. Indeed competing species may be killed off with herbicides or pesticides. In that case the biodiversity would be described as poor and it can result from other causes such as desertification or deforestation. Good biodiversity is usually a sign of a healthy environment.

When one species feeds on another the constituents of the food source are used to provide energy and build new tissue. If a food source is contaminated by a toxin the concentration may not be high enough to cause its death but the one consuming it will also ingest the toxin. If it remains in that species (i.e. it is not excreted or is only slowly) then the concentration can build up as it consumes more food. This is known as bioaccumulation. There is such a dependency among many species in a diverse community. As an example, worms may ingest soil contaminated with mercury. Birds feed on the worms. Animals or other birds feed on them. This is known as a food chain and the mercury moves up the food chain such that the species at the top of the food chain accumulates the highest concentration and may eventually die or suffer other consequences. Birds of prey are very vulnerable to bioaccumulation and the widespread use of some pesticides in the past was blamed for extensive loss of breeding capability and mortality to some species.

- Creation of wild flower meadows through less cutting of grass on roadside verges.
- Reducing impacts on air, water or land and the indirect impacts on ecosystems through better management of the aspects (for example, by introducing new process technology or waste recycling).

3.2.8 Impacts on archaeology and historical structures

Some examples of these have been included above and it is not intended to go into the same level of detail. The impacts are usually detrimental due to damage to sites by excavation or damage to structures by acid rain or other atmospheric fallout. The opportunity to excavate a site scientifically (often a condition of planning approval) could be considered as a beneficial impact although the site is generally then lost forever.

3.2.9 Transport impacts

Again these have been mentioned above. The main impacts are adverse: structural damage to roads, congestion from increased traffic and noise and air pollution from vehicle exhausts. The extent of these depends on the type of transport. A new rail development could have beneficial impacts in reducing car journeys and total emissions to atmosphere but at the cost of visual and other community effects.

Transport developments are an example of the difficulties of impact assessment. The furores created by the proposal to build a new high speed rail link from London to the North West of England and proposals for an additional runway at Heathrow Airport or a completely new airport in the Thames estuary are examples of the difficulty of assessing all of the impacts. They would be large projects covering a big area and involving impacts to all media. The technical issues are difficult enough but the community impacts present a real challenge when emotions run high.

3.3 Nature and key sources of environmental information

Gathering the information to assess impacts may seem a daunting task but there are many sources that could be sought out or consulted. Some of this information will be held by the organisation, especially if it is well established and has been operating to good practice. There are many external sources of information as well and they may be all that are available if contemplating a new development, activity, product or service on a virgin site. It needs to be borne in mind that the worst impacts will occur under abnormal conditions, such as an accidental spill or plant failure, and information on the potential impacts may be less readily available.

3.3.1 Sources of information internal to the organisation

Impact assessment often starts with a look at the aspects so that only the relevant impacts are considered. For example, there is unlikely to be an impact on water if water is not used on site, no discharges are made or no chemicals or oil are stored. The main sources of internal information will be data collected on a routine basis and the results of inspections and audits. These should all be features from the introduction of an EMS or its ongoing implementation.

Raw material usage and supply

This could include materials extracted by the organisation or supplied to them. The raw materials for a process could be timber, ores, metals, plastics, etc. and the utilities used to process them: water, gas, electricity and other fuels. A farm will use seed, animal feed, fertiliser and other chemicals. Box 2.8 gives some examples of indicators to be collected. The organisation should have knowledge of the quantities used and their sources from which an impact assessment can be made of their initial extraction, subsequent processing or production and distribution as appropriate. Some of this information may not be internal and has to be obtained from suppliers as described below.

Waste production

There should also be good information on waste, including hazardous waste: the amounts produced, the amounts recycled and how much went off site to landfill or another route. This information should be available from the records of the organisation and from the information supplied to the carriers (see Chapter 6 for details of transfer and consignment notes). The producer of the waste also has obligations to monitor the ultimate disposal sites and to monitor the amounts of packaging used (again more information is on Chapter 6).

Use of environmental monitoring data to evaluate risk

If the organisation is producing emissions then these ought to be monitored on a regular basis. There may also be monitoring of the local environment to check on the impacts and for noise or other pollution. Looking at the trends of these data will show how close the organisation is to causing a damaging environmental impact either by a gradual deterioration or due to spot increases in emissions. The emissions may be well within any concentration of concern or any limits within a permit most of the time under normal conditions. However, there is the possibility of abnormal conditions resulting from changes in the business or potential incidents that need to be taken into account in assessing the overall risk to the environment.

Audit and investigation reports

Audits and investigations are snapshots in time either pre-programmed or in response to an event. They form part of an EMS as described in Section 2.2.6. Reports will be produced in the event of an audit which may have highlighted issues that could have an impact on the environment. Failures to comply with the EMS or other policy and operational procedures or with legislation and permit conditions are all relevant. Anything that has an environmental impact or potential should have been reported on with recommendations for action to put things right. Similarly, investigations into unusual occurrences as described in Section 2.2.8 may highlight similar issues. An incident such as a serious spill or an enforcement notice from a regulator should trigger an investigation and a report that may also highlight potential environmental risks. Again there should be recommendations to correct any shortcomings.

Maintenance records

Routine maintenance of plant and machinery should be part of good practice to ensure that they are working properly and that faults are found quickly and put right. The routine tasks may include servicing, cleaning, inspection, lubrication and adjustment of machinery; cleaning and calibration of monitoring probes and instrumentation; or the emptying and cleaning of storage tanks or oil interceptors. Non-routine maintenance may be required in the event of breakdown or failure. The records should record when the maintenance took place, by whom, what was found and any remedial actions taken. If all is well there should be little to report but the need for increased maintenance due to breakdowns or calibration drift on monitors and similar departures from normal operation indicate that there is something more fundamental wrong and that the situation could deteriorate and present an increased risk to the environment. An impact assessment needs to take account of these risks.

Inspections

Inspections (Section 2.2.7) are concerned with management checks that all the plant and procedures are operating according to instructions or as expected and finding opportunities to improve. The results of the inspections may identify shortcomings which could be relevant to an impact assessment. Examples could be incorrect storage of chemicals, leaking equipment, poor practices by operators or waste of water or energy. Better still, the inspections may lead to improvements that reduce the potential impact on the environment, for example, by improving storage facilities or providing training to operators.

Job and task procedures and analysis

A job is a description of the role of an operator such as machine tool operator or driver of a goods vehicle. A task is the activity taking place such as machining steel parts using cutting oils or transporting hazardous waste. Organisations may set out instructions or procedures from either point of view of an activity or both. The descriptions should cover what needs to be done, how it is to be carried out, what are the potential hazards (to health and safety as well as the environment), protective measures to be adopted and any special features. Particular attention may be required at start up or shut down as these are the times when problems often occur: plant may not work properly after a long shut down, the temperature may not be optimal at these times or a new shift may not have been briefed on an earlier problem. The impact assessment should take account of all of these types of issues as potential future risks.

Incident data

An incident is an indication that something has previously gone wrong. It could be uncovered as a result of a complaint from neighbours, a report from a regulator or discovered by the organisation itself. All incidents should have been investigated and a report produced explaining as a minimum: what went wrong, what caused it, what action was taken at the time and what needs to be done to avoid a repetition.

Reviewing previous incident data and reports enables the organisation to assess whether there is still a risk. Ideally, any failures have been put right and the process of continual improvement should have reduced the risk but sometimes there may be a time lag or there is little that can be easily done. As an example, a problem caused by flooding of a site may be expensive to remedy for the future and require time and money to resolve. In the meantime, the risk of flooding may remain low but the potential for environmental impact still needs to be considered.

Site history

Records of previous activities on the site, including maps and plans, may indicate that there are redundant structures underground or the possibility of contaminated land. If a site has been used for many years this is a very real possibility as the controls were not in place that apply now.

Employees

Senior managers come and go but many employees spend a lifetime on the same site. They know the history, previous incidents or other problems and where the 'skeletons may be buried'. Ask them!

3.3.2 Sources of information external to the organisation

Internal information should be easy to find and evaluate, especially if records are kept and structured according to ISO 14001. Looking outside the organisation is more complex; knowing where to start and how reliable the information is can present difficulties. The following suggestions are intended as a guide structured around the ease of access and how useful they may be.

Manufacturers and suppliers

Reliable suppliers should provide material safety data sheets (abbreviated to MSDS) for their products if there is likely to be any risks attached to their use (this is also a legal requirement for many products). Some of these will be from the supplier or the supplier may pass on information from the manufacturer. The information should cover health and safety considerations for the users and environmental risks such as the presence of hazardous materials, the safe disposal of the products or their containers and what to do in the event of a spill. This type of information should enable a risk assessment to be made and attention focused on those most likely to cause an environmental problem. If the data are not supplied then it is probably time to change supplier.

Legislation

Referring to legislation is not always easy as seeking out subsequent amendments and trying to make sense of the contents can be a challenge. However, it can be a good source of useful information about regulations that may apply to an organisation. The sources are often available online, especially for more recent acts and regulations and there are organisations that publish compendia of legislation with commentaries on recent changes, although these can be expensive to maintain and may contain some information of little interest to the user. Later chapters in this book refer to specific issues but there will be relevant legislation on emissions to air, water and land, waste management and energy use that will apply to many organisations. The need to consult legislation concerning aspects of less wide application such as wildlife protection or disposal at sea will depend on circumstances and finding clues as to what is relevant may involve trade associations or the use of other routes to find what may apply.

There will also be useful information in any consents, licences or permits already issued. These will have been issued under the relevant legislation and the contents will be based on the risks to the environment. They may contain specific clauses that are relevant to conducting an impact assessment such as the need for an emergency plan or specific preparations that are required. The COMAH regulations in the UK, which feature in Chapter 9, are a case in point (COMAH 2015).

Government and regulatory bodies

The government and regulatory bodies publish a range of useful information. As well as access to original legislation, they may offer guidance on interpretation or compliance and further supporting information and advice. Examples from the UK are included in Box 3.5. The environmental agencies in the UK have offered guidance aimed at specific sectors of the economy as well as general guidance on pollution prevention or waste management, the principles of which should be applicable elsewhere. As further examples, the European Environment Protection Agency and the US Environmental Protection Agency (EPA) have a range of publications which are good for background reading on specific topics and are available through their web sites.

Trade associations

In the UK and elsewhere many businesses have formed trade associations. These usually require membership to access their full range of services. They provide advice on proposed new or changed legislation, compliance, new technologies and other topics of interest and may offer training courses or seminars that are relevant. They also provide networking opportunities to meet others in the same business to exchange ideas.

International, European and British standards

This book has already referred to some standards such as the ISO 14000 series which are relevant to environmental management. There are also many standards for specific products or services that may be worth consulting and may already be available in the organisation if they relate to its activities, products and services. The potential list is long but a web search should help to find them.

IT sources

The use of the internet to access some of the sources above has become the normal way of operation now. It also offers access to further services. A search on a topic will usually bring up thousands of potential links but care is needed in taking all of them at face value. Many are aimed at selling a product or service and the quality of the information may be suspect. Treated with care, though, they may lead directly to useful information or to further links to other web sites. Some web sites offer direct access to electronic advice and information such as legislative changes provided on a subscription basis and kept regularly up to date. This can be more reliable than dated paper publications. Some companies offer this service on CD ROMS. Other IT sources are newsletters offered by e-mail from environmental or trade organisations.

Reference books

There is an enormous literature on all aspects of the environment from textbooks and encyclopaedias with a general coverage to more learned tomes with a very limited scope. Finding the right one with relevant information may prove difficult but advice on suitable texts may be available from professional bodies or review articles. Always check the publication date or that the information is still current.

Box 3.5 UK government and related sources of information

Government

The various government departments have content with links to legislation as well as supporting information used in its preparation (such as consultation responses).

The main access site is www.direct.gov.uk which has links to other government and non-government sites. For England these include the Department for Environment, Food and Rural affairs; the Department of Energy and Climate Change; and the Department for Communities and Local Government. The government web sites and those for the regulatory agencies for the environment and nature in Scotland, Wales and Northern Ireland are given in Box 1.6. Note that the departmental names, responsibilities and web sites change, often with a change of government or as a result of a reshuffle of responsibilities.

Non-government

The regulatory bodies were given in Table 1.6. A joint SEPA NIEA web site 'Netregs' (www.netregs.gov.uk/), a link in the EA web site and one for Natural Resources Wales provide supporting information.

There are other government agencies such as WRAP (for waste at www.wrap.org.uk/) and the Energy Saving Trust (www.energysavingtrust.org.uk/) which may also be of help.

Some of the non-governmental agencies offer regular newsletters or e-mail letters.

As another example, in England the Environment Agency publishes summary data on environmental quality in its public register on the web site www.gov.uk/guidance/access-the-public-register-for-environmental-information.

Old maps and directories

These may provide information on sites (especially previous uses) that could be of interest to an organisation looking to acquire a new location.

Consultants

The recourse for many organisations faced with finding new information is to ask a consultant or a university for advice. The right choice may bring previous relevant experience that will provide a short cut and be good value for money.

Professional organisations

The professional bodies (such as the Chartered Institution of Water and Environmental Management, the Institute of Environmental Management and Assessment, the Chartered Institute of Environmental Health and the Chartered Institution of Wastes Management in the UK) produce journals, books and other useful publications. Membership is available at various grades or levels of experience and offers a network of useful contacts as well as access to conferences and training.

Organisations concerned with the natural environment

There are official regulatory organisations such as Natural England (see Box 1.6) that have a duty to protect the natural environment. There are also local wildlife trusts and national organisations such as the Royal Society for the Protection of Birds that have an interest in specific locations or types of wildlife. They can be useful sources of information for impact assessment.

Pressure groups

Greenpeace, Friends of the Earth or WWF (now the official title – previously known as World Wildlife Fund) are international pressure groups. They have national and local branches and there will be many other local pressure groups: some may target a particular organisation or development. These obviously have a particular view on many of the issues but nevertheless may be useful if for no other reason than to assess the likely public opinion. Public opinion is a factor in assessing the significance of the aspects and impacts.

3.4 Principle and practice of environmental assessments

Earlier sections have identified the reasons for impact assessment: principally in the initial environmental review in preparation for the introduction of an EMS (Section 2.2), in preparing the application for a permit or permission (Section 3.1) or in using LCA. The assessment needs to take account of several factors when collecting the information and then pulling it together. As a reminder, the assessment usually starts with the aspects from the activities, products or services from which the important aspects can be determined. This section develops this approach further.

3.4.1 The circumstances under which operations take place

It is too easy just to take account of what is happening now (the normal conditions) and ignore other scenarios that may occur. Exactly what is relevant depends on the activities, products and services that form part of the operations. The environmental impact of any of these could be independent of other factors but it is necessary to check whether this is the case. Sometimes things happen that were totally unexpected but have a major impact. The tsunami that damaged the nuclear reactors at Fukushima in Japan in March 2011 following an earthquake is a case in point. Japan is known for earthquakes and the nuclear reactors are designed to cope with them but the tsunami damaged the cooling water system and caused a melt down with consequential releases of radiation. So in considering what conditions are relevant it is important to look well beyond the current position or normal operation.

The ones to consider are:

- **Abnormal conditions:** as well as earthquakes in vulnerable areas, this could include extreme weather events such as flooding, storms, drought, extreme heat; disruption to supplies, transport or utilities; failure of or unusual operation of systems or processes; operator error, at start up or close down of a process; in summary events that may or may not be outside the organisation's control but which affect the operation.
- **Emergency situations:** during incidents, accidents and potential emergency situations; there is some overlap with abnormal conditions (such as flooding) but additional examples would include damage to production or treatment plant caused by a fire, a vehicle impact or structural failure; pollution of a river that is used for cooling water; release of unexpected concentrations of contaminants to air, water or land and their consequences (such as fish kill or evacuation of neighbours) or actions required if it becomes clear that something dangerous is about to happen (e.g. deliberate release of gas to prevent pressure build up in a tank that could lead to an explosion).
- **Changes over time:** past activities may have left a legacy of contaminated land or groundwater that is going to become a problem in the future; the current situation may be considered the normal one but there may be plans to change the operations in the future lifetime of the assessment – new products or services or changes in output; the availability or adoption of new technology; market changes driven by fashion or economic conditions; planned new building developments in the vicinity; responses to external pressures such as public opinion or legislation.

Of course not all of these will apply to every situation and there may be others not listed that could be important. In compiling the list of potential impacts it is important to consider the possibilities and rule them out rather than trust to luck.

3.4.2 The source, pathway, receptor model and its application to risk assessment

The actual risk to the environment depends on three factors being in place:

- **A source of risk:** an activity, product or service that has an aspect that has the potential to interact with the environment.
- **A pathway:** for the aspect to get into the environment to the receptor.
- **A receptor:** something that is going to be affected by the aspect.

A source that has the potential to interact to the detriment of the environment would be known as a 'hazard'. (Usage will be familiar to those with a health and safety background. The term hazard is used for something with the potential to cause harm; risk is the probability or likelihood that it will happen.). The hazard could be from a pipe or chimney releasing a toxic discharge or a greenhouse gas. The pathway is self-explanatory; it is usually by way of air, water or land. The new term here is 'receptor'. This could be a watercourse, person, animal, plant; anything that is going to be harmed in some way. Box 3.6 gives some examples which should help explain the concept.

Without all of these being present there is no risk and the potential risk can be mitigated by limiting or removing one of them. Taking the example of the oil spill onto a hard surface in Box 3.6, the risk can be reduced or eliminated by managing the risk of spills by procedures and training or keeping oil storage sites and areas where it is used away from drains (i.e. at source), by having an interceptor in the drainage system (i.e. the pathway) or by discharging drains to foul sewer rather than directly to a watercourse (the receptor). This example and others will come up again in later chapters.

In the context of risk assessment, EIA is about looking at the impacts on the receptors. The model helps to understand the potential risks and how they might lead to environmental damage as well as the opportunities to manage them.

3.4.3 The scope of the environmental assessment

The scope of the impacts has been touched on in previous chapters. The local environment is usually the first to be considered and it is fairly easy to identify receptors at risk: local watercourses, land on site or nearby, neighbours, local flora and fauna or sites with special features (SSSIs, historical buildings, etc.). For some activities this may be all that is required. However, impacts such as climate change, acid rain and depletion of the ozone layer are global in nature and these need to be considered if their causes can be partially attributed (however small) to the aspects of the organisation. Also, if raw material depletion or the geographically remote risks of pollution from extraction and processing of materials and from the transport of materials are issues, they should be included. Waste production and disposal must also be taken into account. Passing waste on to a contractor for removal should not be an excuse for ignoring it, even (or especially) if it goes abroad.

3.4.4 Identifying the receptors at risk

From the analysis so far it should be possible to identify the receptors at risk. Box 3.6 gives examples of the approach but for the organisation conducting the environmental impact assessment these need to be appropriate to all of the activities, products and services, bearing in mind the scope mentioned in the

Box 3.6 Examples of the source, pathway, receptor model

Source	Pathway	Receptor
SO_2 released from chimney	Air	Building or trees – damaged by acid rain
Ammonia in liquid effluent discharge	Watercourse	Fish – killed by the ammonia
Oil from spill on to land	Land	Groundwater
Spill of oil onto hard surface	Surface water drain	Watercourse – contaminated with oil
NO_2 released from chimney	Air	Human with asthma
Methane from land fill	Land	Atmosphere – causing global warming

Note that in these examples the pathway could be also be a receptor depending on circumstances.

previous section. The local flora, fauna, watercourses and residents are the obvious examples but the wider scope potentially brings in indigenous peoples, remote landscapes and damage to the ozone layer or habitats affected by climate change.

3.4.5 Identification of aspects and impacts

Using the information collected so far it should be possible to build up a picture of the organisation's aspects, their impacts or potential impacts and the receptors at risk. The ultimate concern is for the receptors but the control is through the aspects and impacts by managing the sources or the pathways. It would be possible to get carried away and spend all the time analysing the aspects and impacts without actually doing anything with the information. This becomes more of a problem the larger and more complex the organisation.

3.4.6 Determining the significance of impacts

There comes a point when some priorities need to be established and from an environmental point of view this depends on the 'significance' of the impact and thus of the aspects. There could be other points of view that could influence the significance – political, financial and even personal, but some of these can be factored into the determination of significance as outlined below.

The significance is determined by the following factors, most of which are self-evident:

Scale of impact. This can range from on site to global. Something contained on site is of little significance but the wider the scale of impact the more significant it becomes.

Severity of impact. This would normally reflect toxicity but it could be used for noise, visual intrusion or other impact. An event that causes harm to humans will be more significant than killing a few plants nearby, which in turn will be more significant than releasing an odorous substance of no toxicity.

Probability of occurrence. (Also referred to as frequency or likelihood.) This may be difficult to quantify but some assessment can be made subjectively. The likelihood of an earthquake varies from region to region of the world but if the area is of low risk for earthquakes they may be of lower significance than say the risk of disruption due to flooding. There may be information available for the probabilities of these examples but the likelihood of an operator causing a problem by some act or omission is unknown unless there is already a history of failures.

Duration of impact. The impact could be fleeting and of low significance but some chemicals can persist in the environment for years and radioactive substances can have half-lives measured in thousands of years. These are direct impacts that are clearly more significant. A similar argument would apply for an impact other than on the environment, for example on the business where an incident could disrupt the business for a long period. The leak of radioactivity following the tsunami in Japan and the BP oil spill in the Gulf of Mexico are examples of indirect impacts that will go on for years as the damage is repaired, clean-up is completed and insurance claims settled or compensation paid as appropriate.

Sensitivity of the receiving environment. An impact on a nature reserve or an SSSI would be considered more significant than on a piece of land already contaminated. It should be noted that the receiving environment could have varying sensitivities to different hazards. For example, some plants are particularly sensitive to a few herbicides but can tolerate others (this is the basis of action for selective herbicides).

The extent to which the impact is reversible. Some receptors may recover quickly: vegetation killed off on the surface of the soil may regenerate once the cause has been stopped.

The ease of remediation. Clean-up of the spillage of a solid may resolve a problem quickly but land extensively contaminated with pollution that does not degrade (such as a heavy metal) will remain like that for a long time.

Ease of control of aspects. Most organisations expect to have control over their impacts either by managing their aspects or by intervention. However if an aspect can become uncontrollable, such as an explosion, the impact is probably going to be serious.

Legal or contractual requirements. These have to be given high significance as there are likely to be penalties if they are not met.

The concern and importance of interested parties. This is more difficult to deal with. It is a fact of life that some stakeholders have more influence than others due to their status or political influence. Some make more noise in protesting than others. The relative influence of employees, neighbours, pressure groups, regulators and local politicians will depend on circumstances including their perceptions of any previous problems. This has to be taken into account in some way although there is the potential for bad publicity if the influential are seen to take precedence over the vulnerable.

Effect on public image. A small event may well go unnoticed. However, even small events can cause adverse publicity if there are signs of a cover up or incompetence. Public image is also affected if the receptor is vulnerable (children or old people, a rare species of plant or animal, a school or hospital) or if the substance involved is considered dangerous by the public or press even if the amount involved is small (radioactivity for example). So there is always the possibility of bad publicity and assigning significance to the impact needs to take account of this.

Costs. Most organisations would assign a high significance to an impact that could result in high costs. These could be for fines and penalties, clean up or compensation, loss of business or loss of output. The costs of dealing with the aspects to avoid adverse impacts will also be significant to the organisation.

3.4.7 Recording and evaluating information about significant aspects and impacts

The documentation of the assessment of significant impacts and aspects should follow the criteria for the EMS outlined in Chapter 2. Version control and safe keeping are important as the results will need to be reviewed from time to time or consulted to justify some course of action or inaction. The recording needs to start with a list of impacts and then a measure of significance attached to each. It is unlikely that every factor listed above would need to be considered so some may be discounted straight away as irrelevant or of such low impact that the significance does not matter. (Comparing an office with a power station again shows how this would apply.)

A simple scheme for evaluation might just use low, medium and high based on a subjective approach or some quantifiable method if one is available. Sometimes it is appropriate to use a scoring system whereby each significance factor is given a number say from one (low, e.g. non-toxic visible steam) to five (high, e.g. toxic vapour) and the scores added. More complex schemes include weighting each factor or applying probabilities of occurrence by multiplying by one (low probability, e.g. extremely unlikely) to five (high probability, e.g. has happened regularly). The numbers for each factor are multiplied together and then added together to obtain a score for each aspect. Whichever method is used, the most significant impact has the highest score. The associated aspects will also be the most significant. It is important to document the procedure and the rationale whichever methodology is used so that it can be understood if revisited later. It is worth remembering that the value of derived information is only as good as the underlying data. If there is a lot of uncertainty about some of the information (e.g. relative toxicities or the ease of reversing the impact) then calculating a score with high precision is meaningless.

3.4.8 Evaluation of the adequacy of controls

The next step of this approach is to develop as complete a picture as possible of the aspects of the organisation and their impact or potential impact – or at least to go as far as is practical. The aspects are largely under the control of the organisation so the way it manages them affects their potential impact. The assigning of significance has two main purposes: it helps determine priorities for action and focuses attention on the adequacy of current procedures or controls. If resources are limited then it may not be possible to tackle everything that may have come to light at the same time so the most significant should be dealt with first. This will be especially true if large expenditure is required to reduce the risk or significance of a potential impact. Equally, if an impact and thus the associated aspects are significant it is important to consider whether the current way of operation or the controls in place are adequate or if some simple changes to procedures or control systems could reduce the risk or significance. Using the source, pathway, receptor model, managing the risk of spills of oil or chemicals and containing spills or firewater on site are examples of procedural changes that could be adopted early on, as explained in Chapter 5.

3.4.9 Activities of suppliers and influence on product design

The extension of the EMS outside the boundaries of the organisation has been referred to previously, particularly when it involves extracting resources in large volumes in faraway places. Procuring any products or services that have significant impacts on the environment should be considered at the same time as looking internally. Large companies often look to their supply chain to demonstrate that they are not harming the environment by insisting on suppliers being accredited to an EMS and that their suppliers in turn are similarly accredited.

The significance of the impacts should also be used to influence the design of products. The sorts of issues that can arise are using less material and energy in manufacture, less impact on the environment in use and the opportunity to reuse or recycle at end of life. The application of LCA as described above will guide this process. Advances in vehicle design is an example that has resulted in less polluting vehicles with better fuel consumption (through engine design and management systems and the use of catalytic converters), longer life (through the use of alternative materials such as plastics and better protection of steel) and the opportunity to reuse or recycle many components or their materials of construction when they wear out or the vehicle is finally scrapped.

3.4.10 Supplier selection and transport issues

The accredited supply chain has also been mentioned previously. For example, businesses and public may buy only from an accredited source such as timber from sustainably managed woodland or paper that meets certain criteria for recycled content and means of production. The use of EU eco-labels and energy efficiency labels or claims such as 'low carbon', 'carbon neutral' or 'environmentally friendly' are becoming common. For the supplier, these trends mean that environmental impact is becoming a significant factor in consumer choice. Banks and investors are also showing interest. Many will look at the environmental credentials of a business, especially if it is seen to be in an activity with high risk of causing environmental damage. Investors may only build or buy properties or shares that meet their own or other criteria for energy efficiency or other measures of sustainability. Finding the information on a large supply chain is time consuming but organisations such as Sedex (www.sedexglobal.com/) have been set up to share information about common suppliers among their customers.

The means by which goods are transported is also important to consider. Air freight is faster than sea but uses far more fuel for equivalent loads and so produces more impact on the atmosphere. It should ideally only be used for perishable goods or where fast delivery is a necessity. On land, rail is generally better than road transport, at least for large volumes or heavy loads. In some countries rivers and canals are also used for these purposes. Of course the minimum transport impact should arise with the shortest delivery routes arguing for local sourcing wherever possible. The case for applying similar criteria applies to moving people. The most significant impacts arise when one person is in a car. Public transport by train, tram or bus is less environmentally damaging and the least is to walk or cycle. (Further aspects of transport are considered in Chapter 7.)

3.4.11 Reviewing impact assessments

The impact assessments need reviewing just as much as other information collected for an EMS. The information used in the assessment can be subject to change and the factors that were used to decide its significance could also change or become out of date. It is good practice to review such assessment every so often even if there is no immediate awareness of a factor that could require it. Annually may be a good target although for a stable organisation in a stable environment less frequent may be acceptable. Other reasons for review include:

- **Incidents, near misses or complaints:** these question the adequacy of current controls and may be detrimental to public image, both factors that were important in determining significance.
- **New processes or equipment:** these should result in reduced impacts if the right choices were made however, the commissioning stage could be a period of higher risk.
- **Changes of staff:** increase the risk as new staff may be less accomplished at managing the processes or responding to risk situations.
- **Changes in legislation or regulatory regime:** these could introduce new requirements that were not considered in the original assessment.

The reasons for reviewing EIAs apply particularly to the EMS and any activity involving permits to operate. These are continuing activities and circumstances but permits can change. If the EIA is for a new development it is undertaken before work begins and the opportunities for change may be more restricted once it is in place (consider the construction of a railway line or large building). However, it is worth reviewing the actual impact of the completed project with the original EIA to see if the original assumptions were correct and to look for opportunities to improve the current project or the EIA process for future similar projects.

CHAPTER

4

Control of emissions to air

After this chapter you should be able to:

1. Outline the principles of air quality standards
2. Outline the main types of emissions to atmosphere and the associated hazards
3. Outline control measures that are available to reduce emissions.

Chapter Contents

Introduction **66**
4.1 The principles of air quality standards **66**
4.2 The main types of emissions to atmosphere **68**
4.3 Control methods for air pollution **71**

INTRODUCTION

Some of the issues arising from air pollution were described in Chapter 1, principally emissions of carbon dioxide and their impact on the climate, damage to the ozone layer caused by other chemical releases and the formation of acid rain from sulphur and nitrogen oxides. This chapter will extend this to cover a wider range of air quality issues and how emissions can be managed.

4.1 The principles of air quality standards

4.1.1 Units of measurement

The main reasons for having standards for air quality and controls in place to limit emissions are to protect people and the wider environment from harm. Most living organisms that live on the surface of the earth take in oxygen from the atmosphere in order to respire. Pollutants in the atmosphere can cause damage to the lungs, interfere with the processes involved in respiration or get into other parts of the organism such that they cause damage or even death. Before getting involved in the specific problems it is important to understand a bit about the atmosphere and its normal composition.

The atmosphere is generally well mixed as the effects of dispersion, helped by the wind, ensure that the composition of the gases present is fairly constant wherever you are on the globe. The main constituents naturally present are nitrogen (about 78 per cent) and oxygen (about 21 per cent). The remaining 1 per cent consists of small quantities of inert gases such as argon and neon and carbon dioxide. Water vapour is also present in variable amounts as its concentration depends on the location and local temperature. There will also be dust and other solid materials present (again depending on location) and traces of other pollutants.

Measuring the trace constituents requires the use of units relevant to the amounts found. The concentrations are small in comparison to the oxygen and nitrogen and the mass is very small. Concentrations are expressed either as mgm^{-3} (milligrams per metre cubed) or ppm (parts per million). The SI units for measurements are explained in Appendix 1 and mgm^{-3} is an expression of the mass of contaminant (mg) in a volume of air (m^3); this unit is used for most pollutants as they are present in small concentrations. The term 'parts per million' can be used to express the concentration as either a mass of contaminant in a million parts by mass of substrate or a volume of contaminant in a million parts by volume of substrate. In the case of air it is conventional to use ppm by volume and the amount of nitrogen would be expressed as 780,000ppm and oxygen as 210,000ppm. These add up to 990,000ppm and the balance of 10,000ppm is made up of the other constituents. The common pollutants are found in lower concentrations and range from well under 1ppm up to about 400ppm for carbon dioxide.

4.1.2 The effects of poor air quality

Poor air quality can have three main effects. The first of these is toxicity, generally by inhalation. Contaminants get breathed into the lungs and either lodge there or are absorbed into the body. Particulate matter is most likely to remain in the lungs unless the particles are very small. They will cause structural damage to the lung cells and may lead to breathing difficulties such as wheezing or coughing, lung diseases such as emphysema or lung cancer and increased risk of infections. Non-particulate matter and very small particles can be adsorbed into the blood stream and circulate around the body causing acute (short-term) or chronic (long-term) effects. Acute toxicity may arise from exposure to a relatively high concentration and could cause collapse or death due to inhibition of breathing or direct effects on other organs such as the heart. Chronic effects are usually due to long-term exposure to lower concentrations of a substance that can accumulate in the body and reach a critical concentration over time or cause damage by continuous exposure to small concentrations. Lung cancer and emphysema are two such reactions. The toxic effects of air pollution can affect life expectancy; in situations where concentrations of pollutants are too high the oldest and youngest cohorts are the most vulnerable, along with any who may have existing medical conditions such as asthma that are exacerbated. Cigarette smokers are also vulnerable. There is increasing concern that long-term exposure to air pollution in urban areas is shortening life expectancy in the general population and a similar problem may occur with fumes from cooking in confined spaces, especially with wood or animal dung in developing countries.

The second type of effect is more aesthetic. Odours can arise from various sources such as sewage works, abattoirs, farms, chemical factories and waste sites. Although the substances causing the odour may not be toxic – at least at the concentrations detectable by smell – nevertheless, they may be a nuisance (see Chapter 1). Often the odour is due to a mixture of substances rather than one specific chemical. The effect is to disrupt peoples' lives, sometimes to the extent that it is difficult to venture outside or to open windows.

The third effect may be described as secondary. An emission may not be directly toxic or a nuisance but by interacting in the atmosphere with oxygen and other chemicals present or with water vapour some other aspect becomes evident. Acid rain caused by the solution of acidic gases into water vapour which then falls onto the ground or buildings is an example; another would be the interaction of chemicals that cause damage to the ozone layer. These are further described in Section 4.2.

4.1.3 The role of air quality standards

Air quality standards are put in place for ambient (ambient means 'in the surroundings' so here means in the general atmosphere) air quality to limit the concentrations of pollutants that may cause damage. They are expressed in ppm or mgm^3. The standards can be used to set limits in the atmosphere or in an emission from a chimney or vent. The relationship between the two will depend on the dilution as the emission is mixed, initially into the local atmosphere and then dispersed more widely. Calculation of emission standards back from atmospheric standards is a complicated procedure often involving computer modelling by specialists.

Actual standards will depend on the circumstances. Within an enclosed space such as a factory or house, the standard is for the protection of the workers or inhabitants and may be stipulated under health and safety regulations. Once a contaminant has left the factory it will disperse into the local atmosphere and can remain locally for some time until it either disperses further afield in the atmosphere or settles on the ground or is washed out by rain. Some pollutants such as hydrogen sulphide are destroyed fairly quickly in the atmosphere and do not migrate far. Dispersion may be inhibited by local conditions, for example within a sheltered area or under atmospheric conditions in certain types of weather patterns such as inversion (see Box 4.1 for explanation). The local air standards need to reflect these situations and can be a particular issue with pollution in city centres arising from vehicle emissions or the burning of fuels such as coal or wood. In all cases the standards are there to protect the environment as previously defined in Chapter 1.

Air quality standards can be derived from various sources. Some states or communities have their own derived by assessment of toxicity, nuisance or secondary effects. These may be set and enforced by the appropriate regulatory agency. There are international agreements, for example to control the emissions of CFCs – the Montreal Protocol on Substances that Deplete the Ozone Layer (UNEP 2009) – and guidance which may be applied directly or used to derive local standards. For example there is guidance on particulate matter, ozone, nitrogen dioxide and sulphur dioxide from the World Health Organisation (WHO 2006 – currently under review). WHO has also produced guidelines for indoor air quality. The United States, European Union, United Kingdom, Canada and Australia have all produced standards which are readily available online and may change over time.

Because the effects of pollutants are time related (long-term exposure is more damaging than short-term exposure), these are usually expressed as limits not to be exceeded in various time periods such as 24 hours or a year and they may allow exceptions for a limited number of times within a short period or as a percentage of the samples analysed. The arrangements for the management of air quality in the UK are outlined in Box 4.2 in Appendix 4.

The actual pollutants present in the atmosphere will depend on the local sources of pollution. These may be natural but are most likely to be from industrial sources or from the combustion of fossil fuels in heating, boilers and vehicles. The local concentrations of pollutants will depend on the amounts emitted and how well they are dispersed into the atmosphere. This can be influenced by discharge from a tall chimney in the case of a factory

Box 4.1 Inversion

Inversion is the term used to describe unusual atmospheric circumstances.

Normally the atmosphere is warmer at ground level than higher up. This is because the sun warms the earth's surface which in turn warms the air immediately above. As the warm air has the lower density, it rises causing vertical air currents (much sought out by those in gliders and parapونts) and promotes mixing in the atmosphere.

An inversion is caused when, due to atmospheric conditions such as a moving warm air front, warmer air sits above cold air. It can also happen at night due to the earth's surface cooling. This is stable as the cold denser air is now at the bottom layer and no rising air currents occur. The presence of surrounding hills can protect this state from being disturbed by winds. Pollution is trapped under the layer and can build up in concentration close to the ground. Reactions can take place among the pollutants typically creating smog which also builds up.

or power station. Fugitive emissions and those from domestic premises and vehicles cannot be influenced in this way. Other factors such as wind direction and speed will also be important.

Standards for emissions

It is more difficult to set standards for emissions to atmosphere as the dispersion is in three dimensions and is affected by the weather and other factors which are variable. The calculations are complex and involve modelling. This can be done for a point source such as a chimney but not for multiple domestic chimneys or for vehicles. The emission standards for industrial sites (including incinerators and ventilation systems) can be included in the permits.

For domestic and business premises the approach in many countries has been to require the use of suitable solid fuel or to replace coal with natural gas for heating. However, coal is still widely used in those developing countries with large quantities cheaply available. Emissions from vehicles have been reduced by progressively tightening the standards from exhausts in new vehicles. In this case the initial controls apply to the manufacturer of the vehicle rather than the user. As the older vehicles get replaced, the general level of emissions has reduced. In those countries with testing regimes for older vehicles, the exhaust emissions may form part of the test and any failure then does become an issue for the user.

4.2 The main types of emissions to atmosphere

Before considering the main types of emissions to atmosphere arising from human activities, it is worth noting that many emissions also come from natural sources. Volcanoes emit large quantities of particulate matter, carbon dioxide, sulphur dioxide, hydrogen chloride and hydrogen fluoride along with trace amounts of many other substances. Methane can arise from the anaerobic decay of organic matter (see Appendix 2 and below), emissions from natural gas deposits or marsh land and from the digestive systems of ruminant animals such as cattle. Nitrogen oxides and ammonia are emitted by biological processes in soil. Radon is a radioactive gas that is found associated with some geological formations and can concentrate inside buildings built over them. Hydrogen sulphide is found in many areas with anaerobic conditions, including some hot water springs and discharges associated with volcanic activity. Carbon dioxide emitted from a volcanic lake in Cameroon was believed to be responsible for a large number of deaths in 1986 (Baxter et al. 1989).

However the main interest for this book is pollution arising from human activities and these can take several forms. These are broadly identified as:

- *gases* such as the oxides of nitrogen, sulphur and carbon which are in this state at normal temperatures and pressures,
- *vapours* which are derived from volatile liquids such as petrol and solvents the concentration of which may vary with temperature and pressure, and
- *particulate matter* such as dust, soot, grit and fibres which can be of varying sizes.

The presence of these substances may be via emissions from a chimney or extraction vent connected to a process or building. These are easier to monitor and control. Air pollution can also arise from fugitive emissions – those which leak from pipework, process plant and gaps in the structure of buildings and which are more difficult to monitor and control. Industrial polluting substances may also arise in open sites such as mines and quarries, construction and waste disposal sites and contaminated land where again monitoring and control are less easy. The other main sources are from fires and boilers in domestic and business premises and vehicles and static engines.

In the remainder of this section we will look at the main atmospheric pollutants, their sources and effects.

4.2.1 Oxides of carbon

There are two oxides of carbon: carbon monoxide (CO) and carbon dioxide (CO_2). The principal source of both is the combustion of fossil fuels. These are derived from organic matter and consequently have high carbon content (see Chapter 7 for more information). Carbon monoxide is formed when there is insufficient oxygen present for complete combustion and is very toxic. It can be dangerous if gas-fuelled appliances do not have sufficient ventilation and is a regular cause of accidental death. It is also present in vehicle emissions as some of the fuel is not completely burned and is a particular hazard in enclosed spaces. Although it is a potential health hazard in the home or work place, it is less of a problem in the wider environment. It should not be formed if equipment is operated and maintained correctly and it disperses and is eventually further oxidised to carbon dioxide in the atmosphere.

Carbon dioxide is emitted by organisms as they respire. It is also usually the main emission from the combustion of fossil fuels. It is not toxic in the same way as the monoxide but in high enough concentrations can cause asphyxiation, as was thought probable in the Cameroons incident described above. The gas is stable in the atmosphere but is dissolved in rain water and into the surface of water bodies and is slowly removed as it reacts with minerals. It is also removed by photosynthesis – the process by which plants convert carbon dioxide back into oxygen in the presence of sunlight. This produces substances for their cell structure and growth along the way. (Note that plants respire giving off carbon dioxide all the time, they only remove it during daytime.) However, as we burn more fossil fuels and so produce more carbon dioxide, these natural processes are not able to keep up with the growth in emissions. Consequently the concentration has been increasing over the last century. Carbon dioxide is believed to be the principal cause of global warming and climate change as described in Chapter 1.

4.2.2 Oxides of sulphur

There are two oxides of sulphur: sulphur dioxide (SO_2) and sulphur trioxide (SO_3). Together they are commonly referred to as SOX or SO_x (pronounced socks). The dioxide is the commonest pollutant and is formed when sulphur or sulphur-containing compounds are heated in the presence of oxygen. The commonest sources are the processing of metal ore containing sulphides in order to refine the metal and the combustion of fossil fuels – mainly coal and some oils – containing sulphur compounds as impurities. The ore processing industries are more able to contain and control the emissions of sulphur dioxide but the emissions from the combustion of coal and oil products have been more dispersed and have only recently been brought under better control.

Sulphur dioxide is directly toxic and can cause death in high concentrations. Environmental concentrations are usually too low for acute effects but long-term exposure can cause lung damage and also damage to plants and other forms of wildlife. The damage is caused by the SO_2 dissolving in water vapour or in water in cell tissue forming sulphurous and sulphuric acids. These acids are responsible for the damage and for causing irritation to the eyes and nose. Equally importantly, the gas can dissolve in water vapour in the air forming a gaseous acid solution which falls as acid rain. The acid rain then causes further effects: dissolution of metals from soils, acidification of rivers and the sea and damage to structures such as steel and some building stones. The mobilisation of metals can be harmful to plant life and fish and other aquatic organisms. The acidification of water (lowering the pH – see Appendix 2) is also directly harmful to aquatic life and to plant life as described in Chapter 1. Sulphur dioxide is also implicated in the formation of smog as described below.

4.2.3 Oxides of nitrogen

There are three oxides of nitrogen: nitrous oxide (N_2O), nitric oxide (NO) and nitrogen dioxide (NO_2). Together they are known as NOX or NO_x (pronounced nocks). Nitrous oxide is mainly formed by natural processes in the soil whereby microorganisms metabolise nitrate or directly in the atmosphere by electrical discharges such as lightning causing a reaction between the oxygen and nitrogen. It is not normally considered a significant pollutant.

Nitric oxide and nitrogen dioxide are formed in combustion processes, especially the burning of coal and in vehicle exhausts. They dissolve in water in a similar way to sulphur oxides forming nitrous and nitric acids and have similar effects – damage to the respiratory system and wild life and the formation of acid rain. Nitrogen oxides are also involved in the formation of ozone in smog near to ground level. (Remember ozone at ground level is an irritant to eyes and causes lung damage; in the higher levels of the atmosphere it protects against UV radiation.)

4.2.4 Halogens and their compounds

Halogen is the collective term used for the elements chlorine, bromine, fluorine and iodine. Minor concentrations of the halogens may occur naturally in the atmosphere but the main concerns are around the compounds containing them (see Appendix 2 for an explanation of the terminology). Chlorine occurs as chloride salts in coal and is released as hydrogen chloride gas on combustion. This dissolves in water to form hydrochloric acid which is as damaging as sulphuric and nitric acids.

The other main form of pollution is from organic compounds containing halogens. Chlorofluorocarbons (CFCs) are inert gases which are relatively non-toxic and do not cause damage to metals or other common materials. They have been widely used as propellants in aerosols, as refrigerants in cooling and air conditioning and to expand plastic foams such as polystyrene. Chlorinated solvents such as carbon tetrachloride and chloroform have been used in cleaning and degreasing and as solvents in chemical processes, although their use is reducing. These are more toxic than the CFCs and some may be carcinogenic.

The main problem identified from the release of chlorinated compounds, particularly the CFCs is their role in damaging the ozone layer but some are also greenhouse gases. These are described in more detail in Chapter 1. Some halogenated compounds can bioaccumulate in animal tissue.

4.2.5 Metals and their compounds

Metals and their compounds are usually solids and many compounds are soluble in water. Their presence in the atmosphere is rare although they will be associated with dust particles particularly in and around mines and in metal refineries. The most toxic metals include lead, cadmium and mercury but most metals are toxic to some degree if inhaled. Lead accumulates in the body and can cause long-term damage to the nervous system. Lead and mercury can form organic compounds which are volatile. Lead was widely used in the form of tetraethyl lead to increase the octane rating of petrol. On combustion it released lead compounds into the atmosphere and these could be inhaled. It is no longer used for this purpose but lead in dust can arise from old paint coatings. Lead and some other metals can be released in fumes (gases containing very small particles) formed when welding and soldering are carried out. Other metals are released in dust particles during the combustion of coal, incineration and cremation. The metals released depend on the quality of the coal or other material being burned. For example mercury can arise from the cremation of bodies with teeth containing amalgam fillings and there may be metal residues in wastes. The metal compounds in air associated with solids generally concentrate near the source and eventually fall to the surface. Releases due to vehicles are concentrated along main arterial routes but the others are generally a problem local to the source.

4.2.6 Volatile organic compounds

Volatile organic compounds (VOCs) is the term used to describe a range of unrelated solvents and similar products that are liquids at normal temperatures and pressures but which release vapour into the atmosphere by evaporation. Liquid fuels such as petrol and diesel and solvents for paints are the most widely used. Fuel and fuel products are released as VOCs around filling stations and when vehicles fail to fully combust the liquid and they contribute to the formation of smog. Paint solvents evaporate when the coatings are applied and left to dry. Similar products are used in other coatings and adhesives.

Many industrial processes produce or use large volumes of VOCs either as solvents or as intermediates in production processes. Common examples are alcohols (such as methanol and ethanol), organic acids (such as acetic acid), acetaldehyde, acetone, glycols and isocyanates. These are used in the manufacture of plastics, cosmetics, pharmaceuticals, antifreeze, paints and other consumer products as well as industrial intermediates. In these cases the major sources of release are likely to be in the processing, storage and transport rather than the end use, although ethanol is increasingly used as a renewable fuel in vehicle engines. Many VOCs will be slowly oxidised in the atmosphere and diminish in concentration.

Some VOCs have been termed persistent organic pollutants (POPs). Other POPs may not be described as VOCs as they are not that volatile but they still find their way into the atmosphere in very low concentrations by similar routes. The terminology arises from their ability to resist degradation due to sunlight, oxygen, water or microbiological action and they have been found distributed widely. They may accumulate in the tissues of animals and plants (bioaccumulation see Box 3.4) through long-term exposure to low concentrations such that they eventually reach levels where toxic effects may be present. They have been found in areas where they have never been used such as the Polar Regions. Examples are dioxins (see Chapter 9 on the Seveso incident) which are formed in some industrial processes and in incineration, polychlorinated biphenyls (PCBs) which are a range of compounds used mainly in electrical transmission equipment and pesticides such as DDT and dieldrin. These compounds may be distributed through water pathways as well as the atmosphere. Most are toxic either directly or by causing cancers, mutagenicity or problems such as weakening of egg shells or deformities in offspring (teratogenic). Many have been phased out and their manufacture and new use banned or strictly controlled through international agreements. There is a remaining problem due to the residual uses in old equipment and disposal of wastes and their resistance to degradation.

4.2.7 Methane

Methane is the main constituent of natural gas and it often occurs with crude oil and in coal mines (see Chapter 7) and so it is inevitable that some will leak out during abstraction, storage, distribution and processing. It also occurs naturally as described above and is released during anaerobic decomposition of organic matter (see Appendix 2), particularly in waste disposal sites. There is more on this source in Chapter 6.

It is not especially toxic but can cause death by asphyxiation. It is a fire risk and in the wrong concentration mixed with air it can form an explosive mixture: there have been fires and explosions at coal mines and associated with contaminated land and landfill sites. The other property that it has is as a greenhouse gas, more powerful than carbon dioxide (see Chapter 1).

4.2.8 Particulate matter

The term particulate matter can include a wide range of substances that can be present as large particles or fibres visible to the naked eye down to particles so small that they can enter the body through the lungs. They can occur naturally, for example, there are periodic deposits of dust from the Sahara Desert blown by the wind into Europe when the atmospheric conditions allow. The main concerns arise though from human-made sources such as mines, quarries, factories, demolition and construction sites and combustion plants.

Quarries and open-cast mines cause dust emissions due to the blasting, crushing and transport operations. Underground mines confine some of these underground but there will be ventilation shafts that will discharge contaminated air as well as dust from operations on the surface. Most of the particles are relatively large and settle close by.

Factories can release dust and fibres from refining of ores, cement manufacture, wood cutting and manufacturing, general manufacturing operations such as cutting, drilling and grinding of metals and plastics, textile processing, waste processing and similar operations. Construction sites cause dust from demolition, crushing materials, laying solid material for road foundations, cutting and drilling concrete and various other activities. Combustion plants release fine particles such as coal dust and ash as well as carbonaceous solids. Soot is fine solids that collect in the chimney and also get emitted into the atmosphere. The fine solids contain partially burned fuel and adsorbed compounds formed in combustion along with any metals released from the fuel. Smoke is the term applied if the emission contains a high concentration of solids such that it is visible. The fine solids may contribute to the formation of smog described below. Vehicles that use petrol or diesel also release fine particles associated with other chemicals in the exhaust that contribute to smog as well as particles that arise from wear on tyres and brake pads and discs.

The damage caused by the particles is dependent on their size, physical properties and composition. Some will settle on land or vegetation causing smothering or toxicity, if for example they contain toxic metals. Toxicity due to chemicals that do not degrade in the environment (such as metals) can build up in the soil over time. Others can lodge in the lungs of people and animals and cause physical or chemical damage. Asbestos is of particular concern although other fibres can also cause problems.

Small particles are defined by size and are usually referred to as PM_x where x is the maximum diameter in μm (micrometre – see Appendix 1). For example PM_{10} would refer to all particles present smaller than 10μm in diameter. Attention is now focusing on $PM_{2.5}$ (i.e. particle size less than 2.5μm), extremely small particles that are thought to be responsible for most of the damage from such particles. These are a particulate component of vehicle exhaust and diesel fuel is now considered a worse contributor to pollution for this reason compared with petrol or liquefied petroleum gas.

4.2.9 Other emissions

The list of other potential pollutants is virtually endless but it is worth mentioning a few that arise in particular circumstances. Polycyclic aromatic hydrocarbons (PAHs – a range of complex organic compounds), benzene, butadiene and other organic compounds are mainly associated with the combustion of fuels. Their formation is worse when combustion is incomplete such as when the fuel/air ratio is wrong (including start-up) or engines are inadequately adjusted or maintained. They are toxic, some causing cancer, and can remain in the atmosphere for some time and contribute to the formation of smog.

Emissions of metals such as lead, zinc and bismuth arise from welding and soldering. They are known as fume.

Ammonia can be emitted from farming activities, especially storage of animal waste in and at other waste sites. It is more of a nuisance than anything else.

Legionella is a bacterium that occurs naturally in the environment. It can cause a disease – legionellosis or Legionnaires' disease (similar to pneumonia and named after the incident that sparked its discovery) if it is inhaled. The bacterium survives best in warm moist conditions and so the main source of disease is in mists associated with shower heads or cooling towers. There is more on legionella in Box 5.2 in Chapter 5.

4.2.10 Smog

There have been several references in previous sections to smog. The word is a contraction of 'smoke' and 'fog'. It is not a simple pollutant that can be ascribed to one substance or process. Rather it is a result of the admixture of a range of substances that react together in the atmosphere. Particulate matter, NOX, SOX and various organic chemicals react to generate a mist which contains this mixture along with new substances formed by their interaction. The reactions are promoted by UV radiation from sunlight and inversion conditions (Box 4.1) can ensure that the products remain close to the ground. One of the substances formed is ozone. This is not normally present at ground level (not to be confused with the *ozone layer* which is present at high altitude and referred to in Chapter 1). The cocktail of chemicals causes lung irritation, coughing and respiratory problems, particularly in those with pre-existing conditions such as asthma and affects the eyes and nose. Long-term exposure is thought to contribute to mortality; death rates increase at the time of exposure but the mortality rates also seem to be higher in cities that are regularly affected. Many countries now release an air quality index when the problem is anticipated and sufferers are advised to stay indoors.

Smog is a common problem in large cities where emissions from vehicles and burning coal are present. It was a regular occurrence in London until the use of coal for domestic heating was phased out in the 1960s. It is still common in cities world-wide such as Los Angeles, Beijing, Mexico, Tehran and Delhi, principally due to traffic but it can occur in more remote areas due to forest fires or burning wood as a fuel.

4.3 Control methods for air pollution

There is no single method that can control all forms of air pollution; it consists of a potential mixture of solids, liquid vapours and gases that can also react to produce new products. As with other forms of pollution the best strategy is tackle it using a hierarchy of methods:

- Eliminate at source.
- Minimise.
- Render harmless.

Some of the ideas developed below could apply across more than one of these methods: for example, filtration can minimise solids in an air flow or render it harmless depending on the circumstances. The substitution of materials at source (e.g. use of different solvents in reactive chemistry) may result in polluting discharges of a different type, in which case it becomes a balance between competing outcomes. This is the sort of issue for which EIA and LCA, as described in Chapter 3, are appropriate.

4.3.1 Elimination at source

Elimination or prevention at source is a good starting point as it easier when it is concentrated in one place and not diluted with air. This could result in the avoidance of expensive alternative measures. From the previous examples of sources of pollution some options are obvious. Substituting water for organic solvents such as VOCs is becoming normal practice for paints, inks and other products. Many manufacturers have adopted the practice voluntarily but it is becoming a requirement in some jurisdictions such as member states of the EU. Compliance requires changes in product formulation and it has taken some time to develop products with acceptable performance. Some industries, notably vehicle manufacturing, have adopted dipping, powder technologies or electro-deposition rather than spraying as an additional measure to reduce the production of mists containing solvents and paint residues.

Solvents used for cleaning and degreasing are traditionally halogenated solvents or other VOCs. Alternative solvents, water-based agents with detergents or ultrasonics and mechanical processes may be suitable. In all cases it is good practice to enclose the plant to prevent releases or recover the solvent. The use of CFCs as refrigerants is also being phased out to be replaced with ammonia or hydrocarbons. These can also be air pollutants but are considered less of a risk.

Vehicle emissions arise mainly from the combustion of fossil fuels. The use of alternative fuels can eliminate or at least reduce all of the emissions. Electric vehicles emit no exhaust gases directly although if the batteries are charged using electricity generated from non-renewable sources the pollution may just be directed elsewhere. The same argument would apply to the use of hydrogen as a fuel. On balance, petrol is probably better than diesel but gas (LPG or natural gas) is better still. The same principles apply to power generation: renewable energy sources are better than coal, oil or gas as detailed further in Chapter 7. If fossil fuels are to be used (e.g. coal and fuel oil on ships or at some power stations where alternatives are less practicable), then choice of a low sulphur fuel should eliminate emissions of SOX.

4.3.2 Minimise

If emissions to air cannot be prevented then actions or processes may be available to minimise them. Methane from waste landfill sites can be reduced by sending less biodegradable waste to landfill in the first place. Modern sites also trap and collect the gas and use it as a fuel. Waste reduction and recycling as described in Chapter 6 will deliver multiple benefits including reduction of methane.

In the case of dust arising from quarrying and construction, regular removal of solids from roadways and other parts of the site will help to prevent it being disturbed by vehicles and plant and blown into the atmosphere by the wind. Dust suppression by spraying with water will also help, although care needs to be taken that air pollution does not become water pollution.

Process control

Managing the processes that produce the pollution is also important. Operation of plant at optimum efficiency should give a higher yield of product for a given input and a lower level of pollution of all forms. This especially applies to vehicle engines, incinerators and boilers. The use of control systems and regular maintenance ensure optimum efficiency. The type of burner and temperature of combustion can affect the emission of NOX and other pollutants such as dioxins from incinerators. Monitoring of control parameters and emissions will help to check that the operation is not drifting away from that required. If the opportunity arises, the replacement of plant with newer, more efficient models will also be beneficial. A good example is the use of condensing gas-fuelled boilers to replace older versions for space heating and hot water production.

4.3.3 Use of removal technology

The next step in minimising emissions is to reduce the polluting substances by removing them from the effluent stream. If this is totally successful then it could also render the emission harmless. Methods are required for removal of solids, vapours and gases and sometimes all three may be present. The outline explanations of treatment processes in the following sections are presented to explain the principles of operation behind each method. The plant available is proprietary and differs in detail among manufacturers. The choice of which process to use is going to be based on the nature of the air flow (temperature, corrosiveness, etc.), the type of the pollutants present, the effectiveness for the emissions concerned, ease of installation and operation and cost. In most cases the removal still results in a solid or liquid waste that will require safe disposal. No method will work well with all pollutants; more than one treatment method may be required – for example, solids removal prior to activated carbon to prevent the latter becoming blocked.

In order to deliver the polluted air to the treatment plant, it will need to be captured and transferred through a pipe or duct to the inlet. Enclosure of the production stage will minimise the volume of air to be moved and ensure that dilution by clean air does not result in the need for a larger treatment plant. Fans are installed in the ducting to ensure that the flow is sufficient to remove the pollution and to prevent solids settling. The design and material of construction of the ducting and fan needs to be able to cope with the nature of the contents and be designed such that flow is smooth and noise from the motor and ducting is not a problem (see Chapter 8). If the pollution in the treated airflow is sufficiently low, the discharge may be made directly to atmosphere – as is the case with local exhaust ventilation (LEV) in many buildings. LEV may include a filter near to the discharge point but it may be just a simple replaceable filter pad. In all cases (LEV and the final emission from any treatment plant) the outlet needs to be sited where it will not cause a nuisance such as noise or smell to workers or neighbours.

Solids removal by gravity

Solids removal can be achieved by several means. Larger or denser solids may separate by gravity and can even deposit within the ducting and on the fan so the air flow in this stage needs to be maintained sufficiently high. If the air flow velocity is reduced in a large chamber the solids will settle but smaller particles may still be carried through. For large-scale operation the size of plant required makes it an unlikely choice.

Cyclone separators

The forces of gravity can be helped by the adoption of a cyclonic separator. The principles are best seen with the help of Figure 4.1. Air is directed into a vertical tube with a spin so that it rotates as it moves down the tube. The solids are thrown onto the surface of the tube by centrifugal force and then slide to the bottom where they collect in a hopper. On reaching the base, the air flow rises up the centre of the tube and exits at the top. The principle is similar to that used in a bagless vacuum cleaner such as that produced by Dyson®. On a larger scale there is a range of products with various features designed to help the process. Cyclones will separate larger or dense particles but the very fine particles are unlikely to be removed sufficiently and an additional process may be required.

Bag filters

An alternative to settlement is removal with a bag filter – similar in principle to a conventional vacuum cleaner. The problem with simple filters is that they clog up and have to be cleaned or replaced. This becomes impractical for high air volumes or heavy solids loads. Large-scale filters are constructed as bags which can be cleaned automatically and without removal. The filter can be made from paper (not for industrial filters) or natural or synthetic fibres. For a filter to be effective the spaces between the fibres in the bag need to be smaller than the particles to be removed. The choice of material will be affected by the nature of the particles and gas flow. High temperature or a corrosive gas mixture can damage the fabric; high moisture content can cause the particles to stick together and seal it. The design can collect the dust within a bag or on the exterior surface of the bag depending on the direction of gas flow. The solids have to be removed periodically as the cake of deposit builds up causing the rate of flow to reduce. This may be achieved by inverting the bag if the solids are inside, or reversing the flow or shaking the bag if the solids are on the outside. The solids normally fall to a hopper in the base for removal. Figure 4.2 shows the principles of one of these. A bag filter is suitable for dry solids provided that they do not form an explosive dust/air mixture (e.g. some fine metal dusts or flour).

Electrostatic precipitators

Another method of separating solids involves an electrostatic precipitator. The principle behind this can be demonstrated by rubbing a plastic comb or pen your sleeve and then passing it over small pieces of paper. The paper jumps on to the plastic and stays there. This is because rubbing causes a static charge to build up on the plastic and small non-conductive particles will be attracted to the charge. Scaling this up requires that a high voltage (50,000 to 100,000 volts) is applied to a metal plate and the particles are given an opposite charge causing them to be attracted to the plate. The details vary between different designs but a common type, illustrated in Figure 4.3 uses wires or a metal grid to apply the opposite charge to the particles. Fine solids can be removed along with materials in a mist such as sulphuric acid and fumes from welding. The contaminants stick to the plates and need to be removed by switching off the power or shaking them or by mechanical means such as a hammer from time to time. They are widely used to treat large air volumes for dust removal such as in a coal-fired power station. They are relatively energy efficient compared to a bag filter as there is lower resistance to the flow and they

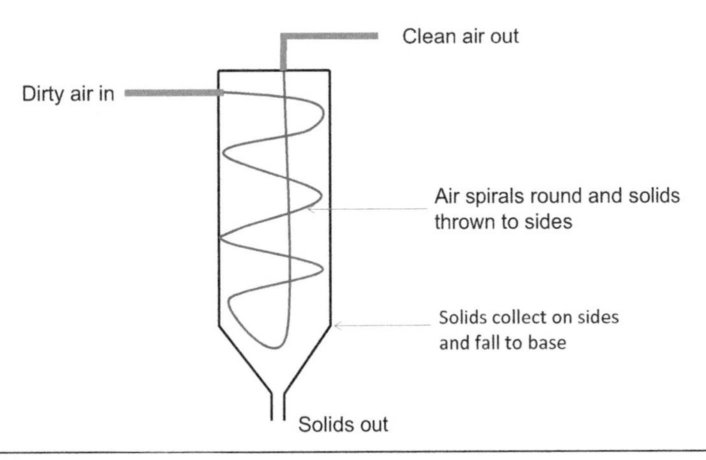

Figure 4.1 Cyclone separator

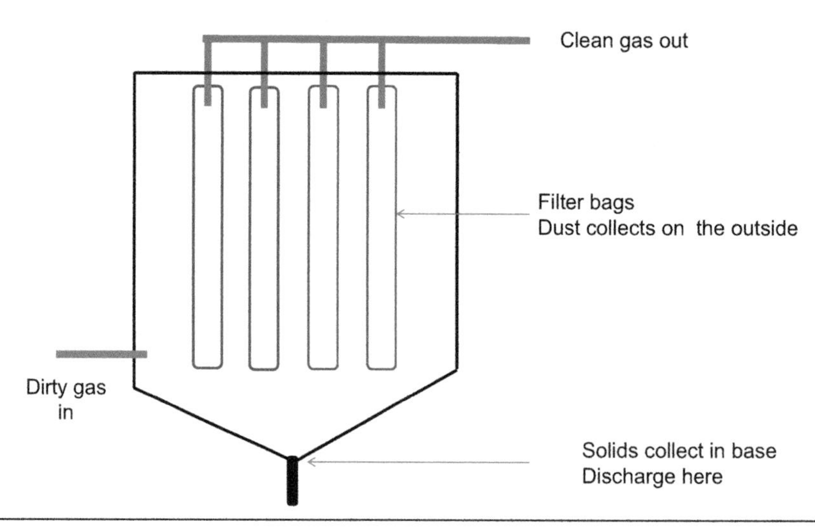

Figure 4.2 Bag filter

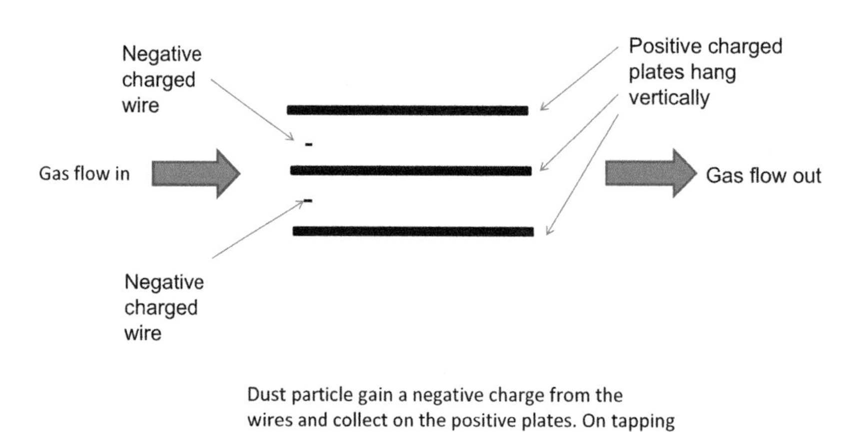

Figure 4.3 Electrostatic precipitator

can cope with high temperatures and corrosive materials given the right materials of construction. The system is not used for very wet gas streams or with potentially explosive mixtures.

A similar process utilises a magnetic separator. In this case the plates are magnetised either permanently or with electromagnets. Iron particles or other magnetic material (such as some metal ores) are attracted to the plate and can be removed mechanically or by switching off the power.

Wet scrubbers

None of the above methods will remove gaseous contaminants. In these cases some form of wet scrubber is used. These operate by passing the gas flow through a wall of water or up a tower with falling water sprayed inside. The gases dissolve in the water along with solids which may dissolve or remain in suspension. The principle is shown in Figure 4.4. The efficiency of the scrubber may be improved by including a packing such as perforated plates or stone or inert plastic shapes with a large surface area that can improve the contact between the gas and water. If the gas is acidic then the addition of an alkali will significantly improve removal. Flue gas desulphurisation (for the removal of sulphur dioxide) at coal-fired power stations uses lime slurry for this purpose and forms calcium sulphate, the chemical name for gypsum. This can be used to manufacture plaster products. Wet scrubbers are also used for flammable or explosive dusts where the solids can be washed out and collected as a safe sludge.

Adsorption on activated carbon

Most organic pollutants are not particularly soluble in water and so the most common method for removing them in low concentrations in a gas stream is to adsorb them on to activated carbon. This is a form of carbon granules derived from various sources such as coal, coconut shell or charcoal that have been treated by heating to high temperature to create an open structure. This

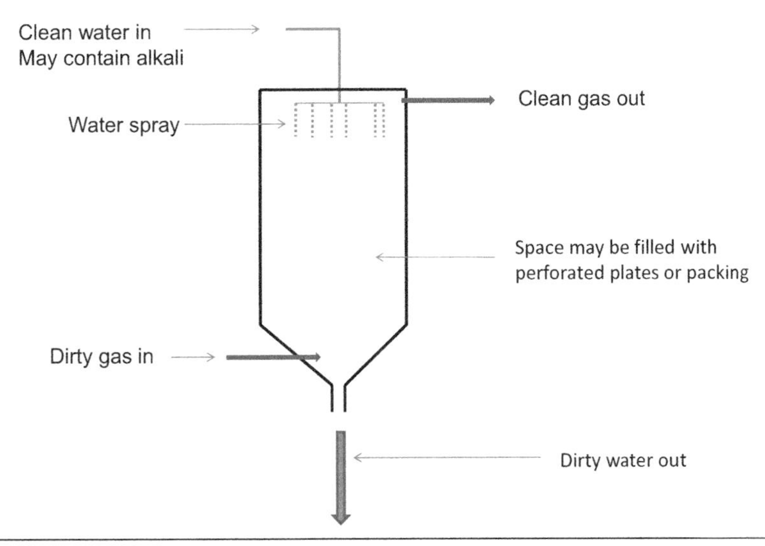

Figure 4.4 Water scrubber

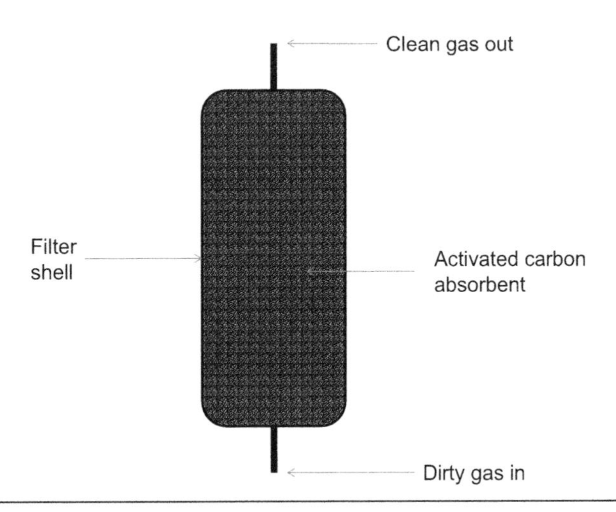

Figure 4.5 Activated carbon adsorber

is packed into a metal container as in Figure 4.5. The surface of the carbon is active in that it adsorbs materials on to the surface. It is particularly effective with organic substances and so will work with CFCs, solvents and substances such as odours not easily removed by other means. As particulates will also settle on to the surface these must be removed first. The carbon eventually becomes exhausted but it can be regenerated, usually by heating again to a high temperature.

Other methods

There are other forms of treatment that may be used for different air pollutants but they tend to be of specialist or limited use. The only one widely used but on a small scale is the catalytic converter that is included in the exhaust stream of vehicles. The catalyst is a precious metal – platinum or palladium – supported on an inert medium in a ceramic honeycomb structure. It works by converting the unburnt hydrocarbons, NOX and carbon monoxide in the exhaust gases from petrol engines into carbon dioxide and nitrogen. They do not remove NOX from diesel engines and this requires more sophisticated measures. Modern large diesel engines may also incorporate a diesel particulate filter to further improve the quality of the discharge.

Carbon capture and carbon storage

These terms are used to describe the capture and safe storage of carbon dioxide formed during the combustion of fossil fuels. It is not in operation on any large scale yet. The idea is to remove the carbon dioxide from the emission stream and then send it underground to an exhausted oil or gas well or similar structure where it can remain indefinitely. This would have to be done on a large scale to be effective. Most proposals use solvents to separate the CO_2 from the other components of the stream. The CO_2 then has to be recovered from the solvent and piped underground. There are a lot of technical problems with several aspects of the proposals and full-scale operation may be many years away.

Choosing and operating the removal process

Choosing which of the mainstream technologies to use depends primarily on which technique is most suitable for the emission. Choice is then refined by other issues such as corrosion and temperature resistance, ease of operation and maintenance, and cost. Power requirements for operation may be a problem in remote sites or regions where power supplies are unreliable. This means that these technologies tend to be reserved to developed countries. In undeveloped or developing regions of the world air pollution continues to be a major problem.

4.3.4 Render harmless

As mentioned above, some of these treatment methods may render a gas flow safe from hazard but it is possible that there may still remain some contamination as removal is rarely 100 per cent. The usual method at the end of the whole process is to ensure dilution in the wider atmosphere so that the remaining contaminants are no longer a threat. Dispersion away from the source will ensure that they are well mixed and diluted and do not fall out or concentrate locally. This can be achieved with a tall chimney, the height of which depends on the topography, wind direction and speed and other factors such as temperature of the gas flow. Complex modelling is required to get the design right.

CHAPTER

5

Control of contamination of water sources

After this chapter you should be able to:

1. Outline the importance of the quality of water for life
2. Outline the main sources of water pollution
3. Outline the main control measures that are available to reduce contamination of water sources.

Chapter Contents

Introduction 78

5.1 Importance of the quality of water for life 78

5.2 Main sources of water pollution 82

5.3 Main control methods available to reduce contamination of water sources 84

INTRODUCTION

About 70 per cent of the earth's surface is covered with water but most of that is ocean and too saline (containing dissolved salts) for many purposes. There is no universal shortage of water, just a shortage of water that can be easily and safely utilised by humans to meet their own needs. The sources of water that are on or under land are generally suitable for human use. These are precious in that they are essential to support the ecosystem and to supply our agriculture, industries and drinking water. This chapter is about protecting water quality to ensure that it is suitable for those purposes. The arrangements for water management in the UK are outlined in Box 5.1 in Appendix 4.

5.1 Importance of the quality of water for life

Water is fundamental for life and its importance and the strains that its availability is under were outlined in Chapter 1. Some plants and animals are adapted to live in an arid environment but most start to suffer and eventually die if dehydrated too much or for too long. They will also die if the water quality is not good enough through acute or chronic toxic effects. An additional problem that occurs in water is that some chemicals known as endocrine disruptors can affect the sexual development of fish and these have been blamed for reductions in fish populations. Various chemicals cause this and their effects on other water users are still under research.

To be suitable for use as drinking water or in agriculture for supply to animals or for irrigation it is important that the quality is such that it will not have toxic effects. The toxic effects may kill or harm animals and kill or damage plant crops or prevent their germination.

5.1.1 Safe drinking water

The definition of safe drinking water is that it does not cause harm to humans. This means that it will not cause immediate harm (acute toxicity) or long-term harm (chronic toxicity). We cannot rely on observing the effects to judge its suitability so in practical terms this means that the water must comply with recognised standards of quality. Most countries have their own standards but they are usually based on those of the World Health Organisation (WHO 2017b) or, within the EU, the drinking water directive (European Commission 1998, under review). The standards set limits for chemical quality and bacteriological quality. Harmful bacteria in water can cause illnesses such as cholera and typhoid. Drinking water is also known as potable water and if it meets the standards it can be called wholesome.

Safe drinking water is not universally available. In 2017 the World Health Organisation estimated that 10 per cent of the population were not using safe water sources (WHO 2017b). The problem is mainly confined to the poorest developing countries and the causes are mostly associated with contamination with bacteria derived from untreated sewage, although industrial pollution may also be a problem. The same source states that over 30 per cent of the world's population do not yet have access to adequate sanitation facilities. In these cases the provision of safe drinking water and sanitation is still at a much lower level of provision than in the developed world. Water supplies may be a shared spring or well but protected from pollution and near to their habitations. Adequate sanitation rarely involves a flush toilet.

5.1.2 The water cycle and sources of water

We are dependent on the water cycle (also known as the hydrological cycle) for our supplies. This is a natural process whereby water is transpired by plants and animals and evaporates from the sea, freshwater sources and land into the atmosphere, helped by the heat from the sun and by the wind. Given the relative surface areas, most water gets into the atmosphere from the sea. Evaporation leaves dissolved salts and other impurities behind and in the higher atmosphere the water condenses into clouds. Rain falls from the clouds and is essentially pure water unless it dissolves gases or picks up solids from the atmosphere. The formation of acid rain is one source of impurity as described in Section 4.2.2.

The rain collects in streams, rivers, lakes and man-made reservoirs as surface sources of fresh water, as shown in Figure 5.1. It can also percolate underground to collect in porous rocks known as aquifers, giving rise to the term 'groundwater', as shown in Figure 5.2. The aquifers are usually in limestone, chalk, sandstone, sand or gravel as these have an open or porous structure. These are not evenly distributed around the globe so that groundwater is not universally available. Groundwater is usually fairly

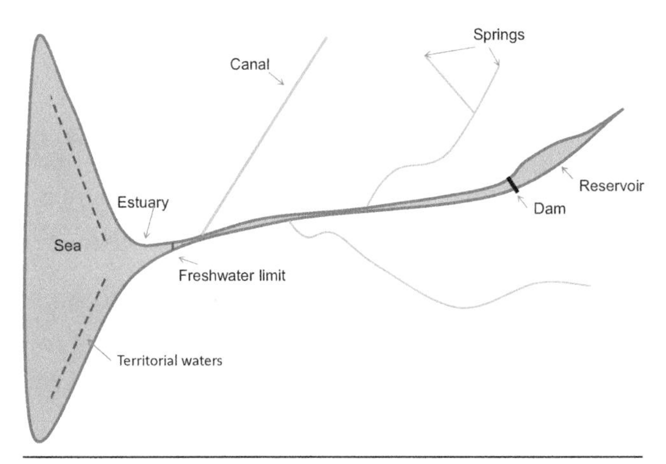

Figure 5.1 Surface water features

pure as a result of the percolation which filters out contaminants or removes them by chemical processes. The depth of the aquifers can vary from just below the surface to hundreds of metres down. The effectiveness of the filtration and purification processes depend on the depth as well as other features but groundwater near to the surface is usually more vulnerable to pollution from activities or pollution on the surface. The rate at which groundwater is recharged with water from the surface depends on the nature of the strata above and the overall depth. In many cases the water can take years to reach the aquifer, leading to the risk of over abstraction if the abstraction rates are not managed. In some parts of the world groundwater is not being recharged at a sufficient rate due to geological or climatic changes and the water that is being abstracted was deposited hundreds or thousands of years ago. It is effectively being mined. An oasis in a desert is usually supplied by groundwater that has travelled a long way or is very old water. Note also that in Figure 5.2 the aquifer continues under the sea. It is possible to draw salt water into the aquifer by over-abstracting the groundwater.

Surface water is more vulnerable to pollution, as we shall see later in this chapter. Man has worked hard to collect surface water and exploit it by installing pumps and building dams and reservoirs. In some regions of the world dams built in the upper reaches of rivers for local supply are taking water that is denied to potential abstractors downstream. This has been a source of conflict and could become more so as demand for water increases.

Surface water has to be treated to make it safe to drink by chemical treatment and filtration to remove impurities followed by disinfection to kill harmful bacteria. Groundwater may just require disinfection to make it safe. In the absence of adequate sources of fresh water, drinking water can be produced from sea water by desalination (i.e. the removal of the dissolved salts). This process involves either distillation or the use of high pressures separating the water from the salts with a special membrane in a process known as reverse osmosis. Both processes use more energy than traditional methods of treatment and so it is an expensive way to produce water. It is only used where there is no alternative (including on ships) or where energy is cheap.

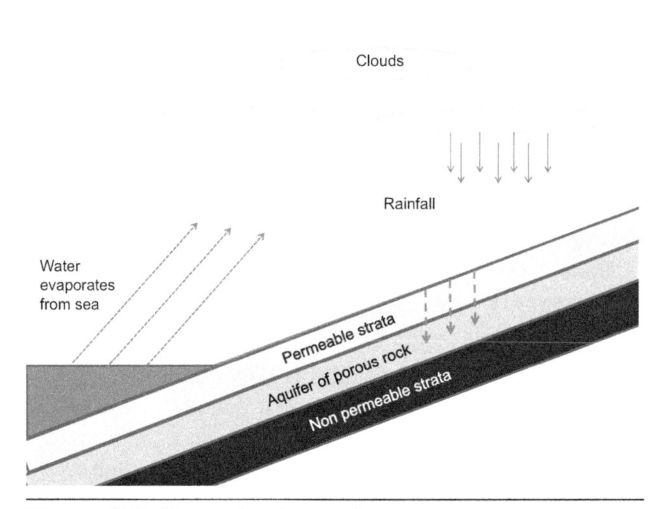

Figure 5.2 Groundwater recharge

5.1.3 Water for agriculture and industry

Water treated to drinking water standard is expensive and it is not necessary for many non-potable uses. Agricultural uses are for irrigation and watering animals and in both instances natural water supplies are used. Many animals drink from streams and irrigation water may be abstracted from surface or groundwater sources. As the need for irrigation water is associated with dry conditions, many farmers have small reservoirs to ensure availability at times of drought. Animals housed under cover may be given potable water and it is also used for washing equipment and surfaces.

Industry uses water for many purposes. Cooling water for power stations is usually abstracted from rivers or, for power stations on the coast, from the sea. The water needs to be screened to prevent fouling of pumps and other equipment but may not have any other treatment. If the water is discharged directly it will be warmer than the river and so may need to be controlled as described later in this chapter. The alternative is to have a closed loop for cooling water and cool the water in cooling towers. These are a feature of many large fossil fuel power stations. The white clouds associated with them under some weather conditions are not pollution in the normal sense but just water vapour.

Cooling water on a smaller scale is used in air conditioning systems for buildings such as offices, hospitals and computer suites. Box 5.2 is about the risk from legionella from aerosols from cooling towers and other sources.

Small industrial users of water may use groundwater if it is available or even potable water. Big users of water such as breweries and food processors may choose to operate on sites that have sources of water suitable for production available nearby. Water used for processes in the chemical and other industries may be taken from surface

Box 5.2 Legionella

Legionella are a type of bacterium present naturally in water. In most cases they do not present a health risk but if inhaled in water vapour or mist they can cause a disease known as Legionnaires' disease or Legionellosis which is a form of pneumonia. It can affect anyone but some groups are more susceptible than others, such as older people, smokers and those with pre-existing lung conditions. Outbreaks of the disease usually affect a few to tens of people and those most at risk may die. The presence of the bacterium in the vapour is most often associated with cooling towers such as those associated with air conditioning systems and shower heads but anywhere that an aerosol is generated needs to be treated as suspect. The bacteria can multiply in the warmer water within the system, especially if organic and other debris is around to act as a food source, and they then may be released in the vapour.

Control measures include attention to the design and cleanliness within the system and managing the temperature regime along with regular flushing and disinfection. There are companies that offer a service to check and maintain systems at risk. More details are easily available, for example, on the web site of the Health and Safety Executive in the UK at www.hse.gov.uk.

water sources or from groundwater or from potable water supplies. Some industrial uses demand water quality that is to a higher standard than for drinking water. The pharmaceutical, semiconductor and dyeing industries are the commonest ones but chemical industries may also have to treat potable water to a higher standard as well, removing the dissolved solids present by ion exchange or reverse osmosis to produce high purity water.

The issues of meeting the demand for water were discussed in Section 1.1.2 including the volumes of water used to produce some common products.

5.1.4 Impact of water pollution on wildlife

Wildlife is just as dependent as humans on water for survival. All need it to drink and some other species live in it so they are even more vulnerable to pollution. Those that just drink water are as prone to acute or chronic toxicity from the same chemicals as humans and will avoid water if they are able to sense that it is polluted by its smell or taste.

The importance of dissolved oxygen

Animals that live in water include fish, amphibians and many species of invertebrates (a very wide range of small animals and the larval stages of some insects such as dragon flies). They need oxygen in the water to survive and are sensitive to the concentration. Under normal atmospheric pressure and ambient temperatures water should contain over 10 mg/l of oxygen to be saturated (the concentration to achieve saturation does vary with pressure and temperature). If it falls below about 50 per cent saturation (5mg/l) then many fish will be in distress; much below that and they will start to die. Invertebrates are also sensitive to oxygen concentrations and the species present and the relative numbers are a good indication of water quality. Reduced oxygen concentrations are usually due to the presence of organic substances that are biodegraded by bacteria and other micro-organisms naturally present in the water and oxygen is consumed in the process. Organic matter is naturally present in surface water from decaying vegetation and run off from fields and is usually not rapidly biodegradable such that it is not normally a problem. If higher concentrations of organic matter are present from sewage or farm slurry or industrial discharges or from spills of organic substances (such as milk from farms or tanker accidents) then the oxygen is rapidly consumed. Under normal circumstances oxygen that is consumed in the water is replaced by exchange with the atmosphere at the surface but in the presence of a high organic load the rate of exchange is too low so the concentration drops quite quickly.

Photosynthesis occurs in plants in water just as on land and this will also replenish dissolved oxygen. Photosynthesis stops at night but aquatic plants continue to respire carbon dioxide; this is the time when water sources are most vulnerable to deoxygenation.

Other pollutants that commonly occur in water and can damage wildlife include:

- **Ammonia:** mainly from sewage, farm slurry and leachate from landfill sites and contaminated land; this is directly toxic to aquatic organisms and oxygen is also consumed as it is oxidised in the water.
- **Suspended solids:** from sewage and industrial discharges, poorly managed waste sites, run off from land and from mines and quarries. The smaller solids can blanket the river bed and interfere with fish spawning, block out light, interfering with photosynthesis and damage the gills of fish. Larger solids such as plastic bags and other articles can be ingested by fish and other sea animals where they may cause internal blockages and death.
- **Nutrients such as nitrate and phosphate:** these may also come from sewage and industrial discharges and from farms due to the application of fertilisers or farm slurry. They cause excessive growth of algae and other water plants (in the same

way that they help farm crops to grow) which can lead to flow inhibition in rivers and a series of water quality problems as the algae 'bloom' and then die when the nutrients are exhausted. The dead algae consume oxygen as they decompose and in deep reservoirs this can result in water becoming deoxygenated. The effects that are due to the presence of excess nutrients are known as 'eutrophication'.

- **Oil:** this is commonly present as a result of contaminated run off from roads and vehicle parking areas and from spills and illegal disposal of oil. It spreads thinly on water surfaces inhibiting the natural aeration of the water underneath and readily sticks to surfaces, fouling the river banks and vegetation. It also fouls the feathers of water birds such as ducks and swans and they are poisoned by ingesting it as they try to clean them.
- **Acidification:** acids can find their way into water from industrial sources and some natural sources but the main problem has been with acid rain because of the scale of impact. This has already been mentioned in Chapters 1 and 4. Wildlife is acclimatised to the water it lives in and the normal pH for fish and other animals to survive is around 7 to 8. Acidification of some waters has reduced the pH to 5 or 6 and this is too low. Acidification in the sea has not resulted in pH values this low but it is thought to be responsible for the damage to some corals.

Bioaccumulation and the progression of pollutants up the food chain have been discussed in earlier chapters and in Box 3.4. The risks are just as prevalent in water. Small invertebrates can ingest metals or pesticides; they in turn are food for small fish which can become food for bigger fish, or birds such as herons or animals such as otters. Water pollution was thought to be one of the reasons for the decline in the otter population in England in the 1960s.

Pollution of watercourses thus can cause extensive harm to wildlife and it is incumbent on organisations that find that they have caused it to report it to the relevant regulatory body and help to mitigate the effects and remedy any damage caused. This is covered in more detail in Chapter 9.

5.1.5 Abstraction and conserving water

Water can be abstracted by water utilities for supply to consumers and industry and by agriculture, power utilities and industry directly. In developing countries water may be collected directly by villagers from streams or springs or from wells and boreholes into the groundwater. In many locations the availability of water is insufficient to meet all the potential needs. In some places there is no regulation of abstraction and there can be disputes as an upstream abstractor on a river or stream denies water to another lower down. In extreme cases a dam on a river can also reduce availability to others downstream as the reservoir is filled. An abstractor can also take too much groundwater and lower the water level, making it unavailable to others unless they can lower their borehole base or pump.

Over-abstraction also has an effect on wildlife. Abstraction from rivers and streams reduces the flow and level, which will have adverse effects on the flora and fauna. It also reduces the dilution available for pollution. Over-abstraction of groundwater can affect flows in streams as the base flow is usually supported by springs which are derived from groundwater. It can also reduce water levels in wetlands, bogs and marshes where the wildlife often contains rare species that are adapted to those habitats.

As these problems become acute the government may set up a regulatory regime involving licences or permits to abstract (see Box 5.1 for the UK example). The regulator has to balance the demands from all potential users including water supply utilities. Determining abstraction rates has to take account of the water quality and how it is affected by discharges of effluent as well as leaving enough water to support the natural ecosystem of the watercourse. There may not be enough water to meet all the potential demands, especially if there is seasonal variation in flow and also in demand due to requirements for irrigation. In these cases a reservoir may be built to balance supply and demand either by a river authority, a water utility, a user or a consortium of users.

A licence or permit to abstract water is likely to set a limit on the daily or hourly rate of abstraction and may also have clauses that enable restrictions to be placed if there is insufficient water available due to drought (in the UK this would be through a drought order).

As the demand for water is increasing and it is becoming more difficult to find and more expensive (see Section 1.1.9), conserving water is good practice from several points of view:

- It is sensible as a business imperative to maintain and reduce costs such as the direct cost of water.
- Effluent charges are also often calculated on water use so they can be reduced as well.
- In-house costs for the capital and operating costs of water and effluent treatment will be lower and expansion for growth in production may be deferred.
- The energy costs for pumping and heating water will be reduced.
- It can help to maintain supply for production in times of scarcity such as drought.
- It can reduce pressure on the environment and in particular on wildlife.
- It will demonstrate responsibility to other users of the water.
- It should feature as part of the EMS, especially for organisations that are large users.

Water is often wasted, sometimes unintentionally:

- Taps and hoses are left running when water is not being used.

- Drips and leaks are not fixed.
- Taps are opened fully when only a little water is required.
- Cooling water is taken at too high a flow rate for the removal of heat.
- Water is not reused or recycled.

Good housekeeping and operator training can address some of these causes at little cost. Beyond that there is benefit in applying a hierarchy of approaches as has been applied elsewhere in this book. The key steps in the hierarchy for water conservation are:

- **Eliminate** demand: for example, by choice of variety or type of crops by a farmer, a change of process in industry or by collecting rainwater off the roof for flushing toilets.
- **Minimise** demand: for example, by introducing more water-efficient processes or plant, drip irrigation for agriculture, or low flush toilets and waterless urinals.
- **Reuse** water: for example, by adjusting the composition of a dye bath rather than discharging the exhausted bath and preparing a fresh one.
- **Recycle** water: for example, by redirecting cooling water into the process stream or slightly contaminated water such as from washing to flush toilets or for irrigation.

As with all such approaches it is important to monitor use by metering and relating the water consumption to output or other relevant measures. Departures from the normal trend should be investigated to see if new leaks have appeared or some changes in practice have crept in.

5.1.6 The potential effects of pollution on water quality

The impact of pollution on wildlife was discussed in Section 5.1.4 but the impacts are wider than that. Pollution may make the water unsuitable for other uses or at least make it more expensive to treat. In addition to the impacts on wildlife, the other sorts of problems that may arise include the following.

- Water may be unsuitable to abstract and treat for drinking water except at excessive cost due to the presence of toxic substances (e.g. arsenic or pesticides).
- Water may be unsuitable for irrigation in agriculture or horticulture due to herbicide residues from their manufacture, storage or use.
- There may be inhibited flow and increased flood risk due to excessive nutrients from sewage or agricultural practices that can cause eutrophication or excessive plant growth in surface waters and from illegally dumped solid waste such as tyres or supermarket trolleys.
- Deoxygenation and anaerobic conditions can kill wildlife (see Section 5.1.4) but also make the water unsuitable for other purposes.
- Water at too high a temperature will not be as effective for another abstractor with a cooling demand.
- Water may be visually unpleasant due to chemical residues that may not be toxic, such as dye residues, oil film, windblown waste or detergent foam.
- There may be damage to pumps or siltation and increased flood risk from excessive solids in suspension.
- There may be odours caused by anaerobic conditions arising from low dissolved oxygen concentrations.

This is not an exhaustive list as local circumstances such as what gets into the water, the concentration or amount and how it may impact are very variable. It should also be noted that discharges of some chemicals in sewage works or industrial effluent treatment plant may cause the plant to malfunction by damaging or killing the micro-organisms that they depend on (see below). This could result in discharges of untreated or partially treated effluent to a watercourse.

5.2 Main sources of water pollution

The previous section was mainly about the impacts of pollution on wildlife and other users of water. This section is about the generic causes which, for an EMS, would be relevant to the aspects. The relative importance will reflect local circumstances. An urbanised or industrial area will have many different potential sources of pollution (such as sewage or industry) to a rural one but the risks in rural areas could be heavily influenced by the activities on the land, ranging from agriculture and forestry to mining and quarrying. Rural sources of pollution can be more serious than urban because they are less immediately obvious in many cases, are out of sight of the main populations, are often tolerated as the price of development and may not be as effectively regulated. This is particularly true of remote rural areas in developing countries. The following sections expand on the main threats to water quality; Figure 5.3 illustrates some of them.

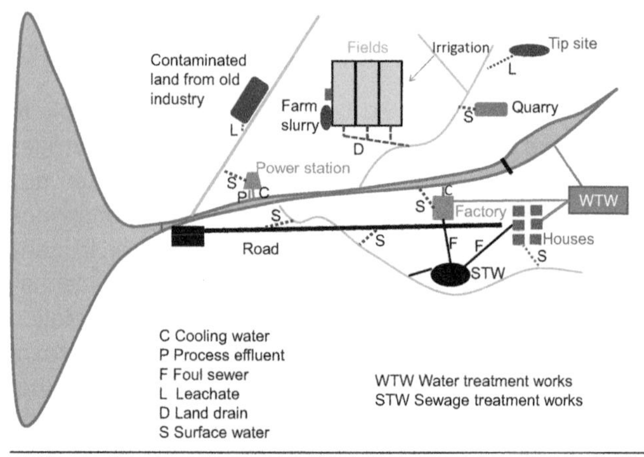

Figure 5.3 Pollution threats to surface water

5.2.1 Surface water drainage

Surface water can find its way into streams and rivers by a number of routes. It can run straight across the land, especially if it is steep or it is impermeable due to soil compaction or artificial surfaces such as concrete. It can also flow into natural or man-made swales and ditches or can run into drains from fields, roads or urbanised and industrial areas that eventually discharge to surface water. It can collect in ponds or similar man-made structures and then be pumped out. Most of the water will arise from rainfall and as it hits the ground it should not be contaminated unless it is subject to air pollution. However, as it flows across the ground it starts to pick up anything that may be present either in suspension or, if soluble, in solution. As the flow increases with heavier rain, or as the flows come together, the scouring effect increases as well such that the amount of suspended material can also increase and new material can be exposed which can also become a problem if it is contaminated. Other flows can arise from liquids that are spilt onto the surface and find their way into watercourses through the drainage systems that deal with rainfall.

The main sources of pollution from surface water drainage are:

- Solids, organic matter, soil or vegetation washed off or scoured from natural surfaces such as fields or mountain slopes.
- Discharges from land drains that collect water from under fields that may be contaminated with slurry, pesticides or fertilisers.
- Surface water carrying solids and other materials eroded from land that has been disturbed by ploughing, timber felling or construction works.
- Solids and contaminants such as oil, combustion emission products and tyre resides washed from roads and vehicle parking and service areas.
- Spillages of oil, fuel and trailer contents associated with transport of the products or road traffic accidents.
- Contaminants from industrial sites such as solids and liquids spilled onto ground as they are used or containers are filled.
- Run off from sites where waste is stored and not properly contained.
- Water that is used for washing down plant, machinery and vehicles, cleaning surfaces and fire water.
- Illegal discharges (for example from process plants or sump oil) into water drains and gullies.

5.2.2 Mining and quarrying

Mining is the term used for extraction of coal or minerals from underground by sinking a vertical shaft and horizontal tunnels. Open-cast mining describes excavation nearer to the surface where top soil and other materials (overburden) are removed to reach the sought material. Quarrying is removal of material such as rock for building materials and limestone for the chemical and cement industries from a vertical face. In each of these cases there is usually a lot of unwanted material: material excavated from shafts and tunnels, top soil and other overburden or materials that are extracted with the rocks or minerals sought but which have no value. In addition there may be sorting, grading and washing of the extracted resources and possibly some purification by chemical and physical processes on site. All of these produce waste material which is collected in heaps on site and poses an environmental threat. Mines and quarries are also dusty sites with fine solids spread on the adjacent land and roads by wind, rain, mining equipment and vehicles. Solids can be washed off by rainfall and there may be leaching of contaminants as water percolates through the heaps. These will find their way into surface watercourses unless intercepted by some form of settling tank or treatment process. Deep mines can also collect water which has to be pumped out to prevent flooding. This can be acidic and contain iron and other metals in solution.

The suspended solids are responsible for the impacts described above. The soluble material, especially the iron salts, can precipitate out and form further fine solids. There is also likely to be copper, zinc, lead, etc. or whichever metal ore is being mined and other associated metals such as arsenic which may be present in the same ores. Most of these will be toxic to organisms which live in water. Mercury is used in the purification of gold at the sites of some mines, particularly in small-scale operations in developing countries and this is released into the environment.

5.2.3 Process water

Many industrial and agricultural sites use water for cooling, washing materials, product or produce, or as part of their process operations. Domestic, commercial, industrial and agricultural wastewater may be sent to a sewage treatment works to be treated or else treated on site prior to discharge. In these cases the treatment process should be to a standard that means that significant pollution of the receiving watercourse is avoided. However, drainage systems often have overflows to cope with high rainfall (a particular problem with combined sewerage systems that receive surface water drainage as well as foul sewage) or there can be leaks or pump or treatment plant failures that can result in untreated wastewater entering the watercourse. Section 5.3.4 outlines the methods available for treatment based mainly on sewage works. It also indicates how they are applied. Similar methods are used for process water.

Sites may also contain disused tanks or pipework which still contain contaminated material that can escape if the containers leak through damage or corrosion. If any of these are underground, leaks will not be obvious and the leakage may also get into groundwater.

5.2.4 Agriculture and horticulture

Sources of pollution from agriculture have been mentioned several times; the risks from horticulture are similar although wastes associated with animal husbandry should not be relevant. In summary the main sources are:

- Run off from fields contaminated with slurry, fertiliser or pesticides and herbicides.
- Slurry stores leaking or, in exceptional cases, collapsing.
- Run off from washing down animal housing and hard surface areas in milking areas and compounds.
- Spillages of milk or washing down dairies.
- Washings from the preparation of farm produce.
- Oil spills and leaks from storage tanks and vehicles.

There are codes of good agricultural practice that are designed to minimise the risks from most of these activities. Various UK codes can be found in several sources but a recent consolidation of the main ones is available on the DEFRA web site (DEFRA 2009a).

5.2.5 Natural sources

There are natural sources of pollution that are the result of the underlying geology. A few examples are:

- Iron and manganese can occur in groundwater that is anaerobic.
- Soluble arsenic (e.g. in Bangladesh) and fluoride (e.g. in India) have caused health problems from the consumption of contaminated groundwater.
- Volcanic regions can produce springs which contain acidic water, and chlorides, metals and sulphides in solution.
- High chloride concentrations can occur due to the infiltration of sea water into groundwater or the deposition of salt deposits underground in the distant past.
- Radon (a radioactive gas) is found associated with groundwater and springs in areas which have underlying granite geology such as Cornwall and Scotland in the UK and the Pyrenees in France.

5.2.6 Contaminated land and other surface activities

The surface soil and the underlying layers often become contaminated as a result of leaching from spoil heaps, industrial activities, coal gasification, spillage onto the surface or waste disposal on or under the surface. Typical contaminants include oil, ammonia, metals, solvents and mixtures of other organic compounds. Water leaches the contaminants out of the land and they pass into solution and down to the groundwater in a similar way to the normal recharge by rainfall as shown in Figure 5.2. Leachates can also find their way into surface water by horizontal migration depending on the topography and the nature of the underlying strata.

The contamination can also occur or be exacerbated by chemical changes underground. Old landfill sites are a particular problem as when the organic waste decomposes it creates anaerobic conditions which help solubilise metals and release ammonia and soluble organic matter into the leachate.

There is more on contaminated land in Section 6.5.

5.2.7 Larger solids

Fine solids in suspension (suspended solids referred to above) cause their own problems. Larger solids such as paper, plastics and the detritus of a developed society (such as traffic cones, large pieces of timber, builders' rubble, tyres and the solid contents of sewage discharges) may appear in watercourses due to deliberate dumping, poor waste management or overflows on drainage systems. This is largely a problem in urbanised areas where urban watercourses may drain poorly managed sites or be hidden from view for part of their route. The solids may not cause harm but they are unsightly and result in watercourses that are objectionable from the perspective of the public.

5.3 Main control methods available to reduce contamination of water sources

The previous two sections have outlined the main sources and potential effects of pollution on water quality. From various perspectives this is something that should be avoided as described in Chapter 3. If there is an EMS in place it is in conflict with the requirements and the concept of continual improvement should be applied to deal with the sources of the problems. Guidance for avoiding water pollution is available for a wide range of industrial sectors on the web sites of the environmental agencies and HSE. A sample is in the list of Pollution Prevention Guidelines (PPG) published by the UK environmental agencies in Box 5.3 in Appendix 4 and their guidance should be of use elsewhere.

5.3.1 Applying the control hierarchy

A control hierarchy has been already introduced in Section 5.1.5 for conserving water. In the case of reducing pollution the details are slightly different:

- **Eliminate** the source of pollution by changing the materials used for those with less polluting potential or by eliminating the potential for loss or discharge.
- **Minimise** the risk of pollution by improved site management or by incorporating new or enhanced effluent treatment processes.
- **Render harmless** any residual polluting material.

Box 5.4 gives some examples of how these steps can be achieved and there are additional details in Sections 5.3.3 and 5.3.4.

Box 5.4 Some examples of applying the control hierarchy to reducing water pollution

Some of the following examples are covered in more detail in Section 5.3 and in other chapters.

Eliminate the source of pollution

- Change the formulation of paints from solvent base to water base.
- Replace cadmium as a plating metal with suitable alternatives.
- Use biodegradable lubricating oils or less persistent herbicides.
- Remove redundant process plant, storage tanks and pipework and materials.
- Contain polluting material safely and removing it from site to be dealt with elsewhere.
- Seal pathways that allow pollution to enter watercourses such as blocking access to surface water drains or applying modern lining techniques to waste landfill sites.
- Seal off or contain and treat leachate from contaminated land.

Minimise the risk of pollution

- Review operating procedures and introduce training programmes for operatives.
- Supervise deliveries and the collection of wastes.
- Enhance the frequency and range of site inspections.
- Plan the site layout to reduce the risk of pollution reaching drains or watercourses.
- Build bunds around storage tanks and store smaller containers on suitable containment trays.
- Build settlement lagoons or oil interceptors into site drainage systems.
- Add additional or improved effluent treatment processes such as nutrient removal at sewage treatment works.
- Improve the monitoring and control of treatment processes with instrumentation.
- Apply animal slurry and fertiliser to land in accordance with good agricultural practice.

Render harmless any residual polluting substances

- Treat effluents to a sufficiently high standard.
- Apply chemical treatment to render polluting substances inert such as precipitation as a hydroxide or sulphate that can be settled or filtered out and safely removed.
- Neutralise acids or alkalis.

5

5.3.2 Regulatory controls to discharges applied at source

The management of water quality is concerned with controlling discharges to water in order to protect the quality and avoid pollution that could harm wildlife or affect the other users of the water. In most countries causing water pollution is a criminal offence so a discharge has to be authorised to make it legal. It remains legal provided that it complies with the terms of the authorisation. Non-compliance with the terms of an authorisation may also be a criminal offence. The authorisation may be known as a permit or consent to discharge. In the UK permits are issued as described in Chapter 1.

Permits to discharge to watercourses

In determining the likelihood of issuing a permit at all and the conditions that it may contain, the regulatory body has to consider the potential impact of the discharge under normal flow conditions and under conditions of low flow in the watercourse. The determination has to protect existing water quality and take account of any objectives that may be in place for improvement. In addition there may be standards for watercourses and groundwater that also have to be met. For example, within the EU there were a number of directives containing standards to protect fisheries and shell fisheries, to protect water for abstraction to provide drinking water, concerning the discharge of dangerous substances and to protect groundwater. These have been replaced by a new directive abbreviated to the Water Framework Directive (WFD), which was phased in over several years to 2015 (European Commission 2000a). By this date inland and coastal water bodies were expected to achieve 'good chemical and ecological status'. These are defined in the directive either directly or by reference to other EU directives. National governments have to establish their own standards in accordance with the WFD.

The calculations of the concentration limits for substances present in the discharge can be complex and involve mathematical modelling to take account of the variations in flow and quality of both the discharge and the receiving water in order to achieve compliance with the appropriate standards for the watercourse. The final permit will contain details of the organisation involved in making the discharge and some or all of the following conditions:

- A description of the trade or process that the permit refers to (changes may mean that a new permit is required as the contaminants present may be different).
- A maximum flow rate (as the flow rate affects the load of polluting substances discharged).
- Concentration limits for a range of parameters (see below).
- Details of the site and representative sampling points.
- Any controls on discharge times or the state of the tide in an estuary (to avoid discharging to an incoming tide for example).
- Any monitoring and reporting requirements.

The parameters (a term used for substances that are subject to control in a permit) will be some or all of:

- Measures of organic load: BOD, COD or TOD (see Box 5.5).
- Suspended solids.
- pH.
- Temperature.
- Other substances likely to be present depending on the trade or process.

Examples of the last of these could be:

- Metals such as copper or chromium from a refinery or metal plating works.
- Organic chemicals such as pesticides or solvents from a chemical manufacturing site.
- Ammonia and nutrients from a sewage works, farm or food processing site.
- Colour, detergents and some metals from a textile or dye works.
- Oil from a refinery, metal processing or vehicle repair shop.
- Anything toxic such as arsenic or cyanide.

The EU Water Framework Directive (see above) identifies the requirement to control a list of priority substances. There was a period of transition from an earlier list of 'hazardous substances' which are toxic, persistent and liable to bio-accumulate and other substances and a list of 'non-hazardous pollutants' (these lists were known as lists I and II of even earlier legislation). These are subject to special consideration and, if present, will be included in a permit to discharge. The new list is in a directive (European Commission 2006b) which is periodically updated. Further information is given in Box 5.6.

Box 5.5 Measures of organic carbon in water

Organic substances present in water may be a mixture of hundreds, even thousands, of different compounds. It would not be possible or even desirable to measure all of them; most are not a problem and the individual concentrations should be low. Those that are toxic such as pesticides may need to be measured individually but for the majority, many of which may be naturally occurring, the interest is on the collective impact on the environment. In the case of organic compounds the concern is the biodegradation of the compounds by micro-organisms (mainly bacteria that occur naturally in all freshwaters) removing oxygen from the water in the process. This can lead to too low a concentration to support fish and other aquatic life. There are three measures that are used for permitting and monitoring: BOD, COD and TOD.

Biochemical oxygen demand (BOD)

This is the traditional measure of organic matter originally used to measure river water quality and the organic load in effluents. It aims to mirror the natural processes in water as the sample is inoculated with a suspension of micro-organisms and incubated for five days at 20°C. The oxygen concentration is measured at the start of the incubation and again at the end. The difference is the BOD: the amount of oxygen (demand) consumed by the biochemical processes. Note that the biodegradable organic carbon present is expressed as an equivalent concentration of oxygen, not as carbon.

The analysis is fairly simple to undertake although care is needed in measuring the oxygen concentrations. As the solubility of oxygen is only about 10mg/l, strong effluents would quickly consume all the oxygen and an accurate measurement at the end would not be possible. Consequently effluents usually have to be diluted with pure water (with no BOD of its own) using a range of dilutions to span the likely range and then the true BOD calculated. The test only measures biodegradable organic compounds (which is what is usually of interest) so the total organic load will be higher than the BOD.

Chemical oxygen demand (COD)

The analysis for BOD takes too long (five days) to be of use for plant control purposes. So another analytical method used to measure organic carbon concentrations in water is COD. There are variations to this method but they are all based on boiling the sample to be measured with an oxidising agent in an acidic solution for a fixed time. The commonest oxidising agent used is potassium dichromate: this will oxidise many organic compounds (but not all) to carbon dioxide. The amount of potassium dichromate left at the end of the boiling period can be measured and the difference from that at the start calculated and converted into the equivalent amount of oxygen, giving the COD. This measure is usually higher than the BOD and the relationship between the two is variable for different effluents although it is usually within a narrow range for different samples of the same effluent.

The result can be available in less than a day so the COD is better for monitoring and controlling effluent plant performance. COD is often used by sewage treatment companies for measuring the load discharged by industry to their sewers and for charging purposes.

Total oxygen demand (TOD)

This test is an even faster method for measuring organic carbon concentrations. Results take a few minutes but the equipment is more expensive and less widely used. This method oxidises the organic carbon by injecting the water sample into a stream of oxygen in an inert carrier gas. The special instrument used can directly measure the amount of oxygen taken to oxidise the carbon. The relationship between TOD and BOD or COD is also variable but within a narrow range again for different samples of the same effluent.

Box 5.6 List I and List II substances as the basis for priority control

This original list is based on Council Directive 76/464/EEC (European Commission, 1976) which has been repealed but elements of it and the daughter directives are still being used.

List I requiring emission limits or water quality objectives

- Organohalogen compounds and substances which may form such compounds in the aquatic environment.
- Organophosphorus compounds.
- Organotin compounds.
- Substances in respect of which it has been proved that they possess carcinogenic properties in or via the aquatic environment.
- Mercury and its compounds.
- Cadmium and its compounds.
- Persistent mineral oils and hydrocarbons of petroleum origin.
- Persistent synthetic substances which may float, remain in suspension or sink and which interfere with any use of the water.

In addition more specific compounds have subsequently been added to the list and daughter directives have been agreed.

List II requiring pollution reduction programmes

The following metalloids and metals and their compounds: zinc, copper, nickel, chromium, lead, selenium, arsenic, antimony, molybdenum, titanium, tin, barium, beryllium, boron, uranium, vanadium, cobalt, thallium, tellurium, silver.

- Biocides and their derivatives not appearing on list I *(NB including pesticides)*.
- Substances which have a deleterious effect on the taste and/or smell of the products for human consumption derived from the aquatic environment and compounds liable to give rise to such substances in water.
- Toxic or persistent organic compounds of silicon, and substances which may give rise to such compounds in water, excluding those which are biologically harmless or which are rapidly converted in water into harmless substances.
- Inorganic compounds of phosphorus and elemental phosphorus.
- Non-persistent mineral oils and hydrocarbons of petroleum origin.
- Cyanides, fluorides.
- Substances which have an adverse effect on the oxygen balance, particularly ammonia and nitrites.

These lists cover a wide range of substances and the details are regularly under review. The latest information is given in European Commission (2006b) but it is important to check the current status if more information is needed.

Discharges to sewer

Discharges to sewer will require authorisation by the sewerage operator in the form of a consent or a legal agreement. The discharge may damage the sewer (for example if it is acidic), cause dangerous conditions within the sewer (such as hydrogen sulphide or cyanide gases or inflammable vapour) or damage or interfere with the operation of the sewage treatment plant due to interference with the processes employed. BOD and solids may still be easily removed but other substances may not and could cause the sewage works to breach its own permit to discharge to a watercourse. The authorisation from the operator will therefore be similar to that for a permit to discharge to a watercourse directly although the contents and any limits applied will differ. Within the EU the operator will also need to be informed of any hazardous substances or non-hazardous pollutants as above and will apply limits. These may force the discharger to sewer to install some preliminary treatment plant on site. In the UK the operator may charge for trade effluent discharges using a formula usually based on volume and load of solids and COD so it is in the interest of the discharger to manage these to minimise costs.

Discharges to groundwater

Because any contamination of groundwater may take many years to degrade or clear, discharges onto land or into groundwater are generally refused unless there is no risk. For example, uncontaminated surface water or the discharges from small domestic sewage treatment plants are often acceptable in the UK. Infiltration from surface activities including leachate from landfill and land disposal of unused agricultural chemicals such as pesticides or sheep dip should also be avoided. Groundwater is included in the Water Framework Directive which is complemented by a directive to protect groundwater intended for abstraction for drinking water (European Commission 2006c) and any discharge involving hazardous substances or non-hazardous pollutants will require a permit and may be refused. The directive also requires that nations protect groundwater; in the UK this is achieved through source protection zones. The Agency has defined three groundwater source protection zones around all water supply boreholes and prescribed the activities that may take place within them. An outline can be found at http://apps.environment-agency.gov.uk/wiyby/37833.aspx.

Monitoring discharges and water quality

Permit conditions often require the discharger to monitor the quality of the discharge and sometimes the receiving watercourse. They may require a continuous monitor for some parameters such as temperature, suspended solids or pH. Even if there are no such conditions, it is sensible to do your own monitoring to check compliance and that everything is working properly.

The first step in monitoring is taking the sample and this is not as easy as it sounds. Getting a representative sample means ensuring that it is taken from a point where the flow is well mixed, which may mean after a pump or valve or well beyond any point where flows come together. As the concentrations to be measured are usually low (in mg/l or µg/l), it is important to avoid contamination of the sample or the container. This can be difficult in dirty or dusty locations where there is potential contamination from the substances to be measured.

The second step is the analysis. Again the low concentrations to be measured mean that contamination has to be avoided in a laboratory which may be mainly concerned with other work. Another issue is the sensitivity needed for the analysis. Many works laboratories are measuring the concentrations in products where the units of measurement are per cent or g/l. The methods and equipment required to measure at much lower concentrations are different and generally more expensive. It may make sense to use an external contract laboratory for analysis and maybe sampling as well.

Instruments can give useful information even if they cannot measure the substances on the permit. For example if the concentration of metal in the effluent is dependent on pH, then monitoring the pH may be enough to assure that all is working. There are also simple colorimetric tests for many substances that may not be as accurate as other methods but suffice for simple checks.

5.3.3 Managing the storage of materials

Spills and leaks are common sources of pollution by liquids (Section 5.2.1). The material finds its way directly into a watercourse or flows there by means of a drain and it can also soak into the ground causing land contamination and posing a risk to groundwater. Causing pollution in this way should be avoided for the reasons previously given. The risks can readily be reduced by site layout and design and by the adoption of good operating procedures. A recommended approach is to use the source, pathway, receptor model to manage the risks. The receptor is the watercourse, land or groundwater. The source is material stored and the storage or potential leak site and the pathway is the surface over which a spill can flow or the drain that may carry it to a watercourse. The risk can be reduced by tackling the source or the pathway. In the rest of this section the material posing a risk is sometimes abbreviated to 'chemical' but it could be oil or fuel and these have some additional features which are dealt with as required.

The quantities held are an important consideration. The greater the volume, the greater the risk and the greater the potential damage from a spill. Chemicals may be held in small containers, standard drums of 205 litres or in bulk storage with thousands of litres. Small containers should be kept in secure cupboards or storerooms where any spills are contained in suitable trays. The management of drums and bulk storage containers and associated pipework is where the main effort should be concentrated.

Storage (and other aspects) of dangerous chemicals is covered in the UK by the COMAH regulations (COMAH 2015) which apply to sites holding more than certain amounts of petroleum products or other listed chemicals (see also Box 9.2). Installations subject to permitting under the IPPC regime may also have requirements in their permit. The PPG series in Box 5.3 covers most aspects of storage. The COSHH (Control of Substances that are Hazardous to Health) regulations, which are concerned with health and safety and reducing the exposure of workers are also likely to apply so the two may be managed in parallel.

Reducing the risk

An obvious initial question to consider is whether it is necessary to hold the chemical that presents a potential hazard on site at all. Substitution of organic solvents with water is becoming common and other examples are in Section 5.3.1 and Box 5.4. If it is unavoidable then minimising the amount held at any time can also help to reduce the risk.

The next consideration would be the location of the chemical storage and any associated equipment. Ideally it should be as far away from groundwater abstractions, watercourses and surface water drains as possible. PPG 2 (Box 5.3) recommends at least 50m from a spring, well or borehole and 10m from a watercourse for oil storage above ground. A part of the site should be selected that is least vulnerable to accidents involving vehicles, fork lift trucks and other hazards such as flooding. Additionally, storage should be as far away as practicable from office accommodation and adjacent housing or business premises.

The storage site also needs to be considered from a safety and security point of view:

- Floors or ground should be level, stable and impermeable.
- Containers should not be stacked in an unsafe manner nor too high.
- Wherever practicable storage should be under cover to exclude rainwater.
- The storage areas should be kept clean and tidy with no trip hazards present and packaging and other waste removed regularly.
- Storage buildings and compounds should be secure with locked and controlled access.
- Crash barriers may be required if site traffic presents a risk.
- Small containers should be kept in a locked cupboard or store room.
- Labels showing common hazard pictograms as in the Classification, Labelling and Packaging (CLP) Regulations should be on the container labels already (European Commission 2008c; these apply world-wide), more information can be found at www.hse.gov.uk.
- Safety signs should be posted indicating the potential hazards (toxicity or flammability for example) and any special precautions needed.

Figures 5.4 and 5.5 illustrate some bad practices for the storage of drums.

Another consideration is the location of storage of individual substances. For example incompatible materials should not be stored closely together, especially chemicals that may react together and create an additional hazard due to a violent reaction or the release of heat or fumes. Inflammable materials and waste should not be stored nearby in case of fire. Particular care is needed with some substances:

- Acids and alkalis can react together releasing heat.
- Concentrated acids and some other chemicals react with water to release heat, sometimes violently.
- Acids corrode iron, aluminium, many other metals and concrete.

- Alkalis corrode aluminium and lead.
- Acids react with chlorine based bleach (solids and liquids) to release chlorine gas.
- Volatile solvents and fuel need to be kept cool and away from sources of heat or ignition.
- Oxidising agents such as sodium chlorate or ammonium nitrate can react with inflammable materials such as paper, fabrics and dusts causing fire or explosion.

This is not a complete list of potential hazards. Suppliers should provide data on safety and storage as well as any environmental hazards. In the UK and many other countries these are referred to as material safety data sheets (MSDS) or safety data sheets. These contain information relevant to safety and to environmental risk and are required by law under the REACH regulations of the EU (European Commission 2006a) for specified chemicals over certain quantities. The packaging should also have relevant information with warning signs as mentioned above. If the supplier does not supply this information they should be avoided.

The choice of container is important. Small quantities of chemicals are often supplied in glass or polyethylene containers. Larger volumes may be in steel or polyethylene drums. Bulk storage can use stainless steel, glass lined steel or plastics reinforced with glass fibres. Each has its advantages and disadvantages related to stability in the presence of the chemicals stored and resistance to external corrosion and accidental damage. The supplier will specify what a chemical arrives in and how it is to be stored. This information needs to be taken into account when transferring chemicals from one container to another. Containers should be protected from damage and inspected regularly for signs of corrosion, leaks or damage.

Figure 5.4 How not to stack drums

Figure 5.5 How not to stack drums 2

Drainage

The site will require surface water drainage to deal with rainfall and there will probably be other drains nearby to deal with sewage and other effluents for discharge to sewer or to a watercourse. Chemicals need to be stored on impermeable ground (such as concrete) to avoid any spilt material soaking into the ground. The risk then is that any spilt chemical may get into one of the drains or be washed in by rainfall. The latter can be avoided by ensuring that all storage areas are roofed to keep rainwater out and uncontaminated. The surface water and foul water drains should be kept separate (i.e. not cross-connected) and clearly marked for identification by colour coding the manhole covers and any visible parts as shown in Figure 5.6. A plan needs to be available showing the layout and all access points, valves, etc. The detailed design features of the storage can minimise the risk of spillage to the impermeable surface (see below) but if it

Figure 5.6 Colour coded drains

does happen then procedures need to be in place to seal off any open gullies, drain covers or other access points and to valve off or block up the drain with a suitable plug (various designs are available commercially) to contain any spilt chemical.

If oil, diesel or petrol is on site (or any other liquid that is immiscible with water and has a lower density such that it floats) then an interceptor can be installed which will separate the less dense material and retain it. This is illustrated in Figure 5.7. PPG 3 referenced in Box 5.3 gives more information. The installation of an interceptor is a common requirement at oil refineries and depots, fuel stations, large vehicle parks and garages, as well as sites that use large quantities of these materials. Its purpose is to capture any minor leaks and spills as well as to deal with a more serious incident. Interceptors need to be inspected frequently and any oil retained needs to be pumped out as, if the oil level builds up inside, it can be flushed out by heavy rainfall. Solids that collect on the bottom will also need to be removed.

Lagoons or drainage ponds are often used on large drainage areas such as motorways or service area car parks which may hold large volumes of fuel and are vulnerable to multiple spills and leaks from vehicles or accidents. There the surface water is collected from the drains and the flow slowed to allow solids to settle as well as holding back any oil. They may be designed with reed beds to help purify the water by removing dissolved organic matter before it overflows to a stream. These sites are often required in areas where drainage is to small streams with little dilution available. Lagoons also help to manage the flow from the site in heavy rain to reduce the risk of flash flooding. They will often attract wildlife such as water fowl so that there is a potential problem if an oil contamination is significant. There are other techniques available for managing water flow and quality from large paved areas. They are known as sustainable drainage systems (SuDS) and more information can be found at www.ciria.org.uk/suds/.

Old tanks, pipework and other equipment should be properly decommissioned by removing any contents and cleaning out any residues. They should be isolated so that they cannot refill with any other material or rainwater. Ideally they will be completely removed. Underground equipment is a particular risk for reasons explained above. Organisations taking over old sites need to be aware of the potential problem and establish their history and conduct suitable surveys. Sites may have been badly managed in the past or abandoned due to company closure and left in a poor state. Any liability resulting from pollution is likely to fall on the new owner as will the costs of dealing with the consequences.

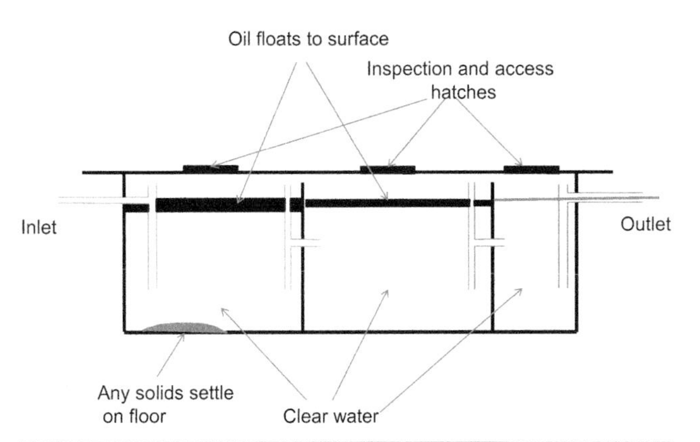

Figure 5.7 Section through an oil interceptor

Storage design and integrity

Some factors that need to be taken into account in locating storage sites were covered at the start of this section. There are several features that can be built into a storage facility itself to minimise the risk of chemicals escaping. Solids are less of a problem unless flushed away by rainwater or deliberate action; liquids need special attention as they will gravitate towards a drain unless intercepted. Keeping rainwater away from potential spill sites is good practice if it can be achieved by covering the site and directing the water to a drain.

The next consideration is the material of construction of a container. It needs to be suitable for the contents, resisting corrosion or other chemical attack and strong enough to sustain handling and knocks without failing. Bulk storage tanks should be protected against external corrosion by suitable surface coatings. The suppliers of chemicals and the containers should provide the necessary information.

Small containers and drums should be stored in plastic drip trays with a lip that enables the tray to hold any liquid that is spilt or leaks. If many containers are together it may be possible to construct a concrete lip around the floor area. The principle is to contain the spill and mop it up before it can escape.

Larger bulk storage tanks should be contained within a bund. Oil storage is a particular concern and in the UK is covered by the Control of Pollution (Oil Storage) (England) Regulations 2001 and similar regulations for Scotland and Northern Ireland. These specify in detail the minimum requirements and non-compliance can result in a fine or penalty. They provide a model of good practice; the key points are:

- They apply to any site storing more than 200 litres of oil above ground.
- Petrol, diesel, vegetable, mineral and synthetic oils are included.
- The secondary container such as bund (or drip tray) should be capable of holding 110 per cent of the contents of an oil tank, bowser or other container.
- If more than one container is present the volume must be the greater of 110 per cent of the largest tank or 25 per cent of the total capacity.
- The walls and floor of the secondary container must be impermeable to oil and water.
- Valves, filters, sight gauges, vent pipes and any other equipment must be within the secondary container except when in use.

- There must not be a drainage valve; any captured liquid including rainwater must be pumped out.
- There needs to be adequate clearance between the tank and the bund wall for inspection and maintenance (75mm minimum is recommended).

This is a summary of the main points and similar principles can be applied to any bulk liquid chemical storage tank. (Similar regulations exist in the UK for storage of animal slurry and oil on agricultural premises.) Figure 5.8 shows a tank without a bund and Figure 5.9 illustrates the main features required to comply with the regulations. More details on the sites included, the exemptions and the details of the design are available in the regulations and from several leaflets on the web sites of the environmental agencies and DEFRA (DEFRA 2001). PPG 2 referenced in Box 5.3 is one. Oil refineries pose a greater risk and they are covered by the IPPC directive (see Chapter 1). Permits will include any requirements for oil or other bulk chemical storage at these regulated sites. Guidelines for petroleum distribution installations in the UK have been produced by the Energy Institute (2015).

Underground storage tanks are a particular hazard as it is not easy to monitor for leaks. The best solution is to closely monitor inputs and outputs and liquid levels inside and account for the contents. Any discrepancies could indicate a leak. Modern storage tanks for underground use have double skinned walls with systems to detect any liquid trapped between them.

Figure 5.8 Where is the bund?

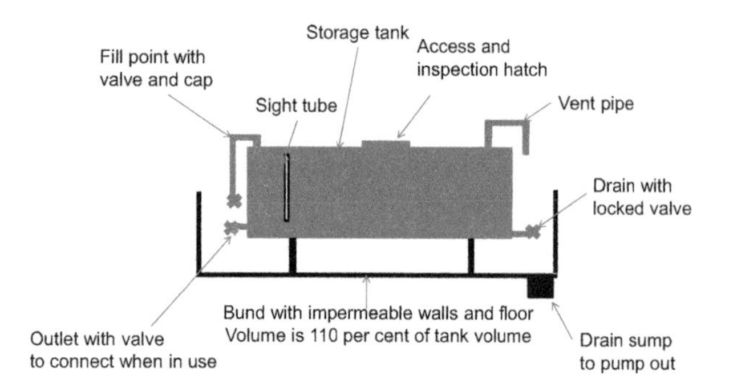

Figure 5.9 Section through a bunded storage tank

Pipework and other features

There will also be risks associated with pipes, valves and pumps that take liquids from storage tanks to where they are used. They may leak and are subject to damage by vehicles and people as well as frost. The routing of pipework needs to take account of these but they should be above ground to enable inspection. Where pipes need to go under a road or elsewhere they should be in ducts that can be easily accessed. Flow, level and pressure meters will show if the system is behaving as expected and alarms can be fitted to indicate a sudden increase in flow or loss of pressure. (This is especially important if pipes are routed underground, for example oil delivery pipes from oil terminals.) All the equipment should be colour coded and labelled to indicate the contents. These sorts of features will be especially important if liquids with hazardous properties are involved and safety precautions will be included as part of the IPPC permit and under COSHH regulations.

Operational procedures

However good the design and construction, many problems are caused by moving and handling chemicals. Much of this content has been covered already in this section but the following additional points need to be made:

- It should be clear who is responsible for site management and safety and what individuals are authorised to do. Authorisation should limit access to the site and to performing operations and require training and certification if appropriate.
- Over-filling of tanks is a regular cause of spills. Deliveries should always be attended; the vehicle driver should be present along with someone from the site. They must not go for tea whilst a vehicle is pumping its contents into a bulk container! Levels inside the tank should be monitored either by dipping or with a sight tube and any alarms responded to promptly. Suitable procedures should be in place to deal immediately with any problems (see below).
- Deliveries of solid materials need to be supervised and any spillages dealt with as below.
- Many chemicals present a personal hazard through inhalation of fumes or skin absorption. Suitable PPE such as safety suits, chemically resistant gloves, eye protection or breathing apparatus must be available, serviceable and used.
- Accidents often occur when transferring chemicals. Decanting from large to smaller containers by pouring risks loss of control of the flow in one way or another. Suitable electric or hand pumps should be used for all but the smallest containers. The transfer of solids also needs care, using appropriate scoops and funnels.
- Sites need to be maintained in a clean and tidy state with clear accesses and good lighting.

- There should be site inspections on a regular and programmed basis. The frequency should depend on the risk based on the hazardous nature and volumes of the substances stored and site knowledge of safety and security or previous problems. They could range from hourly to daily. All inspections should be recorded even with a simple tick sheet and rules established for reporting any problems.
- Supervisors should be regularly conducting their own inspections and observing operations to ensure compliance with procedures.
- Regulatory agencies, such as the Environment Agency in the England, will conduct their own inspections on sites that they consider pose a risk to the environment. If the inspector finds something that they wish to be put right they will raise it either informally or formally depending on the circumstances. If they consider the risk is high they may issue a notice requiring action. For example, the Environment Agency can issue an Anti-pollution Works Improvement Notice or use one of the other sanctions described in Box 1.11.

Dealing with spillages

Chapter 9 is about planning for and dealing with environmental emergencies. This section is about the practical arrangements for dealing with leaks and spills of liquids and solids.

The first action on discovering a leak or spill should be to establish the cause and try to stop it. If a small container is leaking then it can be isolated and the contents transferred. However help must be summoned for leaks from larger tanks, pipework or process plant and equipment. Small leaks may be a sign of a developing problem and they tend to become large ones if left unresolved. Depending on the circumstances, bringing a leak under control may require closing a valve, stopping a pump or closing down a process. This is not something to embark on without adequate knowledge and experience as more severe consequences may result from interfering with a complex process plant. Leaving it for later attention is also a decision that needs to be taken by a suitably knowledgeable and senior person. Emergency plans (see Chapter 9) should cover such events. Temporary sealing may be possible with clamps, tape or special putty that is applied to the site of the leak but a permanent solution should be applied as soon as possible.

Spilt solids can be swept up and, if not suitable for reuse, should be disposed of safely as waste. They may require treatment to render them harmless and this information should be available on site and on the MSDS. Dealing with waste is the subject of Chapter 6. Liquids are not so easy to deal with. Attempts should be made to contain the spread by using sand or a suitable absorbent material (see below). Acids or alkalis should be neutralised and operational procedures should cover the materials to be used and their availability on site. The liquid may be mopped up or pumped or sucked up if suitable equipment is available that is resistant to the chemicals involved. Alternatively it can be mixed with absorbent materials and swept up. Suitable materials are sawdust (but not for incompatible materials such as oxidising agents) or synthetic absorbents. These are supplied mainly for dealing with oil but will also work with many other organic liquids. They come as powders or granules and as sheets and pillows and sausage shaped booms. The booms are designed to place into rivers and streams to hold back and absorb the pollution as shown in Figures 5.10 and 5.11. Their deployment is usually by the environmental

Figure 5.10 Using a boom on a stream

Figure 5.11 Using a larger boom on a river

agencies, especially as handling large booms in wide or fast flowing rivers is difficult and potentially dangerous. The resultant contaminated absorbent materials also need to be disposed of safely as hazardous waste.

Every attempt should be made to avoid material gaining access to drains or watercourses by sealing gullies and covers or sealing the pipes by closing valves or inserting bungs. In the absence of these, sand bags and polythene sheet can be used although not as effectively. Washing away with water should only be used if the properties of the spilt material are non-hazardous and the resulting liquid can be dealt with safely (for example by discharge to foul sewer after consultation and with the consent of the sewerage utility if necessary).

Water used to extinguish fire (fire water) also needs to be considered as it may be of large volume and could be contaminated with chemicals and solids including combustion products from the fire. Careful thought should go into where fire water may occur and where it will drain to. If this is likely to pose an environmental threat then arrangements should be put in place to divert it to a lagoon or waste land where is can be contained and dealt with later. This could involve measures such as building diversion chambers into drainage systems. In many cases no such arrangements have been made and diversion has to make use of sand, sand bags or filled hoses.

Documentation and training

In order to ensure that all the practical arrangements described above work, there is heavy reliance on the people involved to use them properly and react to events in the right way. Operational procedures need to be documented and operatives trained in the safe transport, handling and use of materials. The key information should cover:

- Materials stored on site and their hazards (health, safety as well as environmental).
- Organisational responsibilities and personal authorisations.
- Incompatible materials.
- Maintaining safe storage conditions (site cleanliness, vehicle movements, etc.).
- Deliveries: procedures to be followed.
- Site security.
- Safety equipment and personal protection equipment (PPE).
- Handling drums and small containers.
- Dealing with liquid that has collected in bunds.
- Pipework, valves, etc., locations and operation.
- How to transfer materials between containers.
- Site inspection and reporting requirements.
- Supervision and reporting issues to management.
- Regulatory inspections and notices.
- Alarms and how to trigger or respond to them.
- Dealing with minor leaks and spills.
- Emergency procedures for larger leaks and spills and fire.

In Chapter 9 there is also reference to training to deal with emergencies and the use of practice drills to reinforce and test procedures.

5.3.4 Treatment of wastewater

Wastewater can be treated on site or sent through the foul sewerage network to be treated by the responsible utility. The range of processes available is similar although the scale of operation and the details of the process may vary. The following sections outline the operating principles of the processes available and their use in general terms with comments on their application to sewage treatment. There are proprietary versions of all of them which claim to have special features or to be for particular applications. The choice of process requires detailed investigation and usually field trials. The detailed sizing, shape and flow regime of the stages will affect performance and chemicals may be added to some processes to facilitate solids separation. Many companies treat their effluent to remove solids and BOD prior to discharging to sewer if this can reduce the charge from the utility and may be required to do so to remove substances unacceptable for discharge to sewer.

Screening and solids separation

Large solids (known as 'gross solids' in the trade) gaining entry to a drainage system should not be a problem on a well-run industrial site but leaves and twigs, large bits of wood, sanitary products, plastic bags and other debris can gain access to public sewers and block or damage equipment in a sewage treatment works (STW). These are removed by screens which consist of vertical bars with manual or automated raking or by fully automatic systems to deal with the waste. If the solids are not kept clear they will be forced through as the pressure builds up behind. Stones and grit present similar problems and they are separated by slowing the flow through a larger channel where they settle and can be removed. Both processes are shown diagrammatically in Figure 5.12.

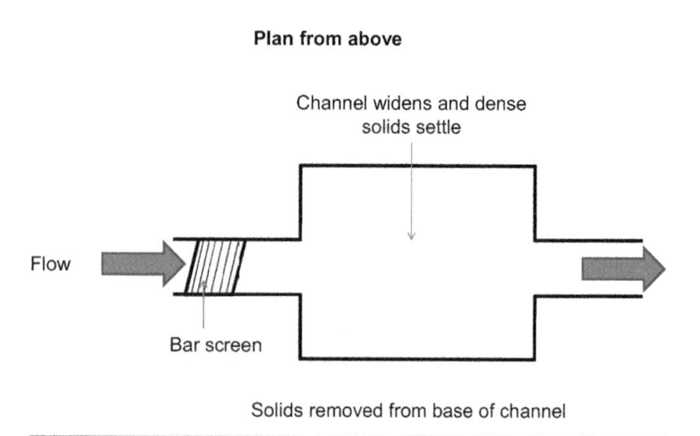

Figure 5.12 Screening and solids separation

Sedimentation

Finer particles, which may contain silt or organic material such as sewage solids or food residues, will only settle if the flow is reduced further. Fine solids may also be formed by chemical treatment of waste to remove metals or other contaminants. Removal can be achieved by using large settling tanks where the flow is horizontal or inverted conical tanks where the flow is vertical and slows as it progresses up the cone. In either case the solids separate by gravity so they need to have a density greater than water. Sedimentation can be helped by the addition of iron or aluminium salts or synthetic polymers (known as coagulants) and these are often used in industrial wastewater treatment. The settled solids can be scraped away or removed hydraulically. Both types of plant are shown diagrammatically in Figure 5.13.

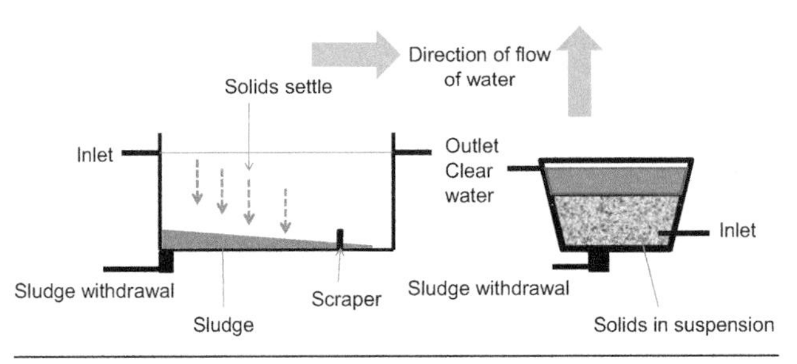

Figure 5.13 Horizontal and vertical sedimentation

Flotation

Fine solids can also be removed by flotation, especially if they have a low density and float naturally. This is a process generally associated with the separation of minerals and industrial wastewater treatment. Air bubbles are attached to solids as they enter a tank by feeding in a solution of air in water under pressure; the pressure is released and fine bubbles form in the larger body of water on the surfaces of the solids. The process may be helped by the addition of coagulants as for sedimentation. The solids then float to the top of the tank (even if the solids themselves have a high density) where they are removed by a scraper as shown in Figure 5.14.

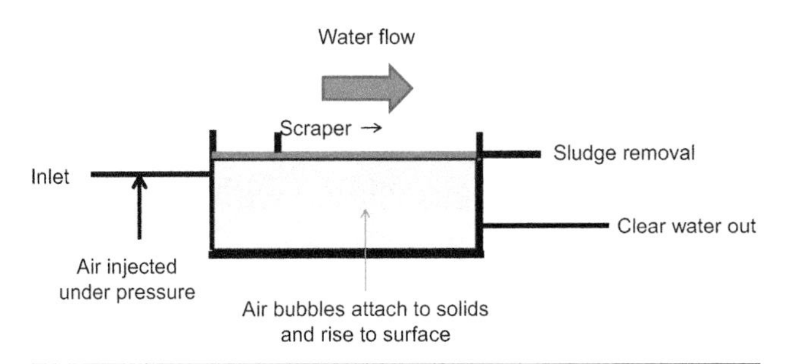

Figure 5.14 Flotation for solids separation

Biological oxidation

The soluble organic matter (BOD) such as found in sewage or from a food processing plant also has to be removed. This is accomplished with biological processes which aim to mirror what happens naturally but under controlled conditions and at a higher rate than in the environment. Two basic approaches are used in both industry and STWs. In one, known as a biological filter, a coarse medium of stones (over 50mm in size) contained in an open structure supports a biological growth of bacteria, fungi and other micro-organisms on the surface. Plastic support media are also used in smaller units, particularly in industry. The effluent is distributed across the surface and trickles down where the micro-organisms consume the organic material as food. This is illustrated in Figure 5.15. A purified effluent emerges from the bottom but may need settlement again to remove solids formed within the filter medium.

The alternative, known as the activated sludge process, is to support the micro-organisms in suspension in a large tank with air blown in or added by surface agitation. The organisms are in high concentration and remove the organic material as food in the same way as the filter. The effluent flows through the tank and is mixed in where the oxidation occurs. Some of the flow is continuously removed to a settlement process to settle the suspended organisms by sedimentation. A proportion of the solids settled out is recycled into the process. This is illustrated in Figure 5.15 and Figure 5.16 shows a plant with surface aeration.

The performance of both of these processes depends on the nature of the effluent and other factors such as flow variability and ambient temperature. STWs use both of these on a large scale. The activated sludge process uses more energy and requires tighter monitoring and control but requires less land so the choice is governed by financial as well as technical issues. There is a variety of proprietary equipment available for biological treatment based on either of these principles suitable for industrial treatment and small STWs.

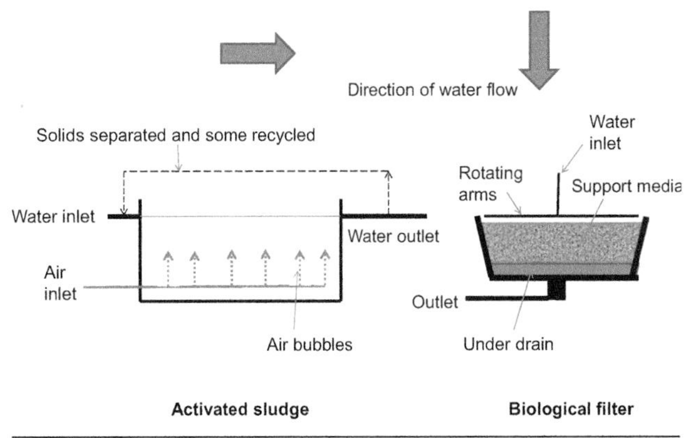

Figure 5.15 Biological oxidation by activated sludge and biological filter

Filtration

Filters can also remove solids and may be added after a sedimentation process to further clean the effluent. They are rarely used in STWs but if they are it is at the end of

Figure 5.16 Activated sludge by surface aeration

the other processes. They usually consist of open tanks or sealed containers with a filter medium such as sand. The effluent flows down through the medium and the solids are retained on the surface and within the medium. This has to be washed periodically by flushing with clean water or filtered effluent. This is illustrated in Figure 5.17.

Filters may also be filled with activated carbon to absorb organic compounds; this is common in chemical plants where trace quantities of residues are not removed by other processes.

Sludge treatment

The processes described above all produce sludges that remain for disposal. Those with a high organic content can be treated further by anaerobic digestion (see below); the rest have to be treated as waste and disposed of by landfill or incineration. Sludges normally require dewatering with a lagoon, a filter press or a centrifuge to reduce the volume for transport and disposal.

Anaerobic digestion

An alternative method of treating wastes with a high organic content is anaerobic digestion. This is a biological process but in the absence of oxygen and is similar to the processes that occur naturally when dissolved oxygen conditions are low, such as a landfill site. Bacteria

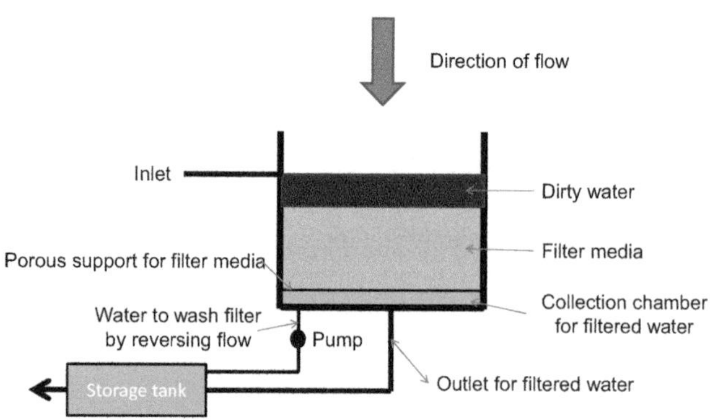

Figure 5.17 Open water filter

consume the organic material and produce methane and other gases; the methane can be captured and used as an energy source for heating or power generation. This process is used in the food and farming industries and for the treatment of sludge in STWs. The remaining solids are reduced in volume and should not cause a nuisance from odour. It is usually spread on land as a fertiliser or soil conditioner.

Centrifugal separation

The separation of solids can also be achieved in a centrifuge in which the flow is spun at high speed and the solids are thrown to the edge of the container where a scraper removes them through one end. This is usually applied to slurries and sludges with high solids content and separation is aided by the addition of synthetic polymers.

Other processes

Discharge to watercourse or sewer may require additional treatment to comply with any permits. The main ones are:

- pH correction to avoid too acidic or alkaline a discharge.
- Nutrient (nitrogen and phosphorus) removal to reduce eutrophication.
- Ammonia removal to reduce toxicity to fish and oxygen demand.
- Special treatments to remove metals or other potential industrial pollutants.

STWs often have to include nutrient and ammonia removal using variants of the biological treatment processes.

5.3.5 Difficulties associated with effluent treatment

Running an effluent treatment plant is not part of the mainstream operation of most organisations and many consider it an unnecessary burden. They may prefer to pass it to the sewerage utility if they can comply with their requirements for a discharge to sewer. The exception may be if the load of solids or BOD (or COD) is high and significant cost savings can be made.

The performance of effluent treatment plants is governed by variations in flow and load as well as the requirements for maintenance and the availability of skilled operatives. Extremes of climate can also affect performance, especially cold weather on biological processes. In some locations remoteness from any other option means that treatment on site is the only possibility. Remote locations, particularly in developing countries, may suffer from variable power availability and the expense and difficulties of obtaining supplies of chemicals or spare parts as well as specialist staff. In these situations it is important to adopt solutions that can be sustained: those that require little power or chemicals and require low maintenance if they will still deliver an acceptable discharge. As land space is not usually a limiting factor, treatment may depend on large settling lagoons for solids removal and rudimentary biological treatment processes if they are required.

CHAPTER

6

Control of waste and land use

After this chapter you should be able to:

1. Outline the significance of different waste categories and the relationship between category and route of disposal
2. Explain the importance of minimising waste
3. Outline how to manage waste
4. Describe the outlets available for waste
5. Outline the risks associated with contaminated land.

Chapter Contents

Introduction **100**
6.1 Waste categories **101**
6.2 Minimising waste **104**
6.3 Managing waste **109**
6.4 Outlets available for waste **113**
6.5 Risks associated with contaminated land **118**

INTRODUCTION

A broad definition of waste includes emissions to air and water which have been covered in earlier chapters. This chapter covers solid waste and contaminated land. Both of these have been mentioned previously and putting them together is justified in that the two are often linked, although waste disposal is not the only source of land contamination; air and water pollution can also be a cause as described in Chapters 4 and 5.

Chapters 1 and 2 described the growing problem of waste and the benefits to the environment, society and to business of minimising it. There are strong environmental reasons for avoiding waste but we should never lose sight of the fact that most items that we consume have waste associated with their production and they eventually end up as waste themselves. All goods have both a cost of waste during production and of final disposal and these have to be paid for by someone. This is usually the ultimate consumer as the costs of production are passed on in prices and the costs of disposal appear as direct charges and taxes or as indirect taxes such as local authority charges for waste collection and disposal. Even illegal waste disposal eventually ends up as a cost to the consumer when the mess is cleared up. There are also the other costs associated with waste such as the impact of the consequential greenhouse gas emissions. The total costs are difficult to estimate but will be measured in billions of pounds in the UK alone. So waste is a cost to us all in the end, the actual amounts depending on the products and wastes in question. The initial EU Framework Directive (see below) listed 15 categories of waste but had a 16th 'any materials, substances or products which are not contained in the above categories' which seems to catch everything else. To gain an understanding of the potential range of wastes produced we can identify the following examples from industrial and commercial activities:

- Spoil associated with the extraction of raw materials.
- Unwanted by-products arising from the processing of raw materials.
- Trimmings from wood, plastics or metal as they are fabricated.
- Excavated materials and waste from demolition and construction.
- Materials over-ordered that cannot be returned or used elsewhere.
- Packaging used to protect materials and products as they are transported.
- Products that are not to specification and have to be disposed of.
- Products such as food or medicines that are past their use-by date.
- Food that is damaged by pests or in storage.
- Products damaged in manufacture, transit or use.
- Products that have reached the end of their life.

Households and businesses also produce waste as part of their daily activities. More examples are:

- Packaging with goods purchased.
- Used paper and items such as old newspapers, magazines and junk mail.
- Waste from food preparation and food not consumed.
- Old clothes and other fabrics.
- Oil and other products from maintaining vehicles and other plant.
- Medical waste from hospitals and doctors surgeries.
- Garden and landscaping waste such as grass clippings, plants and leaves.
- Old furniture, carpets, etc.
- Electrical goods such as old lamp bulbs and white goods that are not repairable.

These lists are not exhaustive and you can probably add more from your home and work environments.

The types and amounts of waste produced are largely determined by the nature of the society. Developed countries produce more waste because they can afford to buy and consume more products and replace them when they are out of date, no longer in fashion or need repair. Poorer societies consume less to start with but are also more likely to repair broken items or, if they are still serviceable but no longer needed, pass or sell them on or find an alternative use. These tendencies are compounded by developed economies having provided means for disposal where poorer societies have not. In poor urban or densely populated areas in particular accumulations of waste can present a health hazard from flies and vermin and a risk to scavengers, especially children searching for items of (usually limited) value. In such societies reuse and recycling are already well established for some items. Sacks, plastic and metal containers and scrap metal are particularly valuable.

Early legislation was mainly concerned with avoiding illegal dumping of waste and introducing controls to regulate the collection, transport and disposal of waste. Until fairly recently landfill was the most common destination of waste in the UK and still is in many countries. Shortages of suitable sites coupled with the realisation that waste disposal is a loss of potential resources has changed the emphasis and progressive regulations are forcing a move away from landfill and towards other ways of managing waste. More recent legislation in Europe has been driven by the Landfill Directive and directives associated with hazardous substances, about which more is below.

6.1 Waste categories

The examples of wastes given above cover a variety of materials with a range of properties. Some may be toxic or carry dangerous organisms such as bacteria and viruses. Some may be dangerous in other ways such as having sharp edges or being radioactive or a fire risk. Some are biodegradable and cause atmospheric or water pollution as they decay. Many are inert but still require space for storage. In an unmanaged system a rubbish dump will contain all sorts of materials, some of which have potential value but become contaminated or difficult and dangerous to recover. In order to establish a management system to control the collection, transport and safe disposal of waste it is essential to categorise it according to its properties so that similar wastes can be kept together and reactive mixtures are avoided (such as acids with sulphides which will release hydrogen sulphide, a toxic gas). Wastes can also be treated if necessary such that they are safer for disposal and, as we shall see later, separation facilitates reuse and recycling of waste.

Countries have adopted different approaches to waste management and categorisation and few of them are simple. The regulations governing the definitions of the types of waste and how they are managed are probably the most complex to understand of all environmental regulations and are regularly subject to change. Expert interpretation is often required and experts may disagree about the finer points. For our purposes we need to keep it as simple as possible, starting with what we mean by waste in Box 6.1.

6.1.1 Hazardous and non-hazardous waste

Within the EU waste management is governed by various directives and these are similar in principle to the legislation adopted in other countries where such legislation exists. The EU Waste Framework Directive (European Commission 2018) clarifies earlier waste framework and hazardous waste directives (European Commission 2008d) and this is the basis for classifying waste and defining hazardous waste within the EU (hence known as 'directive waste'). It also provides a link with the earlier European Waste Catalogue (European Commission 2001b) which set up a list of wastes according to their process, industry or waste type. The latest definitions of waste types can be found at European Commission (2014b). Waste can be categorised using the catalogue (outlined in Box 6.2) and within it the hazardous wastes are marked with an asterisk according to the properties which render them hazardous (outlined in Box 6.3). These properties are self-evident from the definitions but note that these have been simplified from the originals. The full definitions are in the various directives but particular types of waste can be followed up on the web sites of the various government agencies within the UK. Note

Box 6.1 What is waste?

Defining what waste is ought to be simple. If I have something for which I no longer have a use, to me it is waste. The EU (European Commission 2008d) adopts a similar approach and defines waste as 'any substance or object which the holder discards or intends or is required to discard'. So far, so good; but my waste could be of use to someone else even to the extent that they may be prepared to pay me for it or at least remove it at no cost to me. For the new owner it is now a resource so is it still waste? Classic examples are waste oil from vehicle servicing, paper, glass, plastics and metals, which are the basis on which whole recycling industries have been developed. The regulations that govern waste started at a time when waste disposal was mainly to landfill and controlling the collection, transport and storage was about getting it to the right place as quickly and safely as possible. Now the emphasis is on waste minimisation, reuse and recycling it is important that the regulations and the costs and taxes associated with disposal do not inhibit these alternatives. But from a legal point of view, tested in the courts, my waste is still categorised as waste even if it is going to be reused elsewhere.

Recent changes in EU directives have started to change this for iron and aluminium that is destined for recycling and the expectation is these principles will be extended. The rules and regulation that apply to waste as described later in this chapter still apply.

Box 6.2 The EU waste catalogue

The catalogue sets out to give all wastes a six-digit code which is built up systematically starting with the industry responsible for its production. For example, the first few lines of the catalogue are:

01 01	**wastes from mineral excavation**
01 01 01	wastes from mineral metalliferous excavation
01 01 02	wastes from mineral non-metalliferous excavation
01 03	**wastes from physical and chemical processing of metalliferous minerals**
01 03 04*	acid-generating tailings from processing of sulphide ore
01 03 05*	other tailings containing dangerous substances (Mirror)
01 03 06	tailings other than those mentioned in 01 03 04 and 01 03 05

The code 01 03 04 marked with an asterisk (*) is deemed to be 'absolute' hazardous (according to the rules in Box 6.3). The next code 01 03 05* is known as a 'mirror' hazardous code as the hazardous nature will depend on the concentrations present. Thus acid generated tailings from processing sulphide ore are always considered hazardous and will have the code 01 03 04*; other tailings may be hazardous or not depending on content and could have codes 01 03 05 or 01 03 05*.

There are 20 groups of industry- or source-based codes, each with multiple sub-codes with a total of about 800 individual codes of which about half are hazardous. The codes are often referred to as EWC codes.

Box 6.3 Defining hazardous waste

The EU defines waste as hazardous if it meets one or more of the following categories *which have been simplified from the original text*:

H1	*Explosive* when exposed to flame, shocks or friction
H2	*Oxidising* which generate heat when in contact with other substances
H3	*Flammable* depending on flash point or other nature
H4	*Irritant* causing inflammation on skin contact
H5	*Harmful* causing limited health risks if inhaled, ingested or by skin contact
H6	*Toxic* causing serious acute, chronic health effects or death
H7	*Carcinogenic* which may induce cancer or increase its risk
H8	*Corrosive* by destroying living tissue on contact
H9	*Infectious* causing disease from microorganisms or their toxins
H10	*Toxic for reproduction* causing non-hereditary congenital malformations
H11	*Mutagenic* causing hereditary genetic defects
H12	*Releases toxic gases* on contact with air, water or acids
H13	*Sensitizing* causing a sensitive reaction
H14	*Ecotoxic* causing risks to one or more sectors of the environment
H15	*Yielding another substance with any of the characteristics above*

that these are updated from time to time as the regulations change. In particular the hazardous waste regulations are being linked with EU regulations concerned with handling chemicals which are themselves undergoing a series of changes (European Commission 2008c) and subsequent Decisions which can be tracked online. The above is an illustration of the difficulty of keeping track on the legislation!

Hazardous waste cannot be disposed of in most landfill sites. This co-disposal practice was stopped by the Landfill Directive. Now it has to go to sites with permits to deal with it after it has been treated if necessary, for example to neutralise it. Liquid wastes are also no longer accepted at landfill sites.

The implementation of the EU regulations within the UK is expanded further in Sections 6.1.3 and 6.3. Outside of the EU information can be found on the web site of the UN Environment Programme (unenvironment.org) on the practice in various regions of the world and different countries and on plans to improve them.

6.1.2 Inert waste

The EU Landfill Directive was agreed with the aim of reducing the pollution risk from landfill, the waste of resources and the amount of land taken (European Commission 1999). The ramifications of this are explained in Section 6.4 but the acceptance of inert waste at landfill sites should present no pollution risk. Inert waste is defined in the Directive as waste that

> does not undergo any physical, chemical or biological transformations. Inert waste does not dissolve, burn or otherwise physically or chemically react, biodegrade or adversely affect other matter with which it comes into contact in a way likely to give rise to environmental pollution or harm human health. The total leachability and pollution content of the waste and the ecotoxicity of the leachate must be insignificant, and in particular not endanger the quality of surface water and/or groundwater.

Thus inert waste could include broken concrete, glass, bricks and similar materials but would exclude wood, paper, plant material or anything else that degrades.

6.1.3 Other categories of waste

The terms used so far would not be familiar to many people. They would be more likely to refer to household waste, commercial waste and industrial waste reflecting the sources. Many countries use these or similar terms. Broadly, household waste is that produced by domestic premises; commercial waste is that produced by businesses such as offices and shops and industrial waste is from manufacturing processes. Municipal waste is often used to describe waste collected by or on behalf of local authorities and would consist mainly of household waste but may include some from local businesses.

Controlled waste

Under UK legislation the term 'controlled waste' (Environmental Protection Act 1990, Part II) is used for household, commercial and industrial wastes. This Act and subsequent regulations control the collection and management of waste in the UK. The regulations also have to take account of the EU directives which affect the management of hazardous waste and the use of landfill as a disposal option. The term is still used in the UK and has been extended to include mining and agricultural wastes. The latest version is the Controlled Waste (England and Wales) Regulations 2012.

Clinical waste

This is waste from medical establishments that may contain material contaminated with tissue or infective agents. This could include tissue removed in surgery, blood and blood products, bandages and other dressings, hypodermic needles and syringes, paper and fabric bed covers, etc. Within the UK and many other countries these are expected to be kept separate from other waste as they pose a threat to anyone coming into contact with them, including those responsible for the collection and ultimate disposal. The waste is not allowed to be disposed of in landfill. Medical establishments should use appropriate containers with clear labelling. The waste is collected and incinerated either on site in large hospitals or in specialist units elsewhere. Other waste such as paper and food waste from these establishments can be removed and treated as commercial waste.

Radioactive waste

Radioactive substances are used quite widely. The radioactive sources and wastes with the highest activity are associated with nuclear power production and nuclear weapons and are subject to special regulation. However, radioactive isotopes are used widely in research, medicine and industry. Applications include investigating chemical reactions, tracing flows in natural systems, monitoring levels in containers and monitoring and treating diseases. These activities produce wastes not just from the isotopes themselves but from the containers and other materials such as measuring equipment and protective clothing that become contaminated in handling them. The level of activity of the waste depends on the nature of the source and how it was used. The waste can remain radioactive for many years (even centuries) depending on the isotopes used.

Within the UK, most uses of radioactive substances (outside of nuclear power and weapons) are regulated under the Radioactive Substances Act 1993 and may require a permit although there are some exemptions. The regulations are complex and the interpretation is beyond the scope of this text as the need for a permit depends on several factors but the principles behind them are that radioactive sources need to be registered and tracked and that they cannot generally be disposed of with other waste. Radioactive waste has to be kept separate and three levels of activity are recognised: low, intermediate and high level wastes. Exempt wastes may still be classified as hazardous and have to be treated as such. The treatment and storage of radioactive waste is becoming a problem as the amounts increase and discussions on the long-term solutions are continuing: it is a difficult and emotive topic but an underground storage facility is the most likely outcome. The principles of minimising waste covered in Section 6.2 still apply but waste will remain for disposal. Most low level wastes, which consist mainly of the materials used in handling waste and protection of workers, are landfilled subject to the controls on landfill discussed in Section 6.4, but higher level waste, which is of low volume but accounts for most of the radioactivity, ends up in temporary storage.

Classification of waste in the UK

There are two ways of classifying waste in the UK: by type and by source. Box 6.4 in Appendix 4 summarises the definitions.

6.2 Minimising waste

We have already seen that as the population increases and we consume more the amount of waste we produce increases and that this waste represents a loss of resources and a cost for disposal. Part of the solution is to try to minimise the amounts of waste produced in the first place and to manage that which is produced in better ways. This has resulted in the waste hierarchy as a model for tackling the problem, similar to hierarchies referred to elsewhere in this book. The hierarchy is represented diagrammatically in Figure 6.1. Here the order of priority is from top to bottom and the relative sizes of the shapes within the triangle represent the amount of effort that ought to be expected. In summary, more effort should be going into eliminating waste than in planning for its disposal. The extent to which different countries avoid waste varies considerably and it is difficult to get comparable statistics. Definitions vary across countries and over time and comparisons are also compounded by the different stages of the economies. Rapidly developing European economies such as Bulgaria produce a lot of construction waste whereas those that are languishing

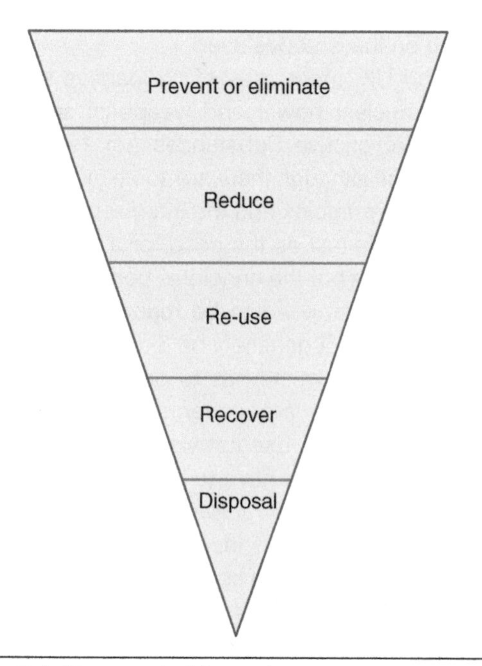

Figure 6.1 The waste hierarchy

do not. Some basic statistics for the UK in Box 6.5 in Appendix 4 help to show the scale of the challenge. These are from reports compiled by DEFRA which analyse waste production by source and type. These data are used by Eurostat (https://ec.europa.eu/eurostat/web/waste) to produce Europe-wide statistics and by the EU to show trends by country and overall (www.eea.europa.eu/data-and-maps/indicators/waste-recycling-1/assessment). These show improving recycling rates over the last decade.

Regulations in the UK (The Waste (England and Wales) Regulations 2011) now require producers of waste to apply the hierarchy and confirm that they have done so on transfer and consignment notes (see Section 6.3.2). This will drive the move to reduce the amount of waste that is not recovered in some way. For example, the Scottish Government is promoting zero waste, linking it to the circular economy referred to in Chapter 1. Dealing with waste according to the hierarchy is already a legal requirement or becoming so over the next few years.

6.2.1 Elimination or prevention of waste

The problems associated with waste disposal can be minimised if less is produced in the first place. Elimination may come down to simple things such as reducing the amount of packaging associated with a product or buying materials in sizes such that less is wasted in subsequent processing or use. More sophisticated approaches may look at the design of a product or manufacturing processes to reduce the amount of material wasted in production. For example, metal parts may now be made by building them up from sintered metal rather than machining them from a larger block. Much food is wasted because of over-ordering, resulting in the excess food being past its best: managing ordering and deliveries better could prevent waste and save money.

At home and in business a lot of waste consists of paper. The use of computers means that paper copies of documents are not needed in many cases; the information can be stored digitally and bills, statements and documents transmitted electronically. (At least that is the theory; experience shows that, without discipline, computers can result in more paper being used!). Where paper copies are required they should be printed on both sides of the paper so in selecting equipment it is important to take this into account. Although double-sided printers cost more, they save paper, storage space and postage costs and should pay for themselves quite quickly.

From the purchaser's point of view, prevention of waste is helped by buying durable products or ones that can be repaired. Many household goods such as kettles and electronic equipment fail this test. Similarly, reusable shopping bags rather than disposable ones eliminate the amount of plastic waste to the extent that most countries have banned the use of the latter or imposed taxes or charges to discourage it.

Reduce waste

Reducing waste means looking at the waste stream and establishing which materials are being thrown away that could be avoided by means other than elimination in the first place. Some waste results from damage to materials so that they are no longer suitable for use. This can be avoided by care in storage, avoiding impacts from people and traffic or the effects of weather. Better control over quality in production could reduce the amount of sub-standard products for disposal. Similarly food should be stored under the right conditions, avoiding damage by pests or other factors and not preparing amounts in excess of those likely to be consumed.

Reuse waste

Some products or materials that are left over may be reused. If the current user no longer needs it, it can be reused by someone else. For domestic consumers, books, clothes, furniture and household goods can be reused by someone else. The rise of e-bay and similar internet sites promote this in the home and charities such as Oxfam may resell donated goods or pass them on to others in need. At the business level similar opportunities exist but may be more difficult to find. There may be local business networks that can work together. However, in many industries there may be internal opportunities to reuse. Bricks and other building materials are recovered for reuse, especially if they are of a type that is no longer produced but of value to those restoring or extending old buildings. Other demolition waste can be used as hard core or crushed and graded for other uses.

At home and in business paper that has been printed on one side may be reused by printing again on the other side or as scrap paper for notes (or shredded and used with packaging – strictly recycling). Packaging received with bought materials may be reused for storage or for sending goods out to customers.

Recover waste

In the waste hierarchy 'recover' includes recycling, composting and energy recovery. The difference between reuse and recycling is not always clear. Reuse implies using something again for the original purpose whereas recycling implies a new use or recovery of the material of construction or fabrication for another purpose, although in common usage the terms may be used interchangeably. So aluminium and steel cans are recycled as they are melted down to produce new metal sheets which may reappear as cans but could also be made into something else. Some glass bottles are reused but increasingly they are collected and the glass recycled by melting it down and making new bottles or other products. Sometimes waste glass is recycled by being incorporated into other materials such as road surfacing. The production of aluminium, steel and glass from raw materials is energy intensive whereas production from recycled materials is less so. The same applies to other metals such as copper and lead. This means that the waste materials have value and, if in sufficient quantity, it can be sold on to a processor.

Used paper and cardboard that cannot be reused can be recycled by sending it back to a processing plant to produce new products. Many plastics can be dealt with in the same way. If separate they can be remade into new material such as polythene film. Plastic drinks bottles can be reprocessed to make new bottles or synthetic fibre such as fleece. Mixed plastic waste is more difficult to deal with. It can be separated into its component materials using mechanical and other means at specialist centres for recycling, or processed as mixed waste to manufacture low-cost items such as garden furniture, street furniture and other products where quality is less critical.

Gypsum (calcium sulphate) produced by the flue gas desulphurisation of the gaseous emissions from coal-fired power stations is used for the production of plasterboard and the ash left from the burning of coal is incorporated into cement and concrete products. Old tyres may be reused either directly or by retreading them or they may be shredded and the steel, nylon and rubber of construction recycled for other purposes. Old tyres may also be incinerated to produce heat, typically in cement kilns.

Composting is another form of recycling. Biodegradable material such as grass cuttings, plants and other organic wastes are composted by keeping them mixed and turned in air such that the natural processes of decomposition are speeded up and the heat produced helps to sterilise the product. The resulting compost can be used as a soil conditioner as it contains nutrients and the remaining organic matter improves soil structure and moisture retention. This has been carried out on a domestic scale (in a garden or small plot) and on an agricultural scale for centuries but now many municipalities offer a service to householders and business to avoid putting biodegradable material into landfill sites. They collect the material (for a fee) and then sell the product as compost.

Energy recovery can also be applied to organic waste. Anaerobic digestion, a process similar to composting but in the absence of air, was described for treating sewage sludge in Section 5.3.4. It produces methane as a by-product: the same process produces methane in a landfill site. Anaerobic digestion is also applied to strong organic wastes such as those from food production and processing and the conditions of digestion help to reduce pathogens that may be present. It can be used for meat and other protein containing wastes which are not composted mainly due to the risk of vermin. Digestion is carried out in a sealed container and the gas collected from the top. The methane can be burnt as a fuel for heating or power generation and the residual product is reduced in volume: it can be spread on land, incinerated or disposed of in a landfill site. Alternatively, waste that has a high organic content can be incinerated directly in a furnace to produce heat which can be used directly or to raise steam for use in power generation. This is considered in more detail in Section 6.4.2. The residual ash is much reduced in volume and it can be landfilled or sometimes it is used in the production of building materials.

Disposal of residual waste

There is always likely to be some waste that cannot be reused or recycled along with the residual matter from some waste recovery processes. This has to go to some form of safe disposal. The previous sections have identified some of these: spreading on to land, landfill sites or incorporation into products used for building and construction. In all circumstances it has to comply with any other relevant regulations; these exist for landfill sites and for the spreading of waste on agricultural land but the waste regulations that govern the other aspects of waste management also apply. The other uses are dependent on the incorporation of the waste not being detrimental to the quality or use of the products accepting it. Landfill sites are considered in more detail in Section 6.4.1.

Zero waste

The concept of zero waste has gathered momentum in recent years. For purists the ultimate aim is that no waste is left as residual waste; it is all avoided in the first place or reused and recycled and incineration is seen as a waste of resources that could be recovered. There is a growing campaign in the US, UK and wider afield, some of it motivated by opposition to alternatives such as incineration. The UK Government adopted a more pragmatic approach in its review of waste strategy which is based on the following characteristics of a zero waste economy:

- Resources are fully valued – financially and environmentally.
- One person's waste is another's resource.
- Over time, we get as close as we possibly can to zero landfill.
- A new public consciousness in our attitude to waste.

This influenced the waste policy review for England (DEFRA 2011b). Scotland is actively pursuing zero waste as outlined above.

This has not stopped some organisations from aiming for zero waste in its narrower sense. For example, Toyota Manufacturing UK claimed zero waste to landfill in its UK car plant in 2003, achieved by a programme of prevention, reuse and recycling (more information can be found at www.toyota.co.uk). Many other companies are aiming the same way and others claim close to zero waste although few achieve it absolutely.

6.2.3 Benefits and limitations of improved waste management

Reducing the amount of waste produced and reusing or recycling waste have many benefits. Some of these have been covered in previous chapters and the sections above but it does no harm to summarise them in Box 6.6. There are many case studies from companies and organisations that justify the need for waste recovery. A good recent starting point is on the environmental benefits of recycling which has used LCA to compare different routes and has been published by Waste and Resources Action Programme (WRAP). Their web site also has much information applicable to various industrial sectors. As a detailed example WRAP estimates that the UK produced approximately 10 million tonnes of food and drink waste per year, 7.0 million tonnes of which was food with a value of over £20 billion a year and associated with potentially 25 million tonnes of greenhouse gases (Waste and resources action programme 2019).

There are also limitations to recycling which are summarised in Box 6.7 and the details of these are considered in the next section.

6.2.4 Barriers to reuse and recycling and how they can be overcome

On the face of it reuse and recycling seem an obvious way forward, with so many benefits so it is hard to see why it has taken so long to become accepted in the UK. Even getting households to cooperate on waste recycling has not been easy, with 45.7 per cent of household waste being recycled in 2017 although this is a big improvement on the 10 per cent recycled in 2000/1. The target is 50 per cent by 2020. In contrast, 91 per cent of non-hazardous construction and demolition waste was recovered in some way in 2016. In round figures, about half of all waste generated is recycled but the statistics quoted above show how difficult it is to find comparable data for each source of waste over the same time periods as data collection and reporting methods change. Some countries, particularly in northern Europe, have a better track record, reflecting tighter regulation and cultural differences. Recycling is also high in the poorest communities in the third world where scrap metals and other materials have high relative values and when wages for scavenging or manual sorting are nil or low. The trend is improving partially in response to government targets for local authorities and the introduction of penalties coupled with increasing landfill tax which hits all producers.

The long-term target in the UK and elsewhere is zero waste, i.e. no waste to landfill or where waste is thrown away as a last resort. This is a challenge as it becomes progressively more difficult to find new ways to reduce this waste. Progress is also slow due to various barriers. The principal barriers and how they may be overcome are discussed in the following sections.

Lack of awareness

This can be at all levels of the business. Directors and managers may not be aware of their legal obligations or the costs that they are incurring. Even if they introduce schemes, the employees may not be aware of

Box 6.6 Benefits of waste recycling

There are many benefits – environmental, financial, legal and ethical – from recycling waste and they also apply to avoiding its production and its reuse. These are:

- Reduced demand for raw materials.
- Conservation of natural resources such as timber and minerals.
- Reduced environmental damage from mining and quarrying new materials.
- Reduced energy required for extraction of raw materials and processes.
- Potential reduced space requirement for storage, treatment or disposal of waste.
- Reduced costs for purchase of materials, packaging, energy, etc.
- Reduced costs for waste removal and processing.
- Reduced taxes for landfill.
- Potential for income if waste can be sold.
- Energy recovery can save money on fuel costs.
- Reduced use of landfill site capacity and land take.
- Reduced risk of environmental pollution from emissions to air and water.
- Reduced damage to ecosystems and loss of biodiversity.
- Reduced emissions of greenhouse gases, especially from methane in landfill.
- Compliance with legal requirements, e.g. for landfill and packaging and the implementation of the waste hierarchy.
- Compliance with the EMS.
- Good publicity.
- Demonstrates commitment to the environment.
- New business opportunities with the potential to create jobs internally or in the wider community.

Box 6.7 Limitations to waste recycling

Successful recycling depends on a number of factors some of which are outside the control of the organisation. The main limitations are:

- It requires leadership and determination to implement and maintain.
- Insufficient space on site to store and handle the different waste streams.
- Having to manage several primary and secondary containers.
- The need to change the organisational culture on handling waste.
- The need to avoid cross-contamination to maximise potential income.
- Inadequate markets for recycled waste materials.
- Concerns for quality of reused goods or recycled materials.
- Lack of suitable contractors to collect the waste.
- The potential cost of multiple collections especially in remote areas.

Some of these barriers are also discussed in more detail in Section 6.2.4.

their obligations and the benefits. Establishing the main costs of waste disposal is usually quite simple as bills are generated by the company that collects it. This will not be the total cost as it excludes internal costs but these will be low if most waste is just mixed into a bin or skip. The collection company will charge based on their costs which may include sorting and recycling at their sites or, if unsorted, the waste will be charged at the highest rate applicable. For example, even if small amounts of hazardous waste are present (such as fluorescent light tubes or tins with paint residues), the whole batch has to be treated as hazardous waste and will cost a lot more than if it were inert or non-hazardous. Establishing the costs of wasted materials that could be avoided in the first place or reused will need someone to examine the waste and estimate the values. Off cuts of metals, wood or plastics and clean packaging materials are good examples of valuable materials that could result in reduced waste disposal costs if reused or recycled, as well as saving on the cost of new materials.

Awareness needs to start at the top of the organisation and that may be the most difficult if their attention is focused on production, product quality or other management issues. Someone needs to recognise the problem and champion a change in attitude. Involving the rest of the employees will be a matter of education and training, explaining the benefits and why their cooperation is important.

Awareness also covers knowledge about what is recyclable and how to manage it. Much information can be found in textbooks and journals or on the internet from trade associations, published texts and the organisations that exist to promote recycling (in the UK WRAP www.wrap.org.uk/ is a good starting point). Even discussions with other companies in the same field of activity or locality and the companies that collect and deal with waste may help to develop a new approach. Training needs to cover the detail such as separation of different types of plastics and metals or different qualities of paper in order to minimise the costs of disposal or maximise any potential income.

Attitudes to reuse and recycling

Even if there is awareness many schemes struggle because it is easier to throw everything into one container than to separate waste and some people may not be convinced that the effort is worth it. Education and training play a part but unless these are followed through with demonstrations of commitment and support by management initial enthusiasm can wear off. The introduction of an EMS with waste as an initial objective could play a part. Managers must show their commitment by example and through such measures as incentive or bonus schemes. These can be extended to the workforce such that savings are shared among the employees or used to fund incentive competitions or special projects such as works amenities or local charities. The use of local champions to spearhead the change and look for new opportunities through involving everyone to come up with ideas can also help.

Attitude is also affected by the ease of doing things. Replacing the mixed waste bins with, say, four different ones will not be welcomed if they are further away, not easy to identify which is which or not emptied regularly. If people are busy they will take the easiest way, so make it easy!

Physical constraints

Some sites may struggle to find space for more containers if one is replaced by several. However they should not need to be so large and the sizes can be matched to the different rates of waste production. Regular emptying or collection will also help but recycling should also be linked to a programme of waste elimination and reduction which can reduce the total volume as well. The collection company is most likely to provide its own containers on a rental basis.

It is important to manage the collections as charges may be based on visits as well as or instead of quantities. Ideally the collection frequencies need to be related to the waste production although in many circumstances this may be variable. Instead it needs a reliable contractor who can respond quickly to requests. This is easier to arrange in urban or industrial areas but remote or rural areas may present a problem as the contractor may have a long way to drive to the production site. However, space is probably more easily found in these sites.

Cost of implementation

There are likely to be costs associated with any change in practice. These could arise from the need to reorganise space, provide collection points and bins, staff training, changing business or processing practices and managing the waste. These will include costs up front before any benefits arise but a well-researched business case should be able to demonstrate a pay back in a reasonable timescale along with any other less tangible benefits connected with image, staff morale or customer requirements. Of course, recycling should not be carried out at any price: the principles of BPEO may need to be followed (described in Section 1.4).

Demand for recycled materials

Some recycled materials have a strong demand but some have a variable demand, which is reflected in the prices paid for recycled waste. This is often linked to the economic cycle. If demand for steel or copper is low anyway, the demand for the recycled metals may also be low. There are sometimes gluts of paper which depress prices. However legislation and government initiatives and taxes are driving the move to greater recycling rates and this is helping the waste collection companies and the recycling companies (often the same) to find additional outlets or generate new markets.

Quality concerns

The reuse of materials or the use of recycled materials in new products often raises questions about the quality implications for the product or the customer. Metals, paper, cardboard and glass have a long history of reuse and recycling, plastics less so. There may be concerns that the material is degraded as it is recycled: paper fibres are broken down, plastics are contaminated or less strong. These concerns are less of a problem now as recycling technology has improved. It is also met by mixing recycled material with new material to form new products or using recycled material to produce different qualities of product. For example, newsprint contains a high proportion of recycled paper as the product has a short life and the quality constraints are less than those for paper to produce books or fine art.

The reuse of products and parts is subject to similar concerns as the customer judges the quality of a product by its performance or life. Some products such as automobiles are increasingly designed and produced with reuse and recycling in mind. Some parts may be refurbished and reused in new vehicles or as spares and other parts are designed for easy separation and recovery of material at the end of the life of the vehicle.

Legal constraints

The example of disposal of vehicles at the end of their life is actually covered by EU legislation aimed at avoiding pollution and promoting recycling as they are considered as hazardous waste. The landfill directive is also forcing the adoption of alternative means of disposal and the hazardous waste directive means that reuse or recycling of hazardous materials makes more sense than other means of disposal. CFCs are recovered from old refrigerators, metals are recovered from liquid wastes and oils and organic solvents are reused or recycled after refining. Even waste vegetable oils from food processing are being converted into fuels. There are other directives, covering packaging and electrical and electronic equipment, which are described in Section 6.3.

It was mentioned at the start of this section that new regulations are forcing changes including the adoption of the waste hierarchy. In the UK companies also need to be aware of the regulations governing the storage and transport of waste and these are covered in the next sections. They apply to their own activities whatever level of recycling is practised. They also constrain companies wishing to work together to manage their wastes collectively. Although there may be advantages in sharing facilities it needs discussion with the regulatory agencies.

Setting targets and monitoring progress

The importance of setting objectives and targets and collecting information to establish whether they are being met was discussed in Chapter 2 as part of establishing an EMS. Regardless of whether an EMS is implemented, changing UK regulations require confirmation that the waste hierarchy has been used. For this reason alone, records of waste produced and where it eventually ends up will help to demonstrate compliance. Setting your own targets will help promote waste recovery and deliver some of the benefits. It will also help to relate amounts of waste to production volumes or other measures that may explain differences in the amounts recorded or recovery rates.

6.3 Managing waste

The legal position that we have seen so far, which has been about definitions of types of waste and regulations that apply, is clear if somewhat complicated. The previous section also looked at the principles behind minimising waste and reuse and recycling. This section develops this into the practical arrangements that have to be in place both to comply with the law and to manage waste cost effectively.

6.3.1 Key steps in waste management

There are some key steps to achieve effective waste management. These apply at most sites from the small office to the large chemical works. The difference is the range and quantities of waste to be dealt with and the increased possibility of cross-contamination or getting something wrong. At all times compliance with the law is essential and Box 6.8 in Appendix 4 briefly summarises the situation where permits are required in the UK. There are differences among the UK administrations and they change over time. The rest of this section is based on UK legal requirements or practice which are generally applicable.

Separation

The first step is to have measures to separate and keep separate the different types of waste according to what is to happen to them. In most circumstances it is easier and cheaper to separate at the start rather than try to sort mixed waste later. This will require different containers at the points where waste is initially produced. This need not mean multiple bins everywhere; most activities are organised around the raw materials in use or the processes being undertaken and there may only be one or two types of waste at any point. The principles apply to hospitals, offices, hotels and shops as much as large industrial sites. They should separate paper, cardboard, plastics and drinks containers in accordance with the specifications of the waste collection company. Food waste may have to go into a general waste container destined for landfill unless a lot is produced in, for example, a supermarket, restaurant or canteen. Special arrangements should be made to collect such large volumes. These premises also need to be aware of the need to keep separate waste such as electrical equipment and concentrated cleaning chemicals.

Separation may focus on waste suitable for reuse or recycling on the site, waste that is to be sent for reuse or recycling elsewhere, hazardous wastes, and waste that will inevitably have to be sent for landfill. In some locations the design of container can help with waste separation. For example at public amenity sites the size and shape of hole in the lid can help direct cans and bottles to the right containers and keep other materials out.

Hazardous wastes need particular attention. Each type must be kept separate and clearly identified. Mixing is generally not allowed and may be dangerous. Remember also that if hazardous waste contaminates other waste the whole batch will have to be treated as hazardous. Many sites do not produce hazardous waste as part of their normal operations but occasional quantities may arise from replacing fluorescent light tubes, computer screens, building and plant maintenance or cleaning.

Employees need to be aware that paint, oil, batteries, electrical equipment, inks, dyes and chemicals need to be dealt with according to the guidance with the products or from the contractors who collect the waste. Asbestos may also be found on some sites, particularly during refurbishment of old buildings but its removal and disposal needs specialist contractors who should deal with the waste appropriately.

Note also that inert waste must not contain any non-inert waste or else it will have to be removed and disposed of as controlled waste with increased costs. Landfill tax in the UK in 2019 for inert waste is £2.90 per tonne; otherwise it is £92.35 per tonne, increasing every year by the retail prices index and each tax subject to 20 per cent VAT on top!

Containers

Containers need to be suitable for the materials being collected. Plastic sacks on a suitable stand may be appropriate for paper in an office or small light items. Larger metal or plastic containers are more usual and the materials of construction need to be inert to the waste being stored and sufficiently robust to withstand rough handling (from heavy items being dropped in to being moved manually or by a forklift truck). The containers need to be easily identifiable, ideally colour coded and labelled in a consistent way throughout. Sizes of containers need to reflect the quantities of waste being produced. If they are too small and quickly filled people will tend to find alternative ways of dealing with the waste.

Figure 6.2 Clear signage is essential

Storage

Storage on site is inevitable as waste is collected from the areas of production and consolidated awaiting further action. There is a continuing need to keep the different wastes separate as it is too easy to tip one initial container into the wrong secondary one. Clear identification with consistent labelling and colour coding help, as in Figure 6.2. As storage is usually away from the main production sites, it needs to be designed and set out to ensure that the waste is kept separate, secure and protected from the weather. If liquids are involved the storage needs to be bunded and managed in accordance with the storage of liquids in Section 5.3.3 in order to avoid the risks of water pollution. Figures 6.3 and 6.4 show two alternative ways of managing waste storage.

The main points to consider are that hazardous wastes are kept separate and care taken to ensure that incompatible materials are also isolated (for example oxidising agents and flammable materials). Containers or storage areas should be covered to keep rain out and to stop the escape of waste and the whole area should be secure against vandalism, unauthorised visitors and, as far as

Figure 6.3 A tidy site with multiple containers

Figure 6.4 Cutting oil escaping from a waste metal skip

practicable, vermin. The location and protection should take account of vehicle movements with sufficient space for safe manoeuvring and protection barriers where necessary.

Storage containers need regular inspection for corrosion or damage to ensure that they retain the waste and will not fail on being removed from site. The storage containers such as skips or large bins are usually supplied by the waste removal contractors so it is necessary to check on them as they are replaced.

Transportation

Some waste will be transported around the site and this is the responsibility of the site operator. The site operator may also remove the waste from site to a recycling plant or to a disposal site. Vehicles transporting waste off site need to meet similar criteria as those for storage: clear labelling, identification and security. The labelling is particularly important and is covered in more detail below. Within the UK persons transporting waste need to meet certain criteria and be registered as waste carriers; there are more details below.

Disposal

If waste is not being reused or recycled its ultimate destination is usually landfill or incineration. The use of landfill sites is regulated in most countries and within the EU is covered by the landfill directive as already explained. Within the UK landfill sites are controlled through the permitting system as described in Box 6.8 in Appendix 4 and in Chapter 1.

6.3.2 Responsible waste management

We have to accept that waste management is rarely the top priority for most businesses so it tends to be tucked away in a corner on a site that is less suitable for productive use and out of sight of the managers of the organisation. The management of waste has to be in accordance with the relevant laws and regulations but the responsibility is on the producer of waste to ensure that it is handled and disposed of responsibly and does not end up in the wrong hands or being fly tipped. In order to achieve this (and go beyond mere compliance with legal requirements as part of an EMS) there are additional points to take into account. Some of these are good business practice but they are enshrined in UK law in the concept of 'duty of care'. This was set out in section 34 of the Environment Protection Act 1990 and subsequent regulations and in statutory guidance from DEFRA. The Waste Duty of care Code of Practice of 2018 is the latest version. The overall aims are to ensure that:

- Waste is not illegally disposed of or dealt with without a licence or in breach of a licence or in a way that causes pollution or harm; waste does not escape from a person's control.
- Waste is transferred only to an authorised person, such as a local authority, a registered carrier or a licensed disposer.

The principles are a suitable model for anywhere. The full details are not included here but the principles are set out in Box 6.9 in Appendix 4 and they serve as a model for good practice elsewhere. Most of the points have been covered in the previous sections but further detail, based on UK practice, is given on the others below.

Regulatory documentation

In most countries that have a regulated waste management system there is a requirement for documentation for describing the waste and monitoring its progress. This section again describes UK (mainly English) practice but the details will vary elsewhere. The description of the waste that accompanies it should be more than the EWC six-digit codes from the EU waste catalogue described in Box 6.2. In the UK it is included in the transfer note (for non-hazardous waste) or consignment note (for hazardous waste) and needs to contain:

- The producer of the waste.
- Details of whom it is being transferred to.
- A full description of the waste which may include an analysis.
- The EWC code.
- The total quantity of waste.

In addition consignment notes for hazardous waste need to include:

- Details of the process that produced the waste.
- Carrier details.
- Whether the waste is subject to the requirements for carrying dangerous goods.
- Where the waste is being taken to.
- Signed declarations by the parties involved.

The notes are produced as copies and accompany the waste with the producer, carrier and receiver keeping copies. All copies should be kept for two years (transfer notes) or three years (consignment notes) as they provide an audit trail. (Note that transfer notes may now be electronic.) Longer periods apply to landfill operators.

Requirement to demonstrate competence for waste operations

Organisations involved in relevant waste operations (defined in legislation) in England will have two particular requirements as a condition of their permit or licence. These will not affect producers but they are relevant for the next sections concerning carriers and organisations that are involved in treating and disposing of waste. The competence of the operating company is measured against a requirement for an EMS. This is required for most waste operations; for the most complex sites it needs to be a formal EMS such as ISO 14001 (see Chapter 2), otherwise an in-house system will be acceptable which may just cover issues such as noise and odour. In addition operators are required to demonstrate technical competence if they are involved in landfill operations or dealing with some types of hazardous waste. Competence is demonstrated by one of two approved certification systems: the CIWM/WAMITAB scheme that has been jointly developed by the Chartered Institution of Wastes Management (CIWM) and the Waste Management Industry Training and Advisory Board (WAMITAB), or the ESA/EU Skills scheme that has been jointly developed by the Environmental Services Association (ESA) and the Energy and Utility Sector Skills Council (EU Sector Skills). There are other requirements for special situations such as radioactive waste.

Waste carriers

It is important that registered carriers or exempt carriers are used to collect and transport waste. Waste cannot be passed on to anyone with a van. Exemptions apply in specific circumstances but it is necessary to check with the appropriate regulator.

Carriers will need to show any relevant Hazchem symbols on their vehicles if carrying hazardous waste (these are signs that show warning symbols for the hazards and use codes to indicate to emergency services, such as fire, the nature of the goods being carried and what action to take) and carry TREM cards (instructions to drivers in the event of an accident). The carrier also needs to ensure that the vehicles are suitable for the wastes being carried, that they are licensed and tested, that their drivers have the correct licences and that the relevant insurances are in place.

Site permits

Most sites that store, sort, treat or dispose of waste require a permit under English and Welsh legislation (a licence in Scotland and Northern Ireland) or, if exempt, may still need to register with the Environment Agency. The permit will specify the types of waste or operations that may be carried out and set out any constraints that may be applied (for example, the requirement to cover waste in landfill at the end of the day). The permit will also specify what competences need to be demonstrated. Details of the types of activity and the requirements in the permit, and the availability of standard permits varies with the different UK administrations and changes over time. More can be found on the web sites of the regulatory agencies.

Management and supervision

The previous sections have set out some steps that are required or help to protect the environment and to ensure legal compliance. In the pressure of business these may slip and so it is important to monitor their continuing relevance and application. It is likely that the regulatory body will inspect the site from time to time. Someone should be given responsibility for managing waste on a day-to-day basis and supervisors should regularly inspect waste collection and storage facilities. They should be checking to ensure that waste is being properly segregated and that no mixing is taking place and that the containers are sound and properly labelled. Equally important is the secure containment and tidiness of all facilities. Overflowing containers may indicate inadequate provision or poor practices, both of which need to be addressed.

Training

Employees may be sceptical of changes, especially if they perceive additional work or frustration. Training can help to overcome this by helping them to understand why reuse or recycling is necessary and why it is important to comply with regulations or regulatory requirements. There are fines for not complying and other potential costs to the business for clean-up or regulatory costs. Vehicles can be impounded if they are used in illegal activity. Training should also cover the processes and procedures being put in place and any changes to current practices that may be required. Training sessions can be good sources of suggestions for further improvement.

6.3.3 Packaging waste

Packaging waste has some rules of its own within the EU. These were designed to reduce the amount of packing used and to promote recycling. Much packaging is only used for one trip and many items come with several layers made up from glass, metals, plastic, paper and cardboard. Some of the packaging materials are made up of layers laminated together which makes them difficult to recycle. Over 11 million tonnes of packaging are put on the market annually in the UK, of which about 70 per cent is now recycled in the UK (Waste data overview (2011b), available at www.defra.gov.uk).

The European Directive on Packaging and Packaging Waste (European Commission 1994 with later amendments) aims to harmonise national measures in order to prevent or reduce the impact of packaging and packaging waste on the environment. It contains provisions on the prevention of packaging waste, on the reuse of packaging and on the recovery and recycling of packaging waste. Member states were required to put measures in place to comply. In the UK businesses that have a turnover of more than £2 million and handle more than 50 tonnes of packaging per annum have to register with the relevant regulatory agency. They then have to meet recovery and recycling targets which are set annually and provide evidence that they have done so. They can do this themselves or join a compliance scheme which operates on behalf of many businesses. Compliance is demonstrated by Packaging Waste Recovery Notes (PRN) and Packaging Waste Export Recovery Notes (PERN) purchased from an accredited reprocessor or the exporter and, as some businesses over-recover and recycle and some fail to meet their targets, these can be traded. Other European countries have approached compliance in other ways, for example by simple regulations.

6.3.4 Electrical and electronic waste

Waste electrical and electronic material is also treated differently. This is because it contains many valuable materials such as copper, gold and silver and some hazardous materials such as mercury, lead, fire retardants and PCBs (polychlorinated biphenyls). The Waste Electrical and Electronic Equipment Directive (WEEE Directive) (European Commission 2003) was introduced to promote the recovery and recycling of WEEE. At the same time another Directive was introduced to promote the replacement of the more hazardous materials in electrical and electronic equipment. These two directives have since been merged into a more comprehensive Directive (European Commission 2018).

In the UK the Waste Electrical and Electronic Regulations 2013 put the onus on manufacturers and importers and some businesses that sell or dispose of equipment to introduce schemes to recover and recycle the waste through an Approved Authorised Treatment Facility which reports on the amounts being recycled each year. In order to meet their obligations, suppliers will take back old equipment from the users. Detailed guidance is available from the Environment Agency (2018).

6.3.5 Waste from construction projects

Waste arising from demolition and construction in the UK is large in volume – estimated as 35 per cent of landfill waste by Sustainable Build (2019) – and traditionally has been badly managed. In 2008 62 per cent was recovered or recycled, whereas other European countries recycled up to 90 per cent of their construction waste. The UK Government introduced the Site Waste Management Plans Regulations 2008, requiring waste management plans to be produced for any project over £300,000 in value. However, these are considered ineffective and were repealed in 2013. Instead reliance is placed on initiatives from within the construction industry. WRAP (undated) estimates that each year the construction industry uses 400Mt of material and produces 100Mt of waste of which 25Mt are disposed to landfill. The reference includes ideas to minimise this waste. The principles of the regulations make good sense as they require planning for waste management to start at the design stage of a project. The waste management plan was established at the same time and covered many of the topics already covered in this book in general terms:

- Designing with waste minimisation in mind (size of materials, etc.).
- Reusing or recycling demolition and excavation waste.
- Managing the quantities ordered and applying security and weather protection so that goods do not get damaged.
- Setting targets and monitoring performance.

Construction sites often rely heavily on contractors so there was emphasis on briefing them and training all workers on their responsibilities. There was also emphasis on complying with the duty of care and ensuring that only registered carriers were used to transport waste. Although the legal requirement for a plan has gone, the principles behind the plan remain good practice.

6.3.6 Domestic waste

Domestic waste accounts for much of the waste produced; 27.3 million tonnes in the UK in 2016. The composition of domestic waste varies as has been previously mentioned depending on economic and cultural differences but consists mainly of waste food and packaging. Waste management practice also varies. The UK is fairly unique in having regular collections from the household. In most of Europe the householder has to take their waste to collection points and in some parts of the world it is just thrown into the street or on to waste land. In Europe householders are required to sort it into different containers for recycling and then most is collected by local authorities or contractors and taken to recycling facilities, landfill sites or incinerators. There is more on UK practice in Section 6.4.4.

Commercial waste from offices, shops and similar locations (i.e. excluding industrial waste) is similar in many ways to domestic waste although the composition will vary. It is normally collected by contractors and disposed of in similar ways to domestic waste with a high degree of recycling of paper, plastics, etc. expected.

6.4 Outlets available for waste

The waste that is recycled will eventually be made into new products. Some, such as paper, metals, glass and plastics, may reappear in forms identical to the original product or may be made into alternative forms (such as plastic bottles being reformulated into synthetic fleece for clothing or glass bottles into construction materials). Wood may be turned into wood chips or pellets and used to make chipboard and other wood products or as fuel. Residual wastes that cannot be recycled have only two legal outlets: landfill (in the EU defined as disposal onto or into land) and incineration. Both should be seen as a last resort. Neither is popular with the public who will object to any proposals for sites near where they live. However, until we are able to achieve zero waste, they will continue to be required.

6.4.1 Landfill

Landfill sites are the commonest form of ultimate disposal worldwide. They range from large unmanaged heaps on the outskirts of towns to highly engineered sites within old quarries that could be seen as a form of recycling in itself. For many years tipping waste into old quarries was seen as simple and safe. Eventually it was realised that liquids disposed of into the sites and created during decomposition of the waste, supplemented by migrating groundwater and collected rainfall, were leaching out of the bottom (hence called leachate) and contaminating groundwater or adjacent land. Since then new rules such as the Landfill Directive, directives to protect groundwater and national laws and regulations have tightened things up. The original Landfill Directive targets for reducing waste to landfill are in Box 6.10. A new proposal (European Commission 2019) requires that only 10 per cent of municipal waste should end up in landfill. Landfill sites that deposit more than 10 tonnes of waste in a day or accept a total of 25,000 tonnes in total (excluding sites that accept only inert waste) are now regulated under the IPPC Directive within the EU.

Modern landfill sites (Figure 6.5) have to meet several criteria, the principal ones of which are:

- Site selection has to be supported by an environmental risk assessment (the granting of a permit has to take account of this: for example, no landfill sites would be permitted within the inner protection zone of a groundwater source).
- Sites are classified as able to accept inert waste, hazardous waste or non-hazardous waste and there are constraints on what each can accept.
- Liquid wastes and some other wastes, such as infectious waste, explosive, corrosive, oxidising, flammable wastes and whole or shredded tyres, are banned.
- The formation of leachate should be minimised by keeping rainfall and groundwater out and any leachate formed should be contained and collected for treatment (e.g. the base and sides should be impermeable, achieved with clay or synthetic membranes).
- The formation of landfill gas has to be controlled and any formed collected and managed by energy generation or flaring.
- Controls should be in place to minimise nuisance such as litter.
- Waste must be pre-treated before acceptance (see below).
- Sites have to be capped when the site is closed and a management system in place to deal with ongoing gas and leachate production.

Figure 6.6 illustrates the main requirements.

There are many other constraints which apply to the designers and operators and information for the UK can be found in the relevant Environmental Permitting Guidance on the appropriate web sites. Practical considerations are that sites are usually managed by filling them progressively in cells and deposited waste is covered with soil at the end of each day to limit vermin and litter. The initial decomposition when air is present (aerobic) releases carbon dioxide. As waste builds and

Figure 6.5 Landfill operations in progress

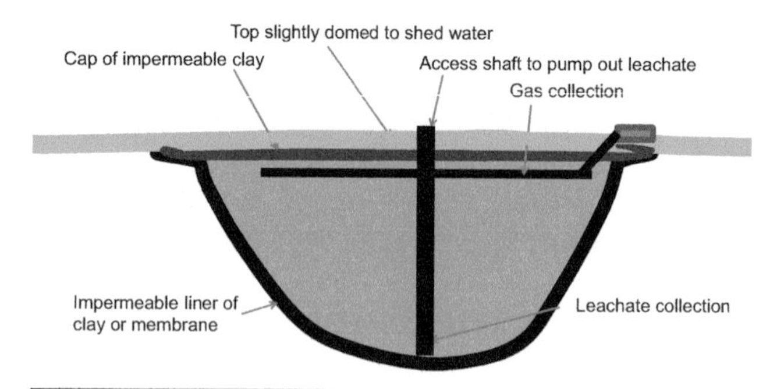

Figure 6.6 Main requirement of a closed landfill site

Box 6.10 EU Landfill Directive targets

The following targets are to reduce the amount of biodegradable municipal waste that can be sent to landfill.

- By 2010 it should fall to 75 per cent of that produced in 1995.
- By 2013 it should fall to 50 per cent of that produced in 1995.
- By 2020 it should fall to 35 per cent of that produced in 1995.

the oxygen is used up the processes become anaerobic and the gas released includes ammonia and other odorous compounds as well as methane. Methane is flammable and can form potentially explosive mixtures so there is a fire and explosion risk that needs to be managed. There have been instances of methane migrating underground into properties. We have also already seen that methane is a powerful greenhouse gas. This means that it has to be either safely vented, or collected by a system of pipes and either used for energy generation or flared.

The leachate has to be collected from the base of the site and contains materials dissolved from the waste (e.g. soluble ions such as chloride) and the liquid products of the chemical and microbiological processes that degrade the waste. It is toxic and has a high BOD so it cannot be discharged directly to a watercourse and would have persisted a long time underground if it had been allowed to escape. Instead it has to be discharged to a sewage treatment works (if available and acceptable to the operator) or treated on site. The amount produced varies with the size of the landfill, the nature of the waste, the degree of filling with waste, the amount of water that infiltrates and the age of the site. The volume and strength builds up over time and then slowly declines after the site is closed. Treatment can take several forms but all rely on further biological treatment (aerobic or anaerobic) as described in Chapter 5.

Note that some businesses operate their own landfill sites rather than rely on contractors to remove and dispose of waste. They are subject to the same controls.

Pre-treatment of waste

It is a requirement of the Landfill Directive that most waste for landfill is pre-treated to meet these three criteria:

- By physical, thermal, chemical or biological processes (including sorting).
- That change the characteristics of waste.
- In order to reduce its volume or hazardous nature, facilitate its handling or enhance recovery.

The exceptions are inert waste for which treatment is not feasible and other waste where treatment would not reduce its risk to the environment or human health.

Segregation at source is considered equivalent to sorting and is acceptable as an alternative but compaction to reduce the volume is not as it would not change the characteristics. If waste is sorted on site there is a presumption that it would be recycled, not just landfilled separately! Other treatments should be designed to reduce the risk to the environment, for example, by reducing the organic content or the hazardous nature. Incineration of organic waste would meet the first objective, leaving a smaller volume of inert solid. Solidification of liquid hazardous waste into a product that is stable under landfill conditions would also qualify.

6.4.2 Incineration

Waste incineration was originally affected by the EU Waste Incineration Directive (European Commission 2000b). This has been replaced by a recast Directive on Industrial Emissions (integrated pollution prevention and control) which consolidates seven directives (European Commission 2010). This change does not significantly affect the position as far as this text is concerned. Some incinerators are already covered by the current IPPC Directive. Current guidance for the UK is found in the relevant Environmental Permitting Guidance for the different jurisdictions. The aim of the directive is to limit the risk to the environment and to human health from incineration and co-incineration of waste. Incineration includes the combustion of the waste whether the heat is recovered or not and includes other thermal treatments such as pyrolysis and gasification (not subjects for this book). Some wastes are excluded from the Directive, principally those involving agricultural and food wastes, wood and fibre. Most of the incinerators included are for clinical and hazardous wastes and for municipal waste. Incinerators require a permit which covers the whole installation including from reception of waste to ash disposal, not just the incinerator itself. They are permitted as a Part A(1) or Part A(2) installation regulated by the environmental regulator or the local authority. The permit will prescribe conditions about the plant, its operation and limits for emissions. Most of the opposition to incinerators focuses on the emissions to air, in particular concerns about contaminants such as dioxins formed in the combustion process, particulates and any possible heavy metals as these have been problems in the past. The permits are meant to ensure that no emissions can occur at concentrations that pose a risk to health and cover a much wider range of substances than dioxins and metals. Combustion of organic matter produces CO_2 so incineration will contribute to climate change but landfill and other options also contribute in different ways. Waste heat is also meant to be recovered as far as is practicable, for example, in local heating schemes or for power generation. For this reason incineration is also referred to as energy recovery and is seen as an alternative to landfill. Its use for energy generation is discussed further in Chapter 7.

Incineration requires that the waste has a sufficiently high calorific value to burn or else it requires the addition of supplementary fuel. Hazardous and clinical wastes have differing qualities and are relatively small in volume anyway so adding fuel is less of an issue. Municipal waste has a much larger potential volume so to be combustible it needs to retain much of the organic matter such as paper and plastic, removing them from the recycling chain. This is another reason quoted by opponents to incineration in the UK. However in countries such as Denmark, Germany and France they are more widely accepted, being built close to urban centres where the energy recovery is most efficient.

6.4.3 Relative advantages and disadvantages of landfill, incineration and reuse and recycling

As alternative choices for the disposal of waste, particularly municipal waste, there is sometimes no competition as suitable sites for landfill may be difficult to find. Where there is a choice there are several factors to take into account. Most of these have been touched on already but they are summarised here as bullet points for ease of access. The benefits and limitations of recycling were shown in Boxes 6.6 and 6.7. Of course not all of these are absolute: for example, the transport issues can be advantages or disadvantages depending on the distance from the sites of production. Waste to landfill can end up travelling large distances if local sites get used up.

Advantages of landfill

- Can be close to the source of the waste so transport is reduced.
- Local waste is dealt with locally.
- Relatively cheap to set up and operate.
- Can accept a wide variety of wastes.
- Only option for some waste (e.g. ash from incineration).
- Potential for restoration of old quarries and other excavations.
- Restored land can be returned to beneficial use such as amenity or agriculture.
- Energy recovery potential from methane (although energy recovered is lower than from incineration) which is considered as renewable energy.

Disadvantages of landfill

- Difficult to find suitable sites in many locations.
- Loss of materials that could be recovered.
- Potential for pollution of air and water.
- Potential to result in contaminated land.
- Fire and explosion risk if not well managed.
- Need to collect and treat leachate even after closure.
- Local nuisance from transport, litter, odour and visual impact.
- Local opposition to new sites.
- Methane emissions linked to climate change.
- Rising taxes increase cost over time.
- Landfill Directive is increasing pressure to avoid its use.
- Ongoing responsibility for sites after closure.
- Some old landfill sites remain a source of pollution to air and water.

Advantages of incineration

- Volume of residual waste much reduced.
- Energy recovery from organic materials.
- Can be sited close to the areas that produce the waste.
- Used for clinical and some hazardous wastes as the only option available.

Disadvantages of incineration

- Loss of materials that could be recovered such as paper and plastics.
- Residual ash still requires disposal.
- Residual ash can be hazardous.
- Potential for air pollution from dioxins and particulates.
- Plant is expensive to build and complex to operate.
- Transport may become a nuisance or add to congestion in an urban area.
- Local opposition is likely to be the most intense (at least in the UK).

6.4.4 The management of domestic waste in the UK

We have already touched on domestic waste in Section 6.3.6 but it has some characteristics that require additional mention. It is high in volume and consists of a lot of material which can be recovered. It can also contain potentially hazardous materials such as batteries and electrical waste. Households also produce waste food, oil and bulky items such as furniture and fridges and garden waste. For many years the smaller items were just mixed together in one bin, collected by the local authority or a contractor and landfilled. Many larger items ended up being fly-tipped as they were difficult or expensive to have removed: sometimes the illegal tipping was carried out by a collection service. The implementation of the landfill directive coupled with government policy and increasing landfill tax is forcing the county councils who have responsibility for waste disposal to make big changes and bring the UK more in line with best practice.

For several years now the householder has been required to separate waste to make it easier for recycling in most areas. The details vary around the country and can be a source of confusion, so attempts are being made to attain some conformity. Most systems expect glass to be kept separate and some expect paper and cardboard to be separated from plastic and metal containers. However others will accept recyclable waste in one container (referred to as 'commingled') with non-recyclable in another and accept batteries in a small bag. If the waste is to be incinerated then separation of combustible material may not be required. Some will collect garden waste separately for composting. They will also collect larger items such as furniture and fridges by arrangement. The alternative is a managed waste centre where householders can take their waste directly. It is difficult to force domestic households to comply with regulations and there have been attempts to use fines but these have now been abandoned. The emphasis is now on education and persuasion.

Once the wastes have been collected by truck it has to be taken for disposal. This may be directly to recycling centres or to landfill or to an incinerator but often these are some distance away. In this case the waste is taken to a transfer station where it may be further sorted for recycling and the waste intended for landfill compressed to reduce its volume before further transport by larger truck, rail or, in the case of London, by barge. Many authorities operate Waste (or Material) Recovery Facilities either at transfer stations or elsewhere. These sort waste for recycling. This could be mixed domestic waste, commingled waste or waste already partially sorted by the householders. The sorting may be by hand but increasingly automated processes are used involving mechanical methods that can separate metals, paper, plastics, etc. into streams that can be further processed by specialist companies. The organic waste comprising mainly food may be sent for biological treatment such as anaerobic digestion. Other organic wastes including garden wastes may be composted. In some cases there is an incinerator on the same site.

There is therefore a variety of ways of managing this waste but they all meet the landfill directive requirement for treatment before disposal. The facilities may also deal with commercial waste from offices, shops, hotels, etc. which is similar in character to household waste. The other provision by local authorities is the managed (or civic) waste centre where householders can take their waste without charge. They are not normally available to businesses and often limit the size of vehicles and the quantity and frequency in which inert waste may be taken. A typical centre may have up to 20 deposit points accepting paper, cardboard, batteries, fluorescent lights, televisions, computer screens, other electrical equipment, refrigerators and freezers, hard plastics, polythene bags, textiles, shoes, oil, garden waste, wood, metals (may be further separated such as drinks cans, copper and other metals), inert waste, paint, and non-recyclable waste. The advantage here is that the waste is sorted already and the risk of contamination is much reduced.

Food waste from households and business premises is largely still sent to landfill, although authorities are starting to arrange separate collections for anaerobic digestion or providing households with special bins that can be used at home if you have the garden space.

All of the facilities provided by local authorities or contractors will require the appropriate permits from the regulatory agencies.

For larger items and many serviceable smaller items there are charities and commercial companies that will collect or accept them for sale or further donation. In addition there are web sites that can be used to sell items (the best known of which is www.e-bay.co.uk) or give items away or exchange them (such as www.uk.freecycle.org which is part of a world-wide movement).

6.4.5 Other issues around waste disposal

The position in the UK is improving in that reuse and recycling are becoming the norm and best practice is being adopted. However, much of the waste that is sorted for recycling is exported. Some of this is to specialist facilities, say in Europe, where it makes sense to consolidate on few sites. However, some is sent to third world countries where it is further processed or just dumped. This includes some potentially hazardous wastes such as electronic equipment, although it should be noted that the Basel Convention does set some controls on the trans-boundary movements of hazardous wastes (the latest version is at UNEP 1998). There is a risk that the waste is often dealt with in ways that would not be acceptable in the UK. It may be dismantled by hand and by young children in conditions that would fail any health and safety legislation. Much of the waste or the residual waste may be deposited in landfill sites that are badly managed, if at all. Some may be incinerated in the open. The net result is that the potential pollution problems are just transported overseas and the resources are not fully recovered. Much of this is illegal as there are laws regulating the export of waste and penalties as well as reputational issues for companies or authorities that fail to comply, but there are profits to be made and some will always try to beat the system. Most countries on the receiving end are now trying to stop this trade and will refuse to accept it or send the waste back to its original source. In the UK the duty of care implies that the onus is on the producer of waste to ensure that it is dealt with legally and safely but if it passes through several hands it is difficult to keep track.

Reference has already been made to the problems of managing waste in developing countries, especially where the legal framework and infrastructure is not in place. Illegal exports just add to that problem. However, the trade is not one-way; western companies have started to offer to buy suitable waste for reprocessing from developing countries, which creates local jobs and income and helps to tackle the growing waste problems as they develop.

6.4.6 The effects of rising costs

The theory is that reuse and recycling saves money overall and in practice this is usually the case with short pay back periods. The costs fall on the producer and the financial benefits may accrue to the producer but also elsewhere. For example, there may be the value of avoided environmental damage caused by mining or greenhouse gas emissions. Many of the changes are now being driven by legislation and financial instruments. The main legislative requirements have been described above and the increasing difficulty of finding sites to accept hazardous wastes is driving up the price. In addition, in the UK, the aggregate levy (tax by another name) is intended to reduce the amount of minerals such as rock, sand and gravel extracted and to encourage their recycling. The long-term effect of this will be to reduce the number of quarry sites available for landfill and this will drive up prices. The landfill tax also adds to the cost of disposal so saving that will be a direct benefit to the producer.

6.4.7 A note on plastic waste

It is impossible to avoid the growing concern about plastic waste in the environment seen in the news (e.g. www.bbc.co.uk/news/uk-39001011) and in television documentaries (www.documentary.org/blog/watch-these-documentaries-about-plastic-pollution). It has been found not only on land, but in the Arctic and Antarctic, the deepest parts of the oceans and in the stomachs and intestines of marine mammals, fish and birds. The visible waste is plastic bags and sheeting, fishing tackle, bottles, pellets for plastic manufacture (called nurdles), etc. Some of this gets into the oceans directly from ships discharging waste or accidentally losing cargo; the rest is carried in rivers or by the wind. Closer study also shows the presence of micro plastic particles and fibres. Part of the problem is that the durability of plastic, one of its positive user benefits, results in waste that does not readily degrade by chemical, microbiological or solar attack – the means by which other wastes are degraded. The small particles are created by abrasion by other materials also present such as other waste or sand. There are also large numbers of microfibers, which seem to come from fabrics, either as waste or as a result of washing clothes. These are so small that they can pass through normal sewage treatment processes. These will be taken in by freshwater and marine organisms.

The wastes are finding their way all over the planet and into the ecosystems. The larger pieces cause the death of species by interfering with their normal behaviour (e.g. by getting caught up in flippers or trapping in nets) or, if they are mistaken for food, by blocking the intestines and causing starvation.

The adverse publicity this has attracted is putting increased pressure on society to reduce the production and waste of plastic. As we saw above, one of the problems with the mixed wastes is that it is difficult to separate the different types of plastic for recycling and the mixed waste is of limited use.

6.5 Risks associated with contaminated land

Contaminated land has been mentioned already in several sections of this book. It can be present due to natural causes such as the presence of some minerals. It is also caused by waste disposal, especially from old landfill sites, contaminated water or fallout from air pollution. In the older industrialised countries there are many sites that have been affected by historic practices that would be illegal now. Wastes from mining were built up on the surface and contaminated drainage allowed to soak into the ground. Waste liquids and solids were tipped at the back of a factory site, many of them from processes such as tanning and metal plating which involve strong acids or alkalis and toxic substances such as chromium, copper or cadmium. Radioactive substances may also be present from times when their use was more widespread and the risks less understood. Further problems arose from leaks and spills that were allowed to soak away, including chemicals and oil at vehicle servicing sites. Drums, old storage tanks and plant were abandoned and slowly leaked. This is still going on uncontrolled in some countries. Figure 6.7 shows a site in the UK that has been left after the demolition of old facilities for converting coal into gas and other chemicals.

Figure 6.7 Contaminated land

The pollution that starts on the surface soaks in and the impact will depend on the nature of the substances, the quantity involved, how far it penetrates or spreads and the site location. Small quantities of biodegradable substances will not be a continuing problem. Metals, pesticides, oil, solvents and paints are examples of substances that do not degrade or do so only slowly. The immediate effects will be on the surface and toxic substances will build up in the soil killing plants and threatening other wildlife and humans who may come into physical contact, causing skin irritation, or ingest the soil or inhale blown dust and any associated gases. If plants are not killed, substances may build up in the tissues and get into the food chain. Over time the pollutants will migrate into the ground threatening groundwater. Figure 5.2 showed how rain recharges groundwater and any liquid that is on the surface can get into the groundwater in the same way. Solids may dissolve in rainwater and also migrate. Once there they will mix with the groundwater and migrate with the underground flow. Immiscible liquids such as oil or solvents will settle in a layer above or below the water depending on their density. Although most oils, pesticides and other organic substances may be considered insoluble in water, in practice most are to a small extent. Thus, even if in a separate layer, they can still contaminate the water and render it unsuitable for providing drinking water without expensive treatment. Pollutants may also migrate sideways into surface water where they will also threaten aquatic life and water supplies for irrigation, animals or drinking water, as shown in Figure 5.3.

Some contaminants such as acids may corrode underground services or building structures and organic solvents have been known to penetrate plastic pipes used to distribute drinking water. If the ground becomes anaerobic methane may be generated, presenting a fire or explosion risk. There may also be odours from the site either from the substances directly or as a result of their degradation. Typical odours include solvents and oil, ammonia and hydrogen sulphide. The sites often attract vermin and in total they become a nuisance to neighbours. Usually the land is left abandoned because it is unsuitable for development without expensive remediation.

6.5.1 The legal position

Within the EU the main legislation is the Directive on environmental liability with regard to the prevention and remedying of environmental damage (European Commission 2004). This defines environmental damage as:

- Direct or indirect damage to the aquatic environment covered by Community water management legislation.
- Direct or indirect damage to species and natural habitats protected at Community level by the 1979 'Birds' Directive or by the 1992 'Habitats' Directive (see Box 1.12).
- Direct or indirect contamination of the land which creates a significant risk to human health.

From the previous section it is clear that all three could apply and, in the case of contaminated land where there is a current threat, the competent national body can require the operator to take precautionary measures or do the work themselves and recover the costs. Where damage has already occurred the operator can be required to take restorative measures or the competent body carries out the work and again recovers the cost. In the UK the relevant legislation is the Environment Protection Act 1990: Part 2A supplemented with the Contaminated Land (England) Regulations 1966 and subsequent 2012 amendment and the other national equivalents. Under this legislation the occupier is the responsible party for cleaning up legacy contamination if they caused (or allowed) contamination to occur, even if they no longer operate it or own the site (these are known as Class A). Otherwise it is the current owner or occupier even if they did not cause the contamination (known as Class B). Permits issued under IPPC may also contain references to avoiding contamination of land.

Responsibility for identifying who is liable usually rests with the local authority but the environmental regulators may also get involved, especially if there is an ongoing threat to controlled waters. If no Class A person can be identified or found, the Class B person may be liable. This means that if purchasing a site it is important to investigate any potential for contaminated land, otherwise there is a considerable risk of suddenly gaining a liability.

As usual, the details are more complicated than this and for more information reference needs to be made to the legislation and guidance available on the DEFRA and the environmental regulators web sites.

6.5.2 Dealing with contaminated land

Given the risks to the environment and to humans, there is the problem of what to do about legacy sites (i.e. from past activities) as well as avoiding the creation of more contaminated sites for the future. The latter is the easiest to deal with: previous chapters have highlighted the risks from air and water pollution and waste disposal and they contain the measures that need to be taken. There is no excuse for further contamination. For legacy sites it is a case of a risk-based approach. A risk assessment needs to be made if investigating a problem or purchasing land that has been previously used. This may be cursory if the history of the site is known, otherwise a detailed site investigation may be needed. If contamination is found a risk assessment will guide the extent and type of remediation needed based on the actual or potential environmental damage, the consequences and the intended use for the land. This would use the source, pathway, receptor model referred to previously.

There are many remediation techniques available. The simplest (tackling the pathway) may be isolation to avoid further spread by inserting an impermeable layer to prevent sideways migration, although this will not stop downward migration unless there is clay or a similar material lower down. Sealing the surface to prevent ingress of rainwater may also be used and this could be appropriate if the site is to be developed as a car park or industrial estate with the surface fully sealed. Alternatives (tackling the source) involve excavation of the contaminated material and its subsequent decontamination or disposal as hazardous waste. In situ techniques are also available but they may need to be continued for many months. These rely on aeration, microbiological methods and water or other solvent extraction. The effectiveness and cost of these will depend on the type, degree and extent of the contamination. The end result has to be to achieve land quality fit for any intended purpose. Guidelines on the choice of procedures are available (Environment Agency 2019).

6.5.3 Potential liabilities from contaminated land

From the above it can be seen that the potential liabilities can be large. If you cause the contamination or even if you did not but were left liable because you now own the land and the person who caused it could not be found, the costs may fall on you. The costs of land and groundwater remediation can run into hundreds of thousands of pounds. Along the way there may be costs for site investigations, sample analysis, advice from consultants and lawyers; none of them renowned for their low fees. There could also be costs for damages or remediation for third parties who suffered damage. In addition, the party causing the contamination may be prosecuted, resulting in fines and additional legal costs. *Caveat emptor.*

CHAPTER 7

Sources and use of energy and energy efficiency

After this chapter you should be able to:

1. Outline the benefits and limitations of fossil fuels
2. Outline alternative sources of energy and their benefits and limitations
3. Explain why energy efficiency is important to the business
4. Outline the control measures available to enable energy efficiency.

Chapter Contents

Introduction 122
7.1 Benefits and limitations of fossil fuels 122
7.2 Alternative sources of energy 124
7.3 The importance of energy efficiency 135
7.4 Putting energy efficiency into practice 136

INTRODUCTION

For thousands of years man used energy derived from local sources to cook and to heat his home and to develop technologies such as pottery and metal working. The source of the energy was initially wood and then charcoal, followed by coal mined close to the surface and water wheels. The development of coal mining into deeper seams and the use of steam engines to power the pumps required to remove water improved the availability of cheap energy and helped to nurture the industrial revolution in the 18th and 19th centuries. Subsequent discoveries and exploitation of commercial quantities of oil and natural gas helped to fuel further development, such that we are now dependent on energy to support our way of life. The GDP (gross domestic product) of a country is closely related to its energy consumption. As the population increases (as mentioned in Chapter 1) and the standard of living of the developing countries improves, it is expected that the demand for energy will continue to rise. The energy use in the developed countries is fairly stable or even in decline but it is more than compensated for by the rapidly expanding economies of China, India, Brazil and other developing countries. Recent growth in world energy consumption has averaged about 2.9 per cent a year (BP 2018). Meeting the demand for energy and the climate consequences of energy use are two of the big challenges for the 21st century.

The problems of energy supply and climate change caused by burning fossil fuels were first mentioned in Sections 1.1.6 and 1.1.7 and these and later chapters made the case for taking action. Reductions in emissions of carbon dioxide are also being promoted through legal measures and agreements and trading schemes which are referred to later in this chapter. The pressure is on energy use. This chapter takes the subject further and looks at the alternative sources of energy that are available and which are expected to become more important in the coming years and how energy use can be reduced.

7.1 Benefits and limitations of fossil fuels

Fossil fuels have been used for centuries and have served mankind well. Now we realise what damage they can cause and action needs to be taken to reduce our dependency on them. However, this is not going to happen overnight: fossil fuels are widely used and still have some advantages over the alternatives but the balance is changing and the limitations of fossil fuels are starting to be taken more seriously. We can start by looking at the relative advantages and disadvantages of business as usual before going on to consider the alternatives.

7.1.1 Sources and uses of fossil fuels

Coal, oil and natural gas are known as fossil fuels. This is because they were formed hundreds of millions of years ago from animal and plant remains. Coal was formed from plants and trees that died and settled in swamps under anaerobic conditions. Over time they were buried under silt and other deposits and the action of pressure and heat converted them to coal (peat is a relatively modern equivalent of dead plant material near to the surface). Oil was formed by heat and pressure acting on micro-organisms that deposited under water and natural gas was formed by higher temperatures and pressures on similar organisms. Oil and gas could rise through porous sediments and rocks and through cracks after they formed until they were trapped and collected under impermeable layers of rock or salt deposits.

Coal is a solid consisting of relatively pure carbon but it may contain other substances such as sulphur or chlorine compounds. The presence of impurities depends on the conditions under which the coal was formed and different sources of coal may produce less pollution than others when they burn. It is the most abundant fossil fuel and is found in forms known as lignite, bituminous coal and anthracite. These have different energy contents and produce decreasing amounts of polluting gases when they burn. Coal is extracted from open-cast mines (large open pits) or from deep mines by sinking a shaft and driving horizontal tunnels. It is found in many locations, the larger sources including parts of Europe, China, India, Australia and Russia. Coal is used mainly for power generation (see Figure 7.1) and in industry as a source of heat and carbon (for refining metals for example). It is also used for heating domestic properties although in many countries it has been replaced by natural gas for this purpose. It is transported by ship, train and lorry from the mines to the customers.

Oil is a liquid comprising a mixture of long-chain hydrocarbons which may also contain sulphur compounds. Uncontaminated oil is made up from carbon and hydrogen

Figure 7.1 Coal fired power station

and should burn just to produce water and carbon dioxide. It is extracted by drilling a well into the space below the layer that traps it. It is usually under pressure and then can get to the surface without being pumped. Eventually the pressure drops and the oil has to be pumped out or be displaced with injected water. Different sources of oil also have different compositions of the hydrocarbons, resulting in some being preferred for refining and so being more expensive. It is found predominantly in the Middle East, Russia, some of the former Soviet republics and the United States, although there are smaller deposits in many countries. As the main sources are exhausted it is being sought in deep sea locations and even the Arctic. Crude oil (i.e. as extracted) is transported to the refineries by sea tankers and by rail or road tankers or by pipeline. The long chain hydrocarbons are broken down in an oil refinery and made into liquid fuels such as petrol, diesel and fuel oil and a wide range of other products that form the basis of the organic chemical industry. These are transported by similar means as crude oil. Oil is the starting point for the manufacture of most plastics, pharmaceuticals, synthetic fibres, dyes, and even for the manufacture of CFCs and VOCs.

Natural gas is often found with oil or it can be found on its own. It is methane gas and is usually relatively pure, although it may contain other gases such ethane and carbon dioxide. Methane is also made from carbon and hydrogen and should burn under the right conditions to produce just water and carbon dioxide. It is also extracted by wells and is collected under pressure. If it comes out mixed with oil the two have to be separated and sometimes the gas is just burnt. It is found in most of the locations where oil is extracted. Once extracted as a gas it can be pumped through pipes or it can be liquefied under pressure and low temperatures and transported as such in large tankers. It is used for domestic, commercial and industrial space heating, for heating industrial processes and for power generation.

The fossil fuels have a long history of use and one of their benefits is that they are well understood and the technology for using them is well established and widely available. They are mainly sourced on a large scale and have been relatively cheap to extract, process and distribute. Coal and oil can be easily stored on site. Gas can be stored under pressure or as liquefied natural gas but this is usually done by the suppler as part of the distribution network, with the consumer taking gas on demand. Prices are rising over the long term, as explained in Chapter 1 (although subject to short-term variations depending on demand and politics). As they get used up the sources are becoming more difficult to find and the extraction processes are becoming more expensive. There is no shortage of coal at present although some of the sources produce coal that is more polluting than others. However many oil wells have become exhausted or too expensive to continue to operate and there is an expectation that sometime this century the consumption rate will exceed the production rate (known as 'peak oil'). This is being deferred as more expensive sources are exploited and oil is extracted from shale deposits (another polluting process in itself). Gas production is still increasing, as is consumption as it displaces coal and oil for some uses. It is a cleaner source of energy (see below and Chapter 4) and easier to handle in many ways but this means that more pressure will come onto supplies. It is also being sought in more remote places and by hydraulic fracturing of potential underground sources as mentioned in Section 1.1.6.

7.1.2 Benefits of fossil fuels

The previous section has mentioned some of the benefits: they are a natural resource with good availability, relatively low price and are relatively easy to extract, transport and use. All are good sources of energy in that the energy content is high (they have high calorific values) and the energy is easy to recover and convert to a usable form. All are burnt to produce heat. Oil products are easy to store and use in business and vehicles. Coal is used to generate electricity which is easy to distribute from the power station to the centres of use along high voltage transmission wires. These power stations have to be located near to a suitable source of cooling water (see Chapter 5) and cannot be stopped and started at will. As coal is also the most polluting fuel when burnt (see below) it is being displaced by natural gas in many locations. Coal-fired power stations generate steam which is used to turn turbines to generate the electricity; the process is not as efficient as using gas in a gas turbine. Gas turbines can be stopped and started as required to meet demand. Gas-fuelled power stations tend to be smaller than coal-fired ones and are more flexible in location. Oil is less used for power generation on account of the fuel cost – except in remote locations or on ships.

7.1.3 Limitations of fossil fuels

Reference has already been made to the formation of fossil fuels over millions of years under conditions that no longer persist. (Note that the origin of fossil fuels from animal and plant material was dependent on solar power long in the distant past). As supplies are used up they are not being replaced: their use is not sustainable (see Section 1.3). Sometimes reference is made to supplies running out in the coming decades. What is more likely to happen is that the price will rise with falling output and what is left will not be used as fuel but be reserved for higher value uses such as chemical production. Alternative sources of energy will be required as described in the next section.

An additional problem associated with fossil fuels is the concern about security of supply. Large quantities of the oil and gas used come from countries that are politically unstable or have used them for political ends. Wars and threats of war in the Middle East and disputes over payment between Russia and its customers have reduced availability and caused significant price fluctuations. The major oil producers also operate a cartel which limits (in theory at least) the outputs from the member states which also affects prices. Variable prices lead to business uncertainty in the consuming countries and is blamed for causing price inflation and reduced growth in demand for other forms of output.

Supplies can also be disrupted by interference with the wells, pipelines and other means of distribution either by natural events or by terrorism or industrial disputes. So much of the infrastructure of developed countries is dependent on a continuous supply of energy that newer forms of energy can offer a local, secure alternative.

Other limitations of fossil fuels are their role in causing pollution. Coal extraction is damaging to the landscape and produces large volumes of solid waste (mentioned in Section 1.1.6). Oil extraction, transport and storage are sources of oil pollution from spills and leaks (again mentioned in Chapters 1 and 5). More importantly, all fossil fuels are sources of air pollution on combustion. Natural gas consists mainly of methane and some escapes from extraction sites and in its distribution: it is a powerful greenhouse gas contributing to global warming. All fossil fuels burn to produce carbon dioxide which was described in Section 1.1.7 as the main source of global warming and climate change. They can also produce other pollutants either from contaminants also present or due to the combustion process not being totally efficient as described in Chapter 4. If sulphur compounds are present then sulphur dioxide will be formed and this is a cause of acid rain. Similarly nitrogen oxides are formed in burning coal and in the internal combustion engine and these react with fuel and other residues in the exhaust to form smog as well as being another source of acid rain (see Chapter 4). The handling of coal produces dust and its combustion produces smoke and particulates; vehicle emissions contain very small particles known as PM_{10} (see Section 4.2.8).

7.2 Alternative sources of energy

Fossil fuels have their limitations as described above and attention is turning to alternative sources of energy. Water power (hydropower) has been used for many generations, initially to raise water for irrigation and to power mills for grinding corn. These relied on water wheels driven by flowing water to turn an axle. Traditional windmills work similarly in that wind causes the sails to rotate which also drives an axle. The rotating axle is then used to drive a means of raising water or a millstone. Wind power was also used for centuries to power ships. These applications have generally been replaced by electricity or fossil fuels driving a motor, except in developing countries where an electricity supply is unavailable or too expensive. In more recent times attention has turned to using alternative sources of energy for heat or to generate electricity. This section considers the alternatives currently available and their benefits and limitations.

Electricity is a form of energy and it can be generated by converting another source of energy such as heat or the movement of air or water. In the first of these, heat is used directly or to generate steam which is used to rotate a turbine which in turn drives a generator. In the others, the force of the movement drives a turbine and generator directly. In most cases the problems have been to capture the alternative forms of energy and use them in an efficient way. Historically they have struggled to be competitive in price with the use of fossil fuels and have required grants or subsidies to help them become viable. The motive is now to tackle climate change and the escalating costs and risks associated with the use of fossil fuels and the expectation is that, as the alternatives become more widely adopted, the costs of construction and operation will fall. In fact by 2019 wind turbines and solar power have become competitive as large-scale installations are developed.

There are some general business and social benefits associated with the move to alternative forms of energy and from improving energy efficiency. The introduction of new technologies and the replacement or enhancement of existing means of generation will create new businesses and jobs. For example, companies have sprung up to install solar power on domestic premises in the UK in the last few years. There are opportunities to manufacture the heavy equipment such as wind turbines but other countries also spot the opportunities and the UK seems to lack a strategy that could take advantage of them. Improving the energy efficiency of homes and other sources of energy waste has also created jobs but they are not necessarily those with high skill levels and pay.

7.2.1 Solar energy

Solar energy is derived (given its name) from the sun. The sun radiates on to the earth more energy than is consumed by all current demands and could provide enough to meet future needs if it could be captured and used economically. It can be captured in two forms: solar heat and solar electricity (or photovoltaic energy). Solar heat causes evaporation in nature and has been traditionally

Box 7.1 Heat exchangers

A heat exchanger is a common feature in energy systems and its use can be demonstrated in domestic hot water systems that use a boiler. The heat source (boiler) heats water which is not used directly. This is because this water is often treated with corrosion inhibitors and other chemicals to protect the boiler and the pipework and, in any event, would be anaerobic, discoloured and unsuitable for most uses. Instead it is kept within a closed circuit known as the primary circuit. In order to be useful the heat has to be transferred to a tank of cold water and heat it up. This is done through a coil in the hot water tank where the water heated by the boiler passes through the coil transferring heat to the tank and then returns to the boiler as shown in the upper part of the right-hand side of Figure 7.2. An industrial heat exchanger works in a similar way although the heat exchange may be between two coils or by other means. A supply of hot water is available for other purposes in a secondary circuit: in the domestic system this is from the top of the hot water tank. In other systems it would be from the secondary coil.

Domestic boilers run at about 80°C and heat the water in the tank to about 60°C. If the primary and secondary circuits are operated under pressure it is possible to run at temperatures above 100°C. When the pressure is then reduced in the secondary circuit steam is formed which can be used to drive a turbine. The steam is condensed after use and the water recycled through the secondary circuit. In the process heat energy is converted into motive energy (rotation of the turbine) which, in turn, rotates a generator which converts the rotary energy into electricity.

used for that purpose in concentrating salt from the sea and for drying food, etc. This is inefficient, if cheap. Buildings also gain heat from the sun (known as 'solar gain') on the south side in the northern hemisphere or the north side in the southern hemisphere. This can be used to advantage in designing buildings and site layout. More recent developments are taking advantage of the sun as a potential source of cheap energy for domestic and business use.

Several projects use mirrors to focus the sun's rays on to a tank containing oil or water and use the heat via a heat exchanger (see Box 7.1) directly or to raise steam for generation of electricity. The trials have been in desert areas where the heat from the sun is strongest. A method of more general application uses the sun's rays to heat water in a flat plate collector (similar to a domestic radiator; see Figures 7.2 and 7.3) or in pipes or a heat conducting rod enclosed in a glass tube. The collectors are usually under glass, rather like a greenhouse on a small scale. Different designs use evacuated tubes to increase efficiency. They are mounted on the roofs of buildings or another suitable site facing the sun and the sunlight heats up the collector which transfers the heat to water in a primary circuit which then heats the water in a tank. A controller ensures that heat is not removed from the tank when the outside temperature is low. The hot water is used directly in houses and small business and hotels, saving the amount of fossil fuel required to heat it up using a conventional boiler. In warmer countries, such as around the Mediterranean, the method is also used to heat swimming pools.

Photovoltaic cells convert the sun's energy directly into electricity and they first became familiar from their use on spacecraft. The cells contain a thin surface of semiconductors which are made from silicon and protected

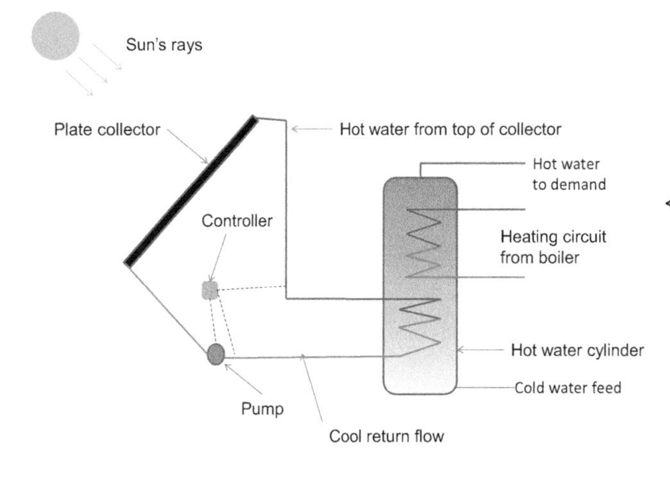

Figure 7.2 Flat plate solar heater

Figure 7.3 Solar heat panels on the roof of a Spanish hotel

under a glass or clear plastic cover. The cells are assembled into a panel and several panels may be installed on a building roof (see Figure 7.4). The panels may stand alone or the cells may be incorporated into roof tiles, glass canopies or even panels on the sides of buildings for different applications. Larger scale generation has also been built on land, particularly in regions with reliable sunlight such as southern Europe or the USA. An internet search for images of solar power will bring up many examples. Much development work is going on to find alternative materials for the cells and to improve the efficiency of conversion of sunlight. Again they work best facing the sun so the aspect and angle of slope are important. The output from each cell is low voltage direct current so many cells are needed to provide enough power for a typical household and it has to be converted to alternating current for most uses. In many countries excess electricity can be fed into the local grid supply and is credited against bills for taking electricity when the demand is too high or the cells are not generating at night.

Figure 7.4 Photovoltaic cells on the roof of social housing

The advantages of solar power are that the sunlight is free once the cost of manufacturing and installing the equipment has been paid for. There are grants and subsidies available in many countries to facilitate the adoption of solar power. Maintenance costs are low and the equipment has an expected lifetime of 20 to 25 years. Both are of particular benefit in locations where centrally supplied energy is not available such as developing countries or remote locations. Many rural communities in developing countries are now using solar cells to power cellular telephones, radios and televisions, making them available for the first time or saving the cost of batteries. Photovoltaic cells are often used to charge batteries that power small public facilities such as weather stations, river level and flow meters or traffic meters. Both forms of solar power do not produce emissions of greenhouse gases or anything else during operation. Emissions are produced during the manufacture and installation of the equipment but this should be lower overall during its lifetime compared with the emissions produced when generating the same amount of energy from fossil fuels.

The disadvantages are that both systems only work in daylight and are most efficient under bright sunlight although there is still some output on cloudy days. The demand for heat and electricity is usually highest at the times of minimum output. There either need to be alternative sources of heat and electricity to cope with the down times or both have to be stored. Heat can be stored as hot water and this is usually part of a domestic hot water system (see Figure 7.2). Electricity can be stored in a rechargeable battery and newer lithium ion batteries on a large scale are becoming more available.

Figure 7.5 Medium sized wind turbines on the edge of a business park

7.2.2 Wind power

The traditional windmill has been displaced by wind turbines. The sails which drove an axle are replaced with solid blades that turn a generator. There are usually three blades mounted vertically like a plane propeller on a horizontal axis. A gearbox is incorporated between the blades and the generator and the whole installation is mounted on top of a column. The column can range from a few metres high, supporting blades of a few centimetres, to towers up to about 100m tall with blades up to 40m long (see Figure 7.5). The outputs range from a few watts for charging batteries up to 7MW in the current largest designs. An alternative design uses vertical blades on a vertical axis. A common design uses blades in the form of an aerofoil (wing cross-section) and twisted at the same time. These are a more recent commercial innovation and not available in larger sizes. They have the advantages that the gearbox and generator are at ground level and easier to service and they produce less vibration, which is important for avoiding structural damage if they are fixed to a building. The differences between the two types are illustrated diagrammatically in Figure 7.6.

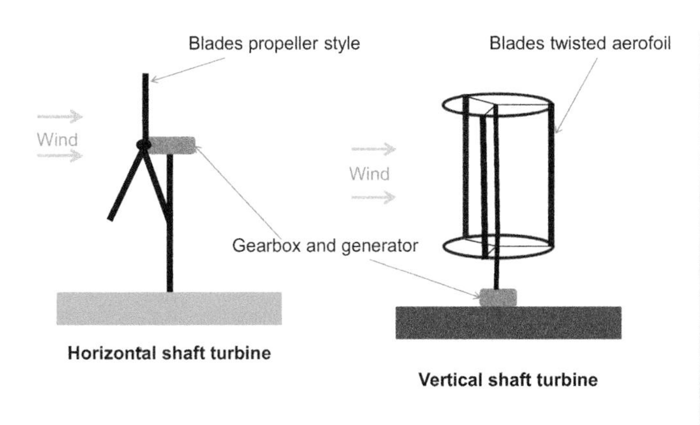

Figure 7.6 Illustration of two designs of wind turbine

Wind turbines need to be sited in locations where they are exposed to wind of sufficient speed and regularity and a vane on the top of smaller structures keeps the turbine facing into the wind for maximum efficiency. If the wind speed is too low the turbines will not work; if it is too high they have to be shut down to avoid damage. Most are sited on high ground or at sea where wind is more reliable and less obstructed by buildings and other structures. One turbine may serve a single site but most of the larger ones are grouped together on wind farms. Wind farms on land may contain just a few turbines; the larger farms are being built in estuaries and offshore. By 2017 onshore wind power was producing 9 per cent of the UK's power needs, enough for 7.25 million homes. Recent wind power has mainly been offshore and by 2020 will produce about 10 per cent of UK demand (www.renewableuk.com/page/WindEnergy).

Wind is another source of free energy and it will not run out. The running costs, including maintenance, should be low. The construction costs can be high especially in remote locations and at sea although most of the units are prefabricated off site. There is also the cost of transmission lines to bring remote power to the centres of demand. These constructions will consume energy (embedded energy as mentioned in Chapter 1) and will be the cause of emissions if this is not from renewable sources.

An advantage of wind turbines over photovoltaic cells is that they can work at night but their main problem is that they are dependent on the wind blowing. Site selection obviously plays a role but wind is unreliable even in the windiest sites. Security of supply means that total reliance on wind power is not sensible where continuity of supply is essential. Alternative means of meeting demand are required for public supplies although remote sites may have to make do or rely on rechargeable batteries.

Another problem is that they can meet with public opposition. If they are on land the best sites are on the skyline which can be unsightly to some people; some consider them quite attractive or, at least, a price worth paying. Power lines on the skyline are also a source of objection (see Figure 7.7). Such sites are often in national

Figure 7.7 Nobody likes power lines

parks or similar relatively unspoilt locations. Turbines can be noisy and will meet with objections if close to habitation. They are also claimed to be a threat to birds and to interfere with radar monitoring low-flying aircraft. These latter concerns would still apply offshore.

7.2.3 Hydropower

Using water as a source of power has a long history, as mentioned at the start of Section 7.2. For centuries the use was for providing mechanical power to drive machines located at the same site. With turbines driving a generator, electricity can be produced and distributed to a wider area. This is also known as hydroelectricity. Hydropower can be on different scales, as with wind power. However, the high cost of construction means that hydroelectric facilities are usually very large. (Farmers may dig small reservoirs on their land to supply irrigation water but they are usually quite small and the engineering problems of construction are not the same.) The emissions of carbon dioxide and other pollutants are nil once built but construction uses a lot of energy, much of it likely to be from fossil fuels. Over their lifetime, which should be decades, they are a very sustainable source of energy.

A turbine can be built into the flow of a river to capture energy but the output will be relatively low and variable depending on the flow. Larger developments store water in reservoirs and release it through the dam or through pipes to drive the turbines which, in turn, drive generators. The energy available is dependent on the flow and pressure at the turbine so a reservoir in the hills can be connected by pipeline to a generator nearer to the centre of demand. The pressure available depends on the height difference between the two ends of the pipe and is called the 'head'. The principles are illustrated in Figure 7.8.

The best potential for hydropower depends on a reliable source of water. A wet climate helps, although dams can be built on large rivers even in regions with low local rainfall (for example the Aswan dam on the River Nile in Egypt which gains its water from rainfall way upstream). Then they rely on sufficient rainfall upstream and are

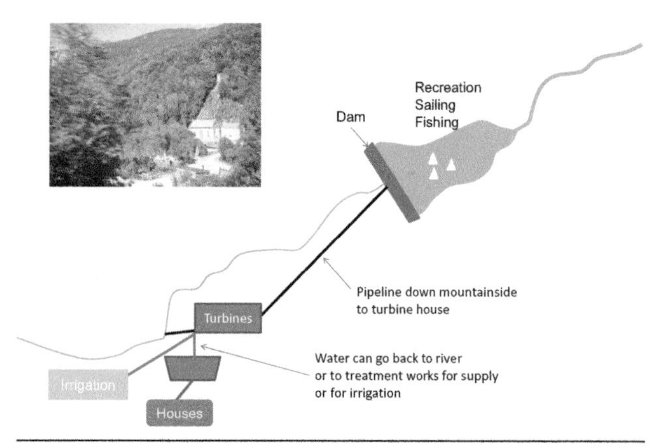

Figure 7.8 Illustration of hydropower

vulnerable to drought. Interfering with the natural flow of a river can lead to disputes with neighbours upstream and downstream over water rights. The water arrives at the river or reservoir by natural means as rainfall by the water cycle referred to in Chapter 5. It is another free source of energy once the infrastructure has been paid for and maintenance costs should be relatively low – although large dams need to be inspected regularly for safety and may require expensive construction works if they are deemed to be at risk from subsidence or other forms of potential damage or risk such as storm flows. Their often remote locations mean that long power transmission lines may be needed.

The main limitation is finding suitable sites for reservoirs. The geology has to be suitable in terms of risk from earthquakes or land slips and the availability of material to construct the dam. The material at the base needs to be impermeable. There has to be sufficient water available but the best sites are in river valleys in hilly areas and, as with wind turbines, there may be objections as these are often valuable sites for nature conservation and for migratory fish such as salmon. Objections may also be made on grounds of disturbance, especially during the construction phase. Loss of amenity may also be claimed but new reservoirs often become popular for recreation, sailing and fishing. In some cases people may be displaced; sometimes whole communities are removed. The construction of the Three Gorges Dam on the River Yangtze in China resulted in over one million people being moved.

The management of reservoirs can be a problem if the water is eutrophic (described in Chapter 5) and they can silt up if the river carries a lot of silt, particularly when the flow is high. On the plus side, reservoirs can serve more than one purpose as the water can also be used for water supply or irrigation after the power has been generated although the balance of the demands is unlikely to be the same.

7.2.4 Wave power

Waves just off the coast are an attractive potential source of free unlimited power as they are formed by the wind and in operation they produce no greenhouse gas emissions. Wave height and frequency varies with location and with the weather but there are sites which have a good prospect of being a reliable source. The challenge is in harnessing the energy on the surface of a body of water as it rises and falls and the difficulty of doing this in a hostile environment such as the sea has meant that wave power has only recently become a serious contender. The devices used mainly work on an oscillating column of water, or on structures floating on the surface that exploit the movement to pressurise air or directly drive a generator to convert it to electricity. In theory this should be simple but many such devices have to be linked together to generate sufficient power and the capital cost and operating and maintenance costs are high. This is due to the need to build in difficult circumstances, the structure having to be able to deal with severe storms, the abrasive properties of suspended sand and the corrosive properties of sea water. Transmission cables are needed to bring the power ashore and these and any visible structures may spark opposition in attractive coastal zones.

7.2.5 Tidal power

The tides offer another attractive potential source of free unlimited power that produces no greenhouse gas emissions. The height of tides is influenced by the moon and, to a lesser extent, the weather so the range between high and low tide varies. Location also plays a part, especially as tides are funnelled into an estuary, such that the high tide is above that on the adjacent coast. Traditional methods of harvesting tidal power to turn a water wheel depended on capturing the rising tide behind a weir and then letting it fall slowly again through a water wheel. Modern methods work in a similar way, such as the French tidal scheme at La Rance, built in 1966. Here the average tidal range is about 8m with a maximum of 13.5m. The water is trapped behind a dam 350m long and then released through 24 turbines to produce an average output of 64MW (www.reuk.co.uk/La-Rance-Tidal-Power-Plant.htm). A similar scheme has been proposed for the Severn Estuary in the UK but ruled out on grounds of the cost of construction, public opposition, environmental concerns (such as fish migration and impact on birds that live on the estuary) and the power supplied not being competitive with other sources. An alternative means of generating power from the tides relies on the presence of strong currents as the tide rises and falls. These can be particularly strong where the water flows in a channel between two islands. Turbines in the water flow can then be used to drive generators.

The problems with tidal power are similar to wave power due to the hostile environment in the sea.

7.2.6 Geothermal power

The core of the earth is molten and has a temperature of several thousand degrees centigrade. The ground gets warmer the deeper you sink a borehole from the surface.

In some locations hot water appears at the surface (in geysers or hot springs), predominantly in volcanic or earthquake zones such as are found in Iceland, Japan and New Zealand. This is already captured for district heating, to supply hot water or to generate electricity. Otherwise a borehole has to be drilled several km down from the surface to find hot water or hot rocks. Hot water can be extracted and used directly on the surface, otherwise cold water has to be pumped down and heated underground by passage through cracks in the rock (see Figure 7.9). It then has to be extracted through a second borehole. The use of the water depends on the temperature. It needs to be hot enough to form steam for generating electricity.

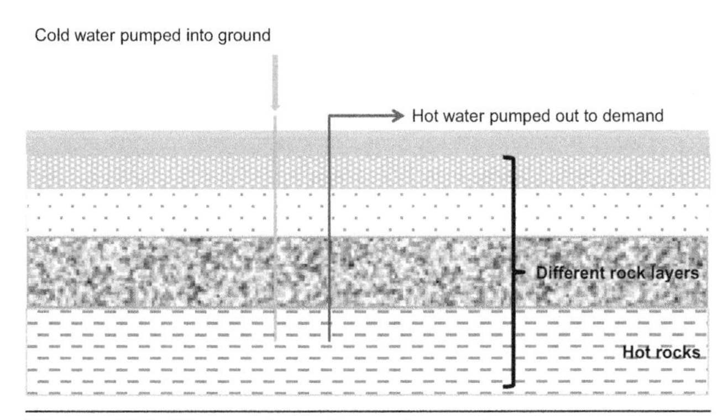

Figure 7.9 Illustration of geothermal power

Geothermal energy offers another route to renewable energy without greenhouse gas emissions but it has not been widely adopted where it is not near to the surface. It is expensive to drill the boreholes. The depth required is uncertain at the start and often the drills hit unsuspected difficult drilling conditions or fail to find a suitable source of heat, so it is a speculative proposition with failure always a possibility. Where it is possible it does not suffer from the objections that other renewable sources can raise.

7.2.7 Nuclear power

Nuclear power could take two forms: fission and fusion. Fission is the process by which the radioactive properties of some elements are used as the unstable nuclei of the elements decay. If the elements are in the right form a chain reaction is set up and heat is released (see Box 7.2 for a simplified explanation of what is going on). This is what keeps the core of the earth molten. This process has been used in nuclear reactors to generate electricity on a large scale since the 1950s and has been adopted by many countries. The main users are France, Japan and the US although many countries, including the UK, have a few reactors and China is proposing to build many in the coming years. They are built to provide power equivalent to the largest coal-fired power stations. The early reactors are now coming to the end of their lives and will need to be replaced.

Box 7.2 The basis of nuclear fission

Elements are the basic building blocks from which all chemical compounds are formed. They are the pure form of the simplest substances such as hydrogen, carbon, nitrogen and iron (see Appendix 1). The atoms of each element are made up of neutrons, protons and electrons and the number of protons determines which element is present: hydrogen has one, carbon six and iron 26, for example. The number of electrons is the same as the number of protons in the pure element in its natural state. Some elements exist as slightly different forms known as 'isotopes'; they are chemically identical but have varying numbers of neutrons. Many of these are unstable and radioactive meaning that they emit energy as various forms of radiation as they decay either to a more stable form of the element or another element entirely. Many radioisotopes, as they are known, occur naturally and radiation is emitted from rocks depending on their chemical composition.

Uranium exists naturally as two main isotopes ^{235}U and ^{238}U (there are six in all). Each isotope has 92 protons but they have different numbers of neutrons; 143 and 146 respectively. Both forms are radioactive but ^{235}U is the fuel choice because it emits neutrons when it decays and its decay is triggered by absorbing neutrons. It is present at less than 1 per cent in natural uranium. In order to be used as a fuel the amount of ^{235}U has to be increased using a process known as enrichment. At the right concentration the rate of radioactive decay and release of neutrons triggers more decay reactions and further release of neutrons. This is known as a chain reaction. By bringing the enriched material closely together it sets up a chain reaction which becomes self-sustaining and the uranium becomes hot. (This is also how an atomic bomb works but the amounts and concentrations involved are different.) In a nuclear reactor the uranium is in the form of rods and the rate of reaction is controlled by the distance between the rods and the presence of control rods which can slow the reaction down (or 'moderate' it) by absorbing neutrons as shown in Figure 7.10. The heat from the uranium is removed by water (sometimes gas or molten salts are used) and used to generate power.

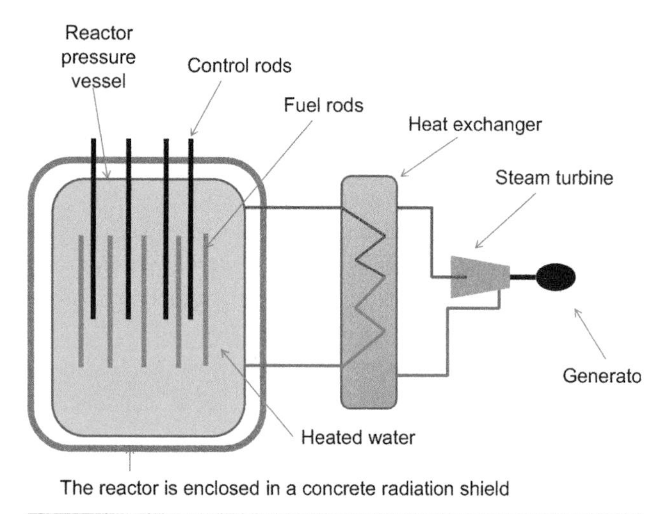

Figure 7.10 Principles of a nuclear fission power reactor

Fusion is the forcing together of atoms to form new elements which also produce heat. The sun produces its energy by this means and it could form the basis of a new generation of nuclear reactors which should be safer to operate as they would produce less radioactive waste. This is still at the experimental stage at the moment and no operational plants exist.

A simplified diagram of a nuclear reactor using fission is shown in Figure 7.10. The reactor is pressurised and contains water which heats up due to the chain reaction. This water is fed through a loop in a heat exchanger and contained within the system. Hot water is taken from the heat exchanger and as the pressure drops it is converted into steam to drive a steam turbine and generator. The nuclear reaction is controlled by raising or lowering the control rods which mediate the chain reaction. The reactor and its contents are highly radioactive and have to be shielded with a thick concrete wall to prevent radiation escaping to the immediate environment. The nuclear process is thermally more efficient than other heat sources such as combustion of coal and it promises very cheap electricity although this has not come about in practice.

Eventually the uranium in the rods becomes exhausted and the rods are replaced. The old fuel rods are refined to recover any useful residual uranium and the remaining fission products which remain radioactive for thousands of years as the residual activity decays. This waste has to be kept contained and stored safely for a long time. Resolving how to do this is still a problem that is becoming more urgent as the quantity of waste builds up as outlined in Box 7.3.

The advantages of nuclear fission are that it generates no greenhouse gases in operation although the initial construction (e.g. from making concrete) and waste disposal do consume a lot of energy and generate greenhouse gases. Supplies of uranium are adequate for now but will eventually be consumed; it is not sustainable in the long term. Its extraction is hazardous and polluting

Box 7.3 Storing nuclear waste

Nuclear waste is a hazard because of the properties of radiation referred to in the main text. There are several sources of waste from a nuclear reactor. Highly active waste is in the form of the material recovered from the exhausted fuel rods from the reactor. A lower level of waste includes the metal casings for the rods and various sludges from water treatment and waste processing and old components of the reactor that have been replaced or arise from decommissioning old plant. The lowest level waste arises from materials used to handle and process the fuel and the waste, protective clothing and laboratory equipment. There are also radioactive wastes arising from non-nuclear applications such as in hospitals and industry referred to in Chapter 6. The fuel rods are usually stored on site under water in large ponds initially to allow the radiation levels to reduce. In the UK waste is sent to Sellafield in Cumbria for reprocessing. This site also accepts waste from outside the UK. After refining to recover useful materials (uranium and plutonium) for making new fuel rods, the residual waste is much reduced in volume and is solidified by fusing it into a glass. This is inert but still radioactive and is likely to end up in the UK in a deep underground storage site, when one is chosen, based on its geology and safety against threats such as earthquakes and leaching by or into underground water. It will have to remain there for thousands of years. Medium level waste is mixed with concrete and stored in metal drums on the surface. Low level waste is compacted and stored in containers and becomes safe in a relatively short time (months or years depending on the source) and most ends up in landfill sites.

The waste problem will become more acute if there is a widespread adoption of nuclear fission to generate electricity and as old reactors are decommissioned. It is all currently stored somewhere although the current volumes are manageable and compare favourably with the volumes of waste that arise from using coal and other activities. One remaining concern is waste (or any other nuclear material) falling into the hands of individuals or organisations that may use it for terrorism or other illegal activity.

and the processing, storage and disposal of materials risk exposure to radiation. The greatest concern, at least of the public, is from radiation leaks arising from an accident or incident. There are also international concerns about the diversion of radioactive materials to make nuclear weapons.

Radiation is dangerous because at high doses it can cause death directly and at low doses it can cause changes in cell structure which can lead to diseases such as cancer and leukaemia. The three most serious incidents in the 60 years of nuclear operation involving leaks of radioactive materials from power plants were at Three Mile Island (TMI) in the US in 1979, Chernobyl (Ukraine) in 1986 and Fukushima (Japan) in 2011. The first two were caused by an initial failure in the reactor or associated plant which was compounded by system and human failures. The last was caused by a tsunami that damaged the cooling system in the reactors. In each case there was a release of radioactive material. At Chernobyl, which was the worst, the radiation spread as far as the UK and can still be traced to this day. At Fukushima an area of about 20km radius was evacuated and work has continued since to prevent radiation spreading into the groundwater and the wider environment. The decommissioning of the site is expected to take 30–40 years. At TMI the radiation was low and the worst was confined to the site.

There have been other accidents at large nuclear reactor sites that did not involve significant releases of radiation and others at sites or on submarines involving small reactors. Generally the danger is closest to the leak. Further away the exposure rapidly drops to be hardly different from the exposure to radiation from natural sources. However, the accident at Fukushima has led to some countries reconsidering their use of nuclear energy. For example in Germany it has been announced that all of their reactors will close by 2022.

This has added to the problems that nuclear energy has always had in terms of public opposition and difficulties in obtaining planning permission. Many are constructed on the coast or on estuaries where water is available for cooling and population densities may be low. However others are constructed inland on major rivers and near to towns. It is a reliable source of energy, not dependent on weather or the regular supplies of distant fossil fuels. The emissions to the atmosphere are low and the running costs are also low relative to the energy output. This means that it ought to be a secure source of energy for the future. Opposition is based on fears of radiation and accidents coupled with the high initial capital cost – some of it necessary to contain these risks and ensure site security. There will be the capital costs for a long-term solution to storing waste and the ongoing cost for waste disposal.

Decommissioning of old reactors will also be expensive and take a long time. Operators are nervous about the financial risks and look to governments for subsidy or guaranteed prices for electricity. Governments look to the electorates.

7.2.8 Biofuels

The use of biofuels can take several forms based either on the direct burning of organic matter or the production of ethanol and biodiesel as liquid fuels. The attraction of all of these is that the source of energy is from crops that are grown to meet the demand rather than the fossilised remains of plants that lived millions of years ago. Although their combustion releases CO_2, an equivalent amount is removed again by photosynthesis when replacement crops are grown. They are renewable sources of energy and in this sense it is sustainable. There are however some associated issues that challenge the sustainability of some sources as described later.

Solid fuels or biomass

The traditional source of fuel for thousands of years was wood but it burned inefficiently and produced smoke and unhealthy fumes. Handling wood for larger burners or boilers was problematic as wood sources naturally occur in various sizes, weights and water content. The calorific value (potential heat content) also depends on the source. Recent developments have resulted in the availability of wood pellets or chips which can be manufactured to standard sizes and water content with improved ease of handling and handling equipment and burners specifically designed to use them. Wood pellets can be made directly from trees but the attraction as a source of fuel ought to be that the pellets can be made from sources of wood that have no other useful purpose. This can include offcuts and wastes resulting from the processing of trees into timber as well as offcuts from the use of the timber in building or manufacturing and from recycled wood from products that have reached the end of their useful life. Wood from some of these sources is also used to make manufactured wood products such as chipboard so there is potential competition for supplies.

The manufacture of the pellets is an industrial process that requires energy and the collection and transport of the wood and pellets also uses energy, which may still be derived from fossil fuels. As woodland and forest is often remote from the centres of demand there is a risk that the energy used to collect, process and transport the wood can be a significant proportion of the energy available from the product and the cost may exceed the value. To be viable there usually needs to be an opportunity that brings together the sources of wood, the processing facility and the demand centres within a small geographical area. The wood may be grown specifically for this purpose using coppiced trees; however this may take land from other productive uses such as farming. Wood burners tend to be on a larger scale than many domestic applications and require large storage facilities for the pellets. They are mainly used for schools, swimming pools, hotels or similar large users of heat.

Waste straw is also used as a fuel and there have been trials with other crops grown specifically as fuels. These are not widely used in the UK but have been trialled as a supplement for coal in power stations. For example, DRAX power station in the UK uses biomass from a variety of sources (for more information visit the web site of DRAX Group plc). Using biomass can be more expensive than coal and to be financially viable depends on the availability of subsidies (see below).

Liquid biofuels

One of the major uses of fossil fuel as oil products such as petrol and diesel is in transport. Ethanol (the sought-after constituent of wine and beer, commonly referred to as alcohol) can be used as a liquid fuel and it is produced by the fermentation of organic matter by yeast. The commonest sources are sugar cane and maize as corn syrup. The fermentation process produces a solution of alcohol that has to be purified and concentrated by distillation. The ethanol can be used directly but is normally mixed with petrol at between 5 and 25 per cent ethanol. The two countries that use it most are the US and Brazil.

Vegetable oils can be burned directly in some engines. The more common use is biodiesel produced from animal fat and vegetable oil by purifying them and reacting them with alcohols. The oils and fats can be waste products derived from cooking for example or extracted from crops grown for the purpose such as rapeseed or sunflower seed. The biodiesel can be used directly but is usually mixed with petroleum derived diesel: in the EU there is now a requirement that all diesel fuel is blended with 5 per cent biodiesel. Similarly manufactured fuels are being used for heating and are being tested in trains and aircraft.

The biofuels have the advantage that they reduce or eliminate the consumption of fossil fuels but a lot of energy is used in their initial production as crops and their fertilising, processing and transport so that the overall energy balance may not be favourable. In addition the major raw materials are food products, so the production of biofuels uses potential sources of food. The land that they are grown on could also be used for the production of other foods but they are sometimes grown in parts of the world where food is in short supply or land is developed by the destruction of forest or taken from indigenous populations. For these reasons, some critics question whether biofuels can legitimately be called sustainable.

7.2.9 Methane from anaerobic digestion and recovery from landfill sites

Natural gas is mainly methane and there are other sources of methane that can be used as a fuel. When organic waste such as that from food processing is digested in the absence of oxygen (a process known as 'anaerobic digestion') the carbon-containing compounds are broken down into methane. A similar process occurs in landfill sites where the organic matter in the waste becomes anaerobic as it is buried. These processes have been covered in Chapters 5 and 6. In both cases the methane can be recovered and burned in a compression engine or a boiler to generate electricity or heat. Waste material is converted into a valuable useful product and methane release into the atmosphere is avoided (remember that methane is a powerful greenhouse gas).

The amount of methane produced depends on the type and quantity of organic matter and is relatively small compared to the availability of natural gas. However, sewage treatment works, farms and food processors take advantage of this process by digesting sewage sludge and organic wastes in large steel or concrete tanks. The methane can be used in a CHP unit (see Section 7.2.11) to provide heat to help the digestion process and for business purposes and also to generate electricity. Large landfill sites have collecting pipes installed which collect the methane for electricity generation for use on site and resale.

The use of organic waste can be considered sustainable at sewage works and food processors as the materials are mainly derived directly or indirectly from crops that regenerate. The collection of methane from landfill is more doubtful. A lot of organic matter that ends up in landfill is waste food and paper, cardboard and plastics which ought to be recycled. Because the targets for reduction of such waste to landfill mean that the amount of methane produced should decline, it should not be considered a long-term source of energy. There is more on this aspect of waste management in Chapter 6.

7.2.10 Waste incineration

This is another route for dealing with waste which has a high organic content including domestic waste and sewage sludges. It is also known as 'energy from waste' or 'waste to energy'. Instead of being disposed of in landfill waste can be incinerated in a large furnace directly to generate heat. Incineration is also used for some hazardous wastes and clinical waste (the waste aspects of incineration are in Chapter 6). Many cities have waste incinerators that provide heat for district heating or similar purposes and the generation of electricity. The waste is screened to remove metal cans and large objects. This outlet avoids the problems of landfill but the gaseous emissions from incinerators can be a source of air pollution. There will be CO_2 from the oxidation of the organic matter but there may also be hydrogen chloride, sulphur dioxide, nitrogen oxides, particulates, heavy metals and trace organic substances such as dioxins. The air pollution from an incinerator depends on what is in the waste to start with, the temperature of incineration and any additional treatment processes such as after burners in the flue gases and filters added to the flue stream. Emissions from incinerators in the EU treating more than 50 tonnes a year were governed by the Waste Incineration Directive (European Commission 2000b) but are now included in the Directive on industrial emissions (European Commission 2010) and most incinerators require a permit. Modern incinerators are claimed to be free from the risks from pollution but their use still meets with local opposition from residents.

The solid residues of fly ash and bottom ash still require disposal. The fly ash is removed from the gaseous effluent and is likely to contain most of the metal content of the original waste and some residual organic substances. Its disposal can be to landfill but at much smaller quantities than the original waste. The bottom ash is inert and can be used in construction projects.

As with waste to landfill, the fuel content can contain materials that ought to be amenable to recycling and some, such as plastics, are derived from non-renewable sources. If they were not present the waste may not be suitable for incineration as the calorific value would be too low. Despite this, waste incineration is regarded as a form of renewable energy.

7.2.11 Combined heat and power

Using steam to generate electricity is inefficient. It is governed by the laws of thermodynamics and some of the heat, typically about half, generated by burning the fuel is lost. In most power stations it goes out in the cooling water or into the atmosphere from the cooling towers. In engines it is also lost to the atmosphere. This represents a loss of a valuable resource made even worse if the fuel is derived from non-renewable sources. The recovery of this heat represents an opportunity to gain more useful energy from the same amount of fuel. The problem is that many power stations are remote from centres of demand and the cost of piping hot water is prohibitively expensive. Where there is suitable colocation then district heating systems could benefit. In some locations the heat is used in greenhouses for food production, fish farming or in local industrial premises.

A simplified diagram outlines the principles of CHP in Figure 7.11. In a conventional power station the condenser is cooled by water that takes the heat away into a river, the sea or the atmosphere. In CHP the heat is transferred to the demand centre through a heat exchanger and the cold water recycled back to the CHP system. There are many variations of detail depending on the scale of operation (some are domestic in size) and the intended use of the heat but the principles of all of them are the same: the heat is removed from the turbine cycle but not wasted. In such systems the overall efficiency of conversion of fuel to useful heat and electricity can be over 80 per cent.

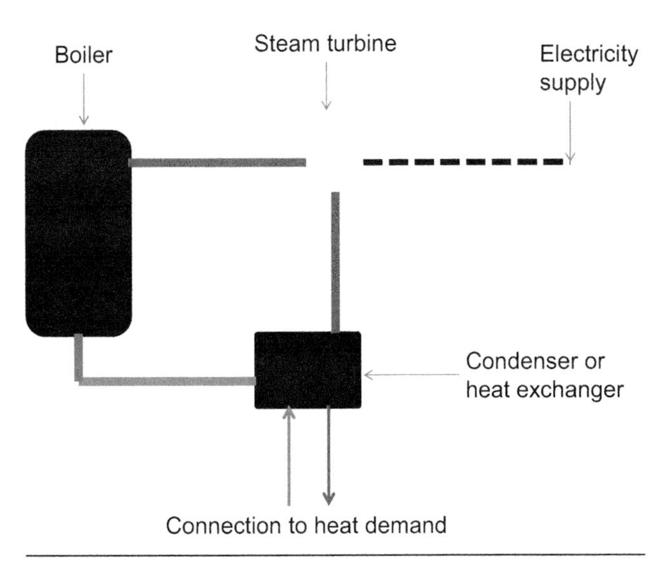

Figure 7.11 Principle of combined heat and power

The principles of CHP are not confined to power stations: they can be applied to most systems that generate electricity from burning fossil fuel, biofuels or waste. Incinerators, compression engines, methane burners, in-house boilers for factories and commercial buildings, even small domestic boilers can be used. In these cases the electricity and heat demands are on the same or adjacent premises and the costs of heat distribution are lower making it financially viable. The use of smaller units for power generation at a local scale can be applied in district heating and electricity supplies on an integrated basis for housing and community buildings. District energy systems have been common in some European countries for many years and are becoming more popular in the UK. Some examples can be found at http://ukdea.org.uk/. The extent to which CHP can be considered sustainable depends on the source of the fuel. Although the overall energy efficiency may be good, if it comes from fossil fuels it can hardly be considered sustainable.

7.2.12 Heat pumps

Heat pumps work by extracting heat from the ground, water or air at ambient temperatures rather than the high temperatures of geothermal power described in Section 7.2.6. They work in a way similar to a refrigerator. A refrigerator takes heat from the cabinet and releases it outside. A heat pump takes heat from the ground, a pond, river or reservoir or the air, effectively cooling them but not to an extent that would be noticed. The heat is used to supply energy to a property. Ground heat pumps usually have a large coil containing water and antifreeze buried just beneath the surface or in a shallow borehole. Coils are also submerged in water in a lake or river and in air a heat exchanger is used. The system relies on a compressor in the same way as a refrigerator and the heat is transferred to another water circuit through a heat exchanger. This circuit is normally used for space heating. Electrical energy is needed to power the compressor but the heat output is higher than would be achieved by the same electricity being used to heat directly. The main limitations are the space required for sourcing the heat and that the temperature of operation is low so that it is only practicable for space heating in a well-insulated building. The heat recovered in excess of the electricity used can be considered as renewable energy but the electricity needs to be from a renewable source for the whole system to be considered fully renewable.

Box 7.4 Principles of operation of a fuel cell

Fuel cells rely on a chemical reaction that takes place in the cell. The cell contains an electrolyte (a solution of salts) and two electrodes (anode and cathode). Chemical reactions take place on the surfaces of both electrodes. One reaction releases electrons to the anode. They flow through the electrical circuit and back to the cathode where the other reaction takes place. The net result is that the fuel is oxidised to water in the case of hydrogen fuel or water and carbon dioxide in the case of hydrocarbons such as methane.

In practice it is a lot more complicated than that. The electrodes are made of catalytic materials which promote the reactions. A membrane is incorporated in the cell to separate the two parts and the materials used mean that the cost is high. The cells can be quite large to deliver higher current but the voltage from each cell is low so for most practical purposes a unit is made up of multiple cells. This means that heat removal is more difficult and the size of a complete unit is large relative to the power output.

The cells generate electricity so can be used in a vehicle to drive a motor. The size of fuel cell required is larger than a conventional petrol engine. The fuel has to be suitable and some experimental vehicles generated hydrogen from fossil fuels on board. This is an additional process which is expensive and takes up more space; it also defeats the objective of reducing dependence on fossil fuels.

7.2.13 Fuel cells and Hydrogen

Fuel cells have only become available as a commercial proposition relatively recently and are becoming more widely used. They work by combining a fuel with oxygen, air or an oxidising agent, generating electricity directly within the cell. Heat is also produced which can be recovered. The original cells used hydrogen as the fuel although alternatives are available that use methane or alcohols for example. The principle of operation is outlined in Box 7.4.

Fuel cells emit no polluting emissions other than CO_2 if hydrogen is not used so are a clean source of energy in use. Their overall pollution impact depends on the source of the hydrogen or other fuel. If they come from renewable sources (e.g. hydrogen from electricity from wind power) then the cells can be seen as clean and sustainable technology. Hydrogen is generated by the electrolysis of water and can be stored as the gas under pressure making fuel cells an alternative to engines burning natural gas or liquid fossil fuel. The cells have various designs and use expensive materials of construction. Their initial use was mainly in space and military situations but now they are available for a much wider range of applications in vehicles, on ships, in buildings and industry. They also offer a route to storing and backing up power as hydrogen can be generated from unreliable sources such as wind.

Hydrogen can also be used directly in a compression engine burning with oxygen in the atmosphere to produce water and no CO_2 emissions. Experimental vehicles have been produced and operated successfully but the storage of hydrogen on board takes a lot of space and the units are heavy. The use of fuel cells and hydrogen directly will develop as the technologies improve and the prices come down.

7.2.14 Summary of the benefits and limitations of alternative sources of energy

The previous sections have presented the alternative sources of energy available and their respective main advantages and limitations. Choice often comes down to circumstances. For utilities the need for large scale and reliability are important and so fossil fuel still tends to be the default option with supplements of biomass to coal in some cases. Hydropower is used where it is available as it is reliable and cheap to operate. Wind power is becoming the next choice but it relies on other sources for back up when wind turbines cannot operate. Nuclear is also reliable but is under scrutiny on grounds of safety and cost. Of the other sources, tidal, wave and geothermal are still in their relative infancy and not practised on a large scale. The uses of methane and waste incineration have their niches but tend to be local to one site or business. Heat pumps are mainly confined to single sites as is solar heat. However, solar electricity may supply one property but larger scale operations are becoming common. Similarly, many CHP installations are confined to one site but their use is increasingly being extended to district- or community-based schemes.

Overall most renewable energy technologies, excepting nuclear fission, have been developed on a scale smaller than conventional coal-fired power stations. The development of large-scale wind farms is changing that and wave and tidal could extend in the future to be comparable in size if they can be demonstrated to be reliable and cheap to operate.

The role of incentives

Many governments have used various incentives to promote renewable energy. The utility companies in the UK have been given targets to use renewable sources such as wind. Domestic, community and commercial properties can qualify for installation grants and ongoing subsidies for using renewable energy; some more information for the UK can be found through the web page of the Department for Business, Energy and Industrial Strategy at www.gov.uk/government/organisations/department-for-business-energy-and-industrial-strategy (previously Department of Energy and Climate Change) with links to other organisations that are involved such as the Energy Savings Trust for householders and the Carbon Trust for business. The details change over time so it pays those interested to check on the appropriate web sites. Renewable heat and generation of renewable electricity qualify under different and changing schemes that include solar, wind, heat pumps and biomass. Some of these schemes have proved so popular that they are being reduced in scope as the cost has become too high. In addition the imposition of carbon emission ceilings and the introduction of carbon trading within the EU are aimed at reducing the use of fossil fuels in power generation and industry. It is notable that this is taken seriously within the EU but the US has resisted taking action at a national level, although some states are going it alone.

7.2.15 Energy supplies in difficult circumstances

The world economy increasingly relies on energy as it is used in industry to replace manual labour and demand increases in homes and for transport. The developed regions of the world have built up supply and distribution systems over many decades that can meet the demand although these can still fail due to exceptional demand or breakdowns in some part. Remote sites and developing economies present challenges as the infrastructure does not exist and is expensive to build from scratch. Many rural communities rely on wood, dung, coal or paraffin for domestic use. These are often difficult to find or expensive. Combustion is inefficient and fumes are a serious health problem indoors as ventilation is often inadequate. In these circumstances electricity is a luxury and batteries are expensive for powering telephones and radios. The education and development of children is limited due to lack of access to good lighting, computers and other gadgets taken for granted elsewhere.

Solar cookers and solar heating are being applied in many developing countries with suitable climates. Renewable energy from solar and wind can help with the supply of electricity if the initial capital cost can be raised. The running costs are low, especially when linked to energy-efficient devices (see the next section) and the benefits potentially huge. They also help in remote locations where finance is less of a problem such as large-scale forestry, mining or agriculture and fossil fuels have to be transported over long distances. The continuing development of the technologies and increased production rates are bringing prices down and bringing them within the availability of quite poor communities. The use of basic mobile phones for market price comparison and money transfers was introduced in some African countries ahead of Europe and has transformed small-scale business. There is an opportunity to avoid the massive infrastructure approach and go for local community schemes which are coming back into the developed world.

7.3 The importance of energy efficiency

However the energy is produced, improving the efficiency of its use will reduce the demand, the cost of supply and any associated emissions. If changing the source of supply is contemplated introducing more efficient use first will reduce the capacity and cost of the replacement. When most surveys show that energy waste in all forms of use is high and can often be reduced by simple and cheap (even cost-free) methods, there is a lot of scope for improvement. Energy efficiency should feature highly in the objectives, targets and continual improvement of an EMS. The main business opportunities are reductions in emissions and costs.

7.3.1 Reductions in emissions

Reducing energy use will reduce emissions directly or indirectly and for increasing numbers of businesses this is becoming a legal requirement. If an organisation operates a boiler or burner or any vehicles it is probably using up some of the earth's resources such as gas or oil and producing emissions of carbon dioxide and possibly other gases such as sulphur or nitrogen oxides. Improving the efficiency of energy conversion and use will reduce all of these impacts. The main opportunity is to reduce the emission of CO_2 from the combustion of carbon-based fuels. This obviously applies to fossil fuels but even reducing the consumption of biofuels and methane reduces the need for land and the other inputs for their production or releases the fuel to displace fossil fuels used elsewhere.

Where electricity is used the savings in emissions are indirect i.e. at the power station. Production of the electricity is outside the control of the organisation but the use is directly under its control. This should still be of concern as it applies equally to the organisation's impact on the environment. It may also feature in life cycle analysis applied to a product, service or activity as described in Chapter 3.

7.3.2 Reductions in costs

The organisation can also gain financially from energy efficiency. There will be a reduced cost for fuel or electricity if usage is reduced as the bills are based at least on the quantity consumed. For larger users of electricity there may be further savings available. Some users may be on tariffs related to peak load or increasing steps in load and the total cost can be reduced by keeping within the peak or avoiding the next peak tariff. Some of these benefits can come just from managing the load without actually reducing the total use but by becoming more energy efficient more headroom is created and there is a wider margin of safety below the tariff change.

As organisations grow they usually use more energy in all forms unless energy efficiency is being pursued. If it is not there will come a time when a larger boiler or bigger electricity supply will be required incurring capital cost as well as increased operating costs. Energy efficiency can delay or avoid those costs altogether.

7.4 Putting energy efficiency into practice

Energy is lost or wasted by many means and improving energy efficiency is aimed at tackling these. Energy efficiency can take several forms: efficiency of production, efficiency of supply and distribution, efficiency of use and the avoidance of loss of energy through improved design, construction or operation of buildings, plant and machinery. These will be considered in turn although there is some overlap and when it comes to making choices more than one may be relevant.

7.4.1 Efficiency of energy production

A lot of energy that is used in the home or business is produced by another organisation, usually electricity from a utility company and fuels from another supplier. However, some businesses generate their own electricity and it makes sense to do this in the most efficient way possible whatever fuel is used. Sometimes a primary fuel such as gas or oil is used or it could be methane from an anaerobic digester. In other locations waste heat or steam from another process may be used. CHP is also worth considering if both electricity and heat are required. Electricity on its own could be a form of renewable energy such as solar or wind. In all of these cases efficient generation should reduce emissions, save money or reduce any need for additional electricity from a utility company. Choice of equipment is the first option, selecting the most efficient that is economically justified and bearing in mind that fuel is free for some renewable energy. It may be that the equipment is already in place and that it is not cost effective to replace it until it is at the end of its useful life. It is always worth checking regularly whether cost-benefit analysis will show that early replacement is justified as fuel prices rise.

7.4.2 Efficiency of energy in use

Energy has multiple uses: electricity for lighting and operation of equipment, gas or other fuel for space heating and in process operation. It is also true that most organisations and individuals waste some of the energy that they consume either by not using it efficiently or by wasting it without using it for any productive purpose. Lights and electrical equipment are left on when not needed. Heating is run when the temperatures are already satisfactory or no one is present. Heat is lost through open doors or poorly insulated buildings, etc. Engines are left running when vehicles are not in motion. The principles apply in the home, commerce, industry and in operations outdoors such as agriculture and forestry. This section will consider the options available for buildings and fixed equipment. Transport is dealt with in Section 7.4.4.

Generation of heat

Most properties have a boiler that is used for heating and for producing hot water that might use natural gas, oil, coal, coke, methane or biomass as fuel. Their relative merits have been discussed earlier in this chapter but whichever is used it makes sense to minimise the amount of fuel used. This will save money and reduce the overall emissions produced by the combustion of the fuel. The factors to consider are similar to those for generating electricity: choice of fuel, choice of equipment; maintenance and boiler controls. Choice of fuel may be limited as with generation but using renewable energy again needs to be taken into account even if it is just using solar power to preheat water. Modern condensing boilers that use gas are more efficient than the traditional models and are now mandatory in domestic situations in the UK. These are rated A (best) to G (worst) and only the better rated ones can be installed now. If the boiler is an older gas model or is using solid fuel such as coal it may make sense to replace it even if that is not yet due.

Maintenance is very important. Regular servicing to adjust boilers and heaters, and responding to faults or changes in fuel consumption should help to keep the plant running efficiently. Descaling of boilers and pipework, chemical treatment to prevent scaling and corrosion and the use of filters will also reduce fuel consumption. In addition controls should be in place on the boiler to regulate the heat output according to the need (such as time of day or building temperature), linked to controls elsewhere in the building.

Use of heat

Once the hot water leaves the boiler there are several ways in which the efficiency of use can be improved. They involve mainly avoiding loss of heat from the system to no purpose and avoiding its use where it is not required. Heat losses can be controlled by the installation of appropriate measures. These are best done at the building or refurbishment stages although some can be retrofitted fairly easily. There will be costs involved but many, such as insulation, will pay back in months. The most used ideas are:

- Insulation of pipework connecting the boiler to radiators, tanks, heat exchangers, and process plant, etc.
- Insulation of hot water storage and other parts of the system where it is not for space heating.
- Insulation of roof spaces, walls and floors to avoid heat loss from buildings.
- Use of double glazing with heat retentive glass on windows and doors.
- Installation of room or radiator thermostats to control room temperature.
- Choice of water and energy efficient equipment such as washing machines.
- Reuse or recycling of hot water.
- Heat recovery from waste hot water or cooling water where practicable.
- Controls to manage the time of operation – start up shortly before people arrive and end just before they leave.
- Installation of monitoring to understand where the main demand is.
- An energy or building management system can take care of managing the heat output from the boiler to the demand and is especially useful for large buildings.
- Testing and maintenance of equipment and control systems.

There may be other ideas that suit a particular building or operation and priorities for investment should start with those activities that use the most heat. In an office or shop this will be space heating but an industrial plant may use most heat in processes. All of the above will incur some cost but there are cost-free or low-cost ways to improve energy efficiency as well:

- Keeping doors and windows shut and sealing draughts, although be aware of the need for ventilation.
- Not running heating with windows open or the air conditioning running to cool the room as well.
- Not heating unoccupied space, although ensure frost protection if required.
- Ensuring that thermostatic controls are at appropriate temperatures for the room use and cannot be tampered with.
- Not wasting hot water by leaving taps running or even leaking.

These are largely about behaviour and supervision and are elaborated below.

Use of electricity

Electrical equipment is everywhere and it is easy to miss or ignore where electricity is being wasted. As with heating there are some ideas which require some expenditure to implement and they may be best timed for a refurbishment programme or at replacement:

- Choosing energy-efficient equipment such as pumps, washing machines, etc.
- Choosing computers and other office equipment that goes into 'sleep mode' or shuts down when not in use.
- Installation of low energy lighting systems or low energy bulbs.
- Use of time switches for equipment or space lighting that is little used.
- Use of proximity sensors to switch off lighting when people are absent.
- Consideration of instant water heaters in kitchens or drinks machines if cost-effective.
- Use of meters to monitor the usage in different departments or areas.
- Linking the information to energy or building management system.
- Testing and maintenance of equipment.

Most domestic electrical equipment has an energy rating in the EU. This ranges from A (best) to G (worst) and the worst rated equipment is no longer available. It covers such items as lamps, cookers, refrigerators and washing machines which are also used in commercial and industrial premises.

Cost-free or low-cost methods include:

- Switching off equipment when not in use.
- Avoiding equipment, chargers, etc. left on standby.
- Switching off the lights when leaving a room or building.
- Not wasting hot water if heated by electricity.
- Only filling kettles with enough water for immediate use.
- Only using air conditioning when necessary and keeping windows closed when it is used.

These are behavioural issues again as discussed later which are just as important as investing in new equipment or upgrading the buildings.

7.4.3 Building layout and design

There have been several references previously to buildings and how the layout and design can affect their use of energy. Opportunities to influence these occur only once and getting it right can have a major influence over the lifetime consumption (and hence cost) of energy. Within the UK many features such as thermal properties and insulation levels and some aspects of energy use are covered by Building Regulations and have to be approved at the design and planning approval stage. The details are complex but should be familiar to the architect. The client needs to be aware of some factors that are influential and may need consideration before an architect is involved – such as site selection. The main ones are:

- Site exposure to wind, sunlight, extreme weather and protection by topography or trees.
- Building location relative to the same issues.
- Building alignment to take advantage of solar gain or to avoid overheating.

- The effect of urban heat islands (overheating in densely built areas).
- Shading by trees or natural features.
- Opportunities for installing renewable energy.
- Choice of materials which affect the thermal properties of the building (insulation, reflection of heat, thermal capacity, etc.).
- The use of natural ventilation rather than air conditioning.

Another feature that is easiest to consider at the design stage is the installation of an energy management system (confusingly, also abbreviated to EMS). This can manage the heating, ventilation and lighting systems and integrate with other building management systems such as security and office space availability to optimise the energy use to occupancy, ambient temperature and energy demand. The complexity of the management systems increases with the size and different uses of a building but the investment probably pays off best in these circumstances.

7.4.4 The human factors in energy efficiency

It is a common experience that spending money on improving energy efficiency does not deliver all of the expected savings. It has been found that households that improve their insulation also improve their comfort by running the house at higher temperatures. Anyone with children will be familiar with the heating at full blast and the windows wide open, lights left on all day and radios, televisions and computers running long after they have left the house. Similar behaviour can negate the best intentions of an organisation. It is not usually done maliciously but is a reflection of lack of appreciation of the issues. So investing in employees can be as effective is as investing in equipment or buildings.

A start can be made with a campaign of publicity. This can be used to explain the issues and why it is important to be efficient in energy use, covering both the environmental reasons and the benefits in reduced costs. A good publicity campaign is helped if it is associated with another event such as the introduction of an EMS, a new building or refurbishment programme or an increase in energy prices but their absence is no reason to avoid it. It will be helped if there are figures to back up the campaign; few employees probably know the size of their employer's energy bills. The figures can be used to highlight any high usages and to set targets for improvement. Employees can help by implementing the ideas above and, by transferring them to their home lives, they should be able to reduce their own energy bills.

Publicity can take the form of briefings, posters in the workplace, 'switch off' labels on lights and equipment, a newsletter, the company intranet, in fact any method of communication currently used, ideally supplemented with something new. Reporting on progress is also important so that those being asked to save energy actually see what has been achieved or how far they are towards reaching the targets. However, involvement can be made more productive. Seeking suggestions from employees can often produce excellent ideas based on their observations. These can be incentivised by offering prizes or a share in any savings either to individuals through a suggestion scheme, to departments or even a charity. 'Energy champions' can be used to seek and promote ideas and encourage changes in practice and, of course, managers and supervisors need to be seen to be putting good behaviour into practice themselves and responding to bad practice.

7.4.4 Transport

Transport can be a major user of energy in many organisations. Goods are received and despatched usually in a van or truck. Employees travel to and from work and between locations or visit clients by car. Vehicles such as forklift trucks or small vans may be used on large sites to transfer goods. Most vehicles run on petrol or diesel fuels which are derived from fossil fuels and they emit carbon dioxide and other emissions as described in Chapter 4. In the UK and many other countries fuel is heavily taxed so there is money to be saved by reducing its use as well as avoiding or reducing the emissions from its combustion. As always there are several ways to approach this.

Choice of fuel

Petrol is generally used for lighter vehicles such as small vans and cars. Diesel is used for trucks and lorries, larger vans and also some models of car. In terms of fuel efficiency expressed as km per litre (or miles per gallon) diesel is slightly more efficient. However it is not as simple as that. Emissions from diesel engines have lower CO_2 for the same distance travelled but emit more small particles and some of the other pollutants. In the UK the tax treatment favours petrol but in other European countries the reverse is true so the financial case also depends on circumstances. Engine technology is advancing rapidly and the use of engine management systems and exhaust filters and catalytic converters keep changing the environmental balance between the two fuels such that the difference is less than just a few years ago. Small diesel vehicles cost more to buy than the petrol equivalent and they are really only cost effective for users with a high mileage. Of course the use of biofuels or fossil fuels with a proportion of added biofuels will reduce the dependency on fossil fuels and save resources as described above but in these cases the emissions will be similar and the fuel costs will be about the same. In practice the initial choice is between petrol and diesel or their derivatives for smaller vehicles but there are other alternatives.

For small vehicles covering a low distance on site, such as forklift trucks inside buildings or for local deliveries with vans, electric vehicles that are driven by a motor that relies on rechargeable batteries have long been a choice. They are limited in range but have other advantages in that they are lighter than vehicles with a combustion engine and have low maintenance costs. Their main advantage is that they have no emissions in operation (important inside buildings) although overall the electricity for charging may counterbalance that if it is not from renewable sources. Cars and vans which are totally electric are now available as the ranges between charges have improved. The better models offer about 500km. The main disadvantage apart from the limited range is that they take several hours to recharge and this is typically done overnight. Battery life can be a problem and they are expensive to replace. Research into battery technology is addressing some of these problems and vehicles continue to improve. They are more expensive to purchase compared with the petrol and diesel equivalents but the electricity cost is a lot less than the cost of fuel. Grants are available to reduce the initial cost in the UK.

If range is a problem then there are hybrid vehicles which run on petrol but have a battery and motor for use at low speeds and to boost acceleration. The battery is of value in towns where speeds are low and emissions are more critical. At higher speeds the petrol engine powers the vehicle and recharges the battery and overall efficiency is improved by techniques such as recovering energy during braking. More recent models use an electric motor to power the car all of the time and a petrol engine cuts in to recharge the battery as needed. Electric vehicles can be recharged from the normal power supply at home or work and many towns are now providing recharge points in parking spaces. Again these vehicles are more expensive to buy but overall they should be more efficient to run and they reduce emissions especially at critical locations.

Other fuels which are based on fossil fuels include compressed natural gas (CNG) and liquefied petroleum gas (LPG). Both of these use gas under pressure and the storage tank takes up some of the space inside the boot or carrying area. The availability of fuel, especially CNG, is variable and poor in the UK but better in many other countries. Most vehicles retain the ability to operate on petrol as the engine is the same. The advantages of these fuels are that emissions of CO_2 are lower and the other emissions non-existent so it is a clean fuel. It is also cheaper per mile and in the UK it gets favourable tax treatment. The fuel savings should compensate for the cost of converting a petrol vehicle to use gas as well, especially for a high mileage.

The choices for larger vehicles are limited. There are versions of trucks and buses that run on LPG or CNG (see Figure 7.12) and experimental vehicles such as buses that run on hydrogen using a compression engine or a fuel cell but at the moment these are not generally available.

Figure 7.12 A bus fuelled with natural gas in Barcelona

Choice of vehicle

Choice of vehicle is clearly related to choice of fuel: petrol, diesel, gas, electricity. Electrical vehicles are small and only make sense for local use. However within the other fuel choices there is a wide range of vehicles available. The larger the vehicle the heavier it is and the more fuel it uses. So carrying capacity will influence environmental impact and fuel costs. Engine size is also important. An engine that is too small for the load will struggle to perform and may well use more fuel than the next size up, especially if driven hard. Various engines using the same fuel and of equivalent output can have different fuel efficiencies due to differences in design and management systems. A more efficient engine may cost more but have lower emissions and be more cost effective. Some overall vehicle designs are also more efficient: the incorporation of designs to smooth air flow over the tops of large lorries being a case in point.

Air transport is also environmentally damaging and the most expensive so should only be used for urgent or perishable goods. The use of rail or ship for delivering goods is a choice that needs to be considered as the fuel use by these larger forms of transport is a lot less than the equivalent loads by road. Both are good for heavy and bulky loads but may be slower and practicality may be a problem. There is a risk of handling packages or containers several times and, if the delivery is by lorry to rail and then from rail by lorry to final destination, for small distances rail may be too slow and expensive. Rail is often used for coal, minerals, steel and similar heavy goods where a branch line can go directly into the production site. Ships are mainly used for long distances by sea with goods in bulk or packed into containers; shorter distances such as from the UK to mainland Europe may be by lorry on a ferry or Eurotunnel. Canals are hardly used at all in the UK for goods but are used along with rivers such as the Rhine in Europe.

Optimisation of vehicle use

For road vehicles, efficiencies can be gained by optimisation of their use. There are several ways to achieve this:

- Match the vehicle to the load.
- Where possible ensure vehicles are fully loaded.
- Look for opportunities to share loads or to carry return loads in a vehicle that would otherwise be empty.
- Choose routes that are optimal in terms of distance travelled or least fuel used.
- Avoid busy roads or travelling at peak traffic times.
- Use traffic warning systems and satellite navigation to avoid hold ups and take advantage of alternative routes.

Some of these involve investment in new technology such as route planning software and on-vehicle navigation. Others may involve working with other local companies to share vehicle carrying capacity. If the use is generally for small numbers or for small packages, the use of a courier or transport firm may make more sense. Another consideration is local sourcing of as many goods and services as possible. Purchasing is often just on price which may rule out a local supplier (consider the example of buying books at the local bookshop as against online) and of course this works both ways in that distant customers may be thinking of the same strategy.

Managing the vehicle and the driver

Despite the best investment in vehicles, choice of fuel and route optimisation, fuel can still be wasted by poor maintenance and driving. Poorly maintained engines consume more fuel and tyre pressures make a difference. Hard acceleration and braking waste fuel; fuel consumption rises with speed. Getting the best out of the investment in vehicles means:

- Ensuring vehicles are maintained to schedule.
- Encourage reporting and fixing of faults as they arise.
- Regular (e.g. daily) checks of oil levels and tyre pressures along with safety features.
- Observance of speed limits and driving at optimum speed and gear selection.
- Speed limiters fitted to lorries.
- Switching off the engine at all stops; some vehicles do this automatically now.
- Use of in-vehicle tracking and monitoring of speed, route, etc.
- Monitoring of fuel consumption by each vehicle, benchmarking against other vehicles or manufacturers' claims and responding to anomalies.
- Setting fuel consumption targets for drivers.
- Driver training on safe and efficient driving.
- Management and supervision of vehicles and drivers.

Travel by employees

Employees travel to and from work and, when they are at work, between locations or to visit clients, customers, etc. They have the same vehicle and fuel choices as the employer and should be encouraged to make the most environmentally sound ones. If the employer provides vehicles it is easier to manage that choice and within the UK there are differences in the tax treatment of provided vehicles. For commuting there are other choices available: walking and cycling are healthy alternatives but weather and distance may present obstacles. Car sharing among employees that travel similar routes at least reduces the number of vehicles on the road and hence emissions and fuel consumption. Some organisations and local authorities have schemes to promote car sharing among their employees. Public transport using bus, tram or train or 'park and ride' help reduce overall pollution especially in town centres. Again local authorities and other employers may develop 'green transport plans' to promote the development and encourage the use of these alternatives. Choice of location by the employer is a factor in these alternatives; out of town business parks are usually difficult to reach except by car. Finally, working from home is possible for an increasing number of employees. Many jobs just require a computer, internet access and a telephone so location is immaterial. Organisations with several offices can accommodate employees dropping in to use a 'hot desk' if they have to travel around and access to the office is required.

The employer can offer facilities and incentives to encourage some of the changes: secure cycle parks and showers, subsidised season tickets or loans for their purchase, incentives such as free or reduced parking charges or even a guaranteed parking space for car sharing. If fewer employees use cars there will be reduced pollution and fuel consumption as progress is made towards the targets in the EMS and the employer may need to provide fewer parking spaces.

For business use some of the same choices may exist but sometimes public transport is inconvenient or too slow. For long distances rail is a good choice not just on environmental grounds as employees can do other work on the train and should arrive in a better frame of mind. Other options to avoid travelling to meetings such as video or telephone conferencing can be very effective. The most basic personal computer can work with video conferencing for two or three people but for more some bespoke equipment is better. The initial equipment cost is usually quickly recovered in savings in travelling cost and time and meetings tend to be shorter. Telephone conferencing over the normal telephone network or by internet-based services is also effective for modest numbers of people.

7.4.5 A footnote on technology

The increasing use of computers, the internet, cloud computing and intelligent devices at home or work (automation or 'the internet of things') is increasing the consumption of electricity such that it could negate the efforts to save energy elsewhere. Estimates are difficult but are suggested to be up to 20 per cent of total use in a few years' time. Battery-powered devices still need charging and some energy is lost each time due to inefficiency. Large data farms consume energy not just to power the computers but also to cool the buildings to remove the heat produced. Each mobile device may consume a small amount of energy but collectively, measured in their billions, this is significant. So energy efficiency is important for the selection of computers, mobile phones, and all the gadgets that are 'necessary' for modern life. In addition, the larger IT service companies are increasingly using renewable energy to power their operations.

CHAPTER

8

Control of environmental noise

After this chapter you should be able to:

1. Describe the potential sources of environmental noise and their consequences
2. Outline the methods available for the control of environmental noise.

Chapter Contents

Introduction **144**

8.1 Sources of environmental noise and their consequences **144**

8.2 Methods for the control of environmental noise **146**

INTRODUCTION

It is a common complaint in modern societies that noise is everywhere and it is virtually impossible to find a location for true peace and quiet. This is not a statement that stands scrutiny as the same person may consider that the sound of birds singing in a wood is not noise as they mean it; what they are complaining about is unwanted sounds. So if I am having a picnic in a field I don't complain about the birds but I would complain about the outdoor pop concert nearby. Yet if I am at the pop concert I have paid to hear the music, in some cases the louder the better! If I return to my garden to sit quietly with a drink and read the paper I may not appreciate the neighbour cutting his lawn or listening to a loud radio. So dealing with noise is subjective and a difficult topic. It is often the cause of falling out between neighbours or disputes with other local businesses and may even end up in court as a civil claim for nuisance as described in Boxes 1.2 and 8.1 in Appendix 4.

8.1 Sources of environmental noise and their consequences

In approaching the subject of noise from the point of view of environmental management we need to understand what makes noise a nuisance and how we can identify and manage the sources or their impacts.

8.1.1 The characteristics of noise which lead to it being a nuisance

The definition that noise is unwanted sound is a good starting point for identifying which aspects of noise make it unwanted. Most complaints arise because the noise has some of the following characteristics which should be familiar from personal experience.

- It is too loud: this may be judged against expectations given the location, etc.
- It is intermittent and unexpected: a sudden noise such as quarry blasting is more disturbing than a continuous one and the sirens on emergency vehicles are meant to catch attention.
- It is persistent continuing for hours or days.
- It is repetitive such as a pile driver.
- The frequency (or pitch) is annoying: this may be because it is too high (such as a screeching sound) or too low (such as a throbbing sound).
- It is unfamiliar: a new sound is more likely to be noticed than a background noise that is always there.
- It is intrusive in that we try to listen in or interpret: a loudspeaker or tannoy (or perhaps the neighbours gossiping).

Other factors that can affect whether noise is likely to become a nuisance are:

- It is uncharacteristic for the neighbourhood: expectations of the types of sounds are different in rural and industrial locations and the levels of background noise make a difference.
- The sensitivity of individuals: the level of hearing and the sensitivity and attitude to noise varies among people.
- Closeness to property: especially a problem if a factory is surrounded by houses.
- It is disturbing given the time of day: what may be acceptable during the day is not at night.
- It is happening too often: once or twice may be acceptable but if a party takes place every night it is not (unless you are invited of course).
- Weather, especially wind direction.

A couple of these points need expansion.

Frequency

Note that the term 'frequency' can have two meanings. The common usage is how often something occurs and this is relevant for noise as included in the list above. The less familiar use is to describe the pitch of a sound. Sound travels through air in a wave form and the frequency of the wave, measured in cycles per second, determines the pitch: high frequency results in high pitch and low frequency results in low pitch. Figure 8.1 shows this diagrammatically. The samples shown would represent pure sounds (or a single pitch) as would come from a tuning fork. In practice most sounds are a mixture of a wide range of frequencies. Humans can only hear these frequencies within a narrow range from about 20 to 20,000 cycles per second (referred to as Hertz or Hz) and the sensitivity to high frequency in particular declines with age. Some animals can detect noises outside of the range of humans. Humans may detect very low frequency sounds as much by feeling as hearing: stand in front of a bass speaker with loud music to sense this. Low frequency sound also travels further in air and can also travel through solid matter more easily. It can arise from heavy machinery, large

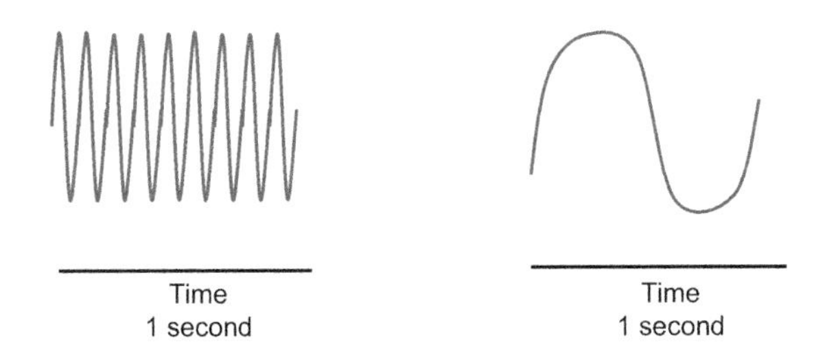

Sound travels as a wave and the wave form affects the pitch as heard.
High pitch has a high frequency as on the left.
Low pitch has a low frequency as on the right.
The height of the wave is a measure of the amplitude or loudness.

Figure 8.1 Frequency and noise

turbines and engines. It may be a nuisance in itself but can also cause secondary noise from vibration in a neighbouring property.

Background noise

People tend to become accustomed to background noise unless it is too loud. A railway, busy road or industrial site may result in continuous or frequent background noise but the immediate neighbours ignore it. A visitor is more likely to notice and say something. This is also relevant to the time of day. During daytime there tends to be more background noise so the odd intrusive one is less of a problem. At night, when it is generally quieter, the distant rumble of traffic or factory operations suddenly become noticeable. Noise can also travel further at night due to different atmospheric conditions.

Loudness

The loudness (you may also come across the terms amplitude or intensity) of a sound is related to the heights of the peaks in Figure 8.1. The unit of measurement of sound is the decibel (dB) and the scale is logarithmic. This means that a change of 3dB represents a doubling of the intensity (strictly sound pressure or amplitude). Noise is measured with a sound level meter and the UK standard for noise measurement is British Standard 4142 *Method for rating and assessing industrial and commercial sound* (British Standards Institution 2019; this replaces the standard dated 1997; there are ISO and other national noise standards relating to a range of equipment and activities). This standard describes how to measure noise and interpret the results. The methodology is quite complex but the main point of interest is that an increase of around 10dB (which represents a tenfold increase in noise levels) would be expected to lead to an increase in complaints. No noise at human threshold is 0dB and the sound intensity in quiet locations would typically measure less than 50dB. Factory locations can be up to 100dB; hearing protection is required at 85dB under UK legislation – see Box 8.2.

8.1.2 The effects of noise

The effects of noise can go beyond mere annoyance. Very loud noise can cause pain and, if persistent, can damage hearing. This is less likely in the wider environment than the work place. Low frequency noise can damage structures but again it would need to be at high intensity. The main environmental impacts for people are:

- Stress which can result in high blood pressure, anxiety and mental problems.
- Loss of sleep especially if at night, or during the day for shift workers.
- Nuisance – see Section 8.2.5 below and Box 8.1 in Appendix 4.

Property owners may also be concerned that noise could affect property values.

Noise can disturb wildlife such as interfering with nesting birds. Remember also that animals such as dogs can hear and react to noise frequencies that humans cannot hear.

Box 8.2 Noise as nuisance

The law as it applies to the tort of nuisance was introduced in Chapter 1 in Box 1.2. The main feature that applies for a civil court case is that to be a nuisance to an individual, the cause has to interfere with the reasonable enjoyment of the land of the person affected. Noise would be a good example, provided that it was excessive by some of the criteria in the main text and is causing harm such as stress or loss of sleep. Temporary or transient noise would not be considered a nuisance, nor would building operations unless they were conducted in an unreasonable manner. A case for public nuisance could be taken by several complainants in which case they would not have to demonstrate the proprietary interest. In either case they would have to show that the defendant had not taken reasonable measures to avoid causing the nuisance. The court could award damages or compensation and issue an injunction or an abatement order requiring measures to be put in place to avoid the nuisance.

If noise affects a wider area or many residents they could complain to the local authority who could issue an abatement notice. They could also class the noise as a statutory nuisance and may go to court under criminal law to seek fines and an injunction. Under these circumstances it is possible for a business to defend itself by demonstrating that it had used 'Best Practicable Means' to minimise the noise.

8.1.3 Sources of industrial environmental noise

There are many sources of noise in industrial settings with the potential for nuisance varying from industry to industry. The common generic ones are:

- Machinery such as presses, machines for cutting, hammering and drilling, ventilation fans, compressors, turbines, electric motors and diesel engines.
- Vibration caused by worn or damaged equipment or by loose parts in ducting, protective shields and casings, or set up by a motor speed causing resonance in related parts.
- Tannoys and public address systems, radios (music or person to person) and sirens.
- Transport such as delivery vehicles and loading and unloading goods, employees' cars, reverse warning alarms.
- Movement of heavy material such as the clashing of steel parts.
- Agricultural noise from heavy machinery such as tractors and combine harvesters, ventilators on intensive rearing sheds and bird scarers.
- Construction plant such as earth movers, pile drivers, pneumatic drills, erection and removal of scaffolding, demolition of buildings, plant for breaking and grading construction materials.
- Quarrying and mining with explosives as well as the plant and machinery above.
- Worker activity such as hammering, drilling and shouting.

Outside of industrial situations, common causes of complaint are:

- Road works and works on services.
- Demolition and building works in or near to residential areas.
- Music and people at pubs and clubs (often late at night) and associated vehicles.
- Fairgrounds, fireworks, open air concerts and similar events.
- Intruder and vehicle security alarms.
- Neighbours shouting or playing loud music.
- Small wind turbines on buildings and large turbines near to property.
- Transport noise from roads (caused by road and engine noise, especially at night), railways and aircraft taking off and landing.

The extent to which any of these is a problem depends on the factors in Section 8.1.1.

8.1.4 Regulation of environmental noise

The main responsibility for regulating environmental noise rests with local authorities in the UK. The Department of Transport regulates noise from vehicles and small civil aircraft. The Health and Safety Executive regulates noise in the work place from the health and safety point of view.

8.2 Methods for the control of environmental noise

Initial controls may start with an application for planning consent for a new construction or building or an activity such a waste site. In the UK the local authority may require an EIA and place conditions in any issued consent controlling the nature of any activities that take place and such issues as time of working. The operator is then not liable to legal action if the conditions laid down are followed. Planning permission is also required for some noisy activities such as shooting and landing helicopters if more than 28 days in a year.

In looking for control measures, the source, pathway, receptor model can be used again and measures could be applied anywhere. However, the pathway is through air (or ground in rare cases) and in all directions, although it may be influenced by wind direction. The receptors may be many and widely spread. In terms of ease of application and cost it is normally best to start with the source, which could be a piece of plant or machinery, and its immediate surroundings such as case or enclosure.

8.2.1 Noise control at source

The principal sources of noise are from vehicle engines, vibration, items knocking or scraping together and air emissions under pressure. This can be down to poor design, poor maintenance, mechanical failure or the activities of the users. It is often a sign that something is wrong with a piece of machinery or a practice, although sometimes it is an inevitable consequence of an activity such as hammering. The options available for control at source are outlined in the following sections. Some of the methods available for fixed sites are illustrated in Figure 8.2.

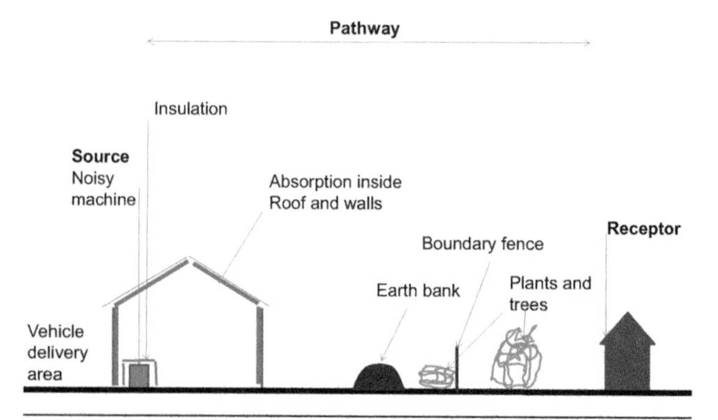

Figure 8.2 Some methods to control noise

Choice of equipment

If you can afford it, a Rolls Royce car is quieter than a model costing a fraction of the price. We are probably familiar with this in other items of everyday use that we are more likely to experience such as refrigerators. This is down to design, fine attention to detail in the components and construction along with inbuilt sound insulation. Even such things as the speed of rotation of a motor may be important. So choices of equipment at purchase or renewal can make a difference. In some cases in industry, electrical motors could be used instead of diesel engines: forklift trucks and some hand tools are examples. Quieter models may cost more but will be better for worker protection and will reduce the cost of dealing with complaints and the need for additional measures. If they are less noisy due to better components and construction they may also be cheaper to maintain in the long term. The noise ratings should be available from the supplier; if not it probably means that the item is best avoided anyway.

Choice of method of working

Often there will be little choice in methods of working due to the nature of the task being carried out or some factor such as ground conditions. However alternatives should be considered: can piling be replaced with drilling; can fixing with nails be by nail gun or can they be replaced with screws or adhesives; can cutting operations be avoided or reduced by better selection of sizes of materials? The decision will have to reflect such factors as safety, cost and the suitability of the method to ensure the quality or endurance of the finished work.

Maintenance

Mechanical plant wears as it is used and tends to become noisier. Regular maintenance should help to control this. Lubrication reduces wear so regular checking and changing oil or greasing parts will delay that process. It may be that, despite this, parts still need to be replaced eventually as bearings and other moving parts wear or fail but, as well as reducing noise, this should extend the life of the equipment. Maintenance checks also include ensuring that insulation and any other features are intact.

Damping

Damping refers to mounting equipment on material which absorbs and reduces vibration. This is usually rubber mats or feet although other materials and mechanical means may be used. Rubber matting can also be used to absorb the impact of dropped items such as rubble into a skip.

Escaping or moving air

Air under pressure and escaping from leaks is noisy and the noise gets worse as the leak gets worse: it needs fixing. An exhaust system on a vehicle or static engine is another example. Air exhausting from ducts on ventilation systems or movement caused by rotating blades (e.g. fan blades or wind turbines occasionally) is less easy to deal with but the design, layout (e.g. avoiding tight bends in ducting) and sizing of the parts will have an effect as will the location of the final vent relative to potential receptors.

Loudspeakers

This includes tannoys and similar equipment whose purpose is usually to convey a message to those nearby and to radios used for on-site entertainment, or as part of a communications system (and to annoy customers in supermarkets). They may need to be there so control comes down to volume control, direction and the use of more than one speaker. A good example is tannoys found on railway platforms or airports: they should point along any platform or be directed at the receptors. There should be more than one at low volume rather than turning a single one up to maximum. Quality of speaker may also help and make the messages more intelligible if the volume is lower.

Reversing warning alarms

These are the ‘beepers’ fitted to trucks and other large vehicles to warn people behind a reversing vehicle whom the driver may not be able to see. It may be possible to use a ‘banksman’ to aid the driver or rely on flashing lights instead if circumstances allow. It is possible to get alarms that link to the vehicle lights and reduce in volume at night.

Silencing

Vehicle emissions (including fixed engines such as compressors) rely on a silencer in the exhaust system to reduce noise. Some are more effective than others and they need to be replaced if corroded, broken internally or damaged.

Isolation (Location)

There may be little choice sometimes in where a piece of plant is located; the decision is determined by its relationship to other pieces of equipment or space considerations. Where possible it makes sense to isolate noisy equipment away from sensitive sites or to locate it where it is shielded by buildings or other structures. Location is important outside for vehicle movements such as delivery areas and siting of waste collection points where there may be a lot of activity and banging of doors or containers.

Insulation

Some items such as ducting, or equipment such as motors or engines may be insulated as part of their construction or it may be possible to put an enclosure round the item with insulation material included.

Outsource noisy work

At a manufacturing plant or construction site it may be possible to arrange for some work to de done elsewhere less sensitive and bring in prefabricated parts to avoid noisy operations at all. This is often the practice for steel work or roof support timbers and steel panels are often cut and pressed at another site before being taken to the assembly site.

8.2.2 Controlling noise on the pathway

Considering previous comments about the nature and extent of the pathway, it makes sense to implement pathway controls as close as possible to the source. The principles of isolation and insulation apply here also by intervening in the pathway and reducing the spread of the noise.

Absorption

This is the use of sound absorbent materials around a large piece of equipment or on the walls and ceilings of the containing building. The material absorbs the sound waves and reduces reflections into the building space (hence also benefiting employees) as well as reducing that which escapes, especially from modern steel or lightweight clad buildings.

Isolation or containment within the building

A lot of noise escapes from buildings through open doors and windows. It may not always be possible to keep them closed for access and comfort reasons but then the emphasis has to be on other measures. Where possible, and especially at night, they should be kept closed.

Screening

As an alternative where noise is still escaping or cannot be easily contained such as on an open site, some form of screen will help to attenuate the noise. Smaller permanent sites and temporary construction sites may have fencing or wooden sheets about 2m high fixed around the whole area: this has other benefits such as reducing the spread of dust and site security. Temporary screens may also be used. For larger construction sites and quarries it may be feasible to build an earth bank. Planting trees and thick bushes also help but these will only be of use if the activity is going on for a long time. All of these help by various mechanisms including absorption and reflecting or deflecting the sound away from potential receptors.

8.2.3 Operational procedures

On a working site there are other things that can be done to avoid causing noise in the first place or to minimise its impact. Some of these are influenced by management decisions and others may be related to employee behaviour.

Hours of working

Night times are sensitive as would be weekends and other holidays. Schools may be particularly concerned about noise during examinations and other infrequent activities such as an outdoor play or concert could benefit from a period of reduced noise. Liaison with the neighbours (see below) will help identify the issues and agreement reached to curtail certain activities at times. It may just be a case of avoiding deliveries at night, putting automatic closures on doors or postponing a particularly noisy activity such as piling to a date that causes the least inconvenience.

Vehicles

Apart from delivery times it may be possible to reduce the impact of vehicles by choice of access route to the site or, on site, by ensuring that routes, collection and delivery areas and parking areas are sited away from neighbours. The steepness of haulage routes in quarries and elsewhere may also be a factor. Engines should not be revved and should be switched off when not in use, also helping to save fuel.

On-site works and construction

A lot of nuisance noise is caused by inconsiderate use of materials, plant and machinery. It may be possible to relocate some operations such as cutting and drilling without interfering with the work; items should be lowered or passed through chutes, not dropped; rubber matting can be used on the ground or in skips and dumper trucks; choice of method and tools can avoid noisy hammering or pneumatic hammers; if several operations are necessary are they all better done in one short annoying burst or can they be phased such that at any one time the noise is acceptable? Anything not in use should be switched off; this also saves energy.

Operator behaviour

The other aspect that needs to be under control is what the operatives do: shouting to each other; noisy meetings in the smoking areas; leaving doors to buildings open; slamming doors on buildings or vehicles and using the wrong or damaged equipment. Manufacturers' instructions for equipment should be followed for operation and maintenance and regular checks made to ensure that insulation or silencers are intact. Employees should be required to carry out basic checks and service equipment at the start of each day or period of use and not to use and to report any that is damaged or malfunctioning.

Supervision

The onus is on the managers and supervisors to provide the right equipment and to put the procedures in place to avoid causing excessive noise. It is also their responsibility to check that the right equipment is being used, that it is being serviced and that the operators are following instructions. They should also respond promptly to operators' concerns so that they have confidence that if they report a fault it will be dealt with.

Monitoring for noise

Part of the regular checks and inspections should include monitoring for noise. At the most basic level this can be observed by listening but for noisy sites or changes in operations the use of a noise meter may be required. This can be bought in as a service. It should include noisy plant and locations but also extend to the site boundary and beyond, especially if there have been complaints. Keeping records of the noise levels may be helpful in the event of any disputes with neighbours or the local authority.

Training

All of the above needs to be supported by training. Operators are more likely to cooperate if they understand the concerns of the neighbours, how their actions may be having an impact and how they can contribute. The topics will benefit by being reinforced at regular site meetings where problems are discussed and resolved. It offers everyone an opportunity to raise problems with the equipment or procedures and to make suggestions for further improvements.

8.2.4 Controlling noise at the receptors

If all else fails or is impracticable this should be the last resort as it is most likely to be the most expensive unless there are only one or two properties or locations involved. Any proposed solution also has to be acceptable to their owners and the only one likely to be accepted is insulation at the house or office block by installing double or triple glazing. This has other benefits to the property owner such as reducing energy loss. It has been adopted around airports where noise from aircraft taking off and landing is difficult to control by any means other than the adoption of quieter engines.

Liaison with neighbours

The neighbours are likely to be critical and many may have objected to a new factory, quarry, waste site or construction project at the planning stage. Whatever the history, once the site is established, good relations need to be encouraged. These will help to identify problems early on so that they can be headed off or actions agreed to mitigate them. Neighbours need a route to bring their concerns directly to the site manager or a nominee and regular meetings with site visits will help to build trust.

8.2.5 Noise and the law

UK regulatory agencies concerned with law were covered in Section 8.1.4. In the UK and elsewhere there are acts and regulations that are relevant to noise under specific circumstances. Local authorities have powers under planning law and other laws to regulate new sites, construction sites, some premises and noise at public events and public places. Vehicles are governed by regulations from the Department of Transport. These are all specialist topics outside of the scope of this book but if you think that you have or may have an issue you need to find out more. A good starting point is the web site of Environmental Protection UK at www.environmental-protection.org.uk. This gives an overview of the topic. Noise is also covered in detail in their guide *Essential Environment* (2019).

CHAPTER

Planning for and dealing with environmental emergencies

After this chapter you should be able to:

1. Explain why emergency preparedness and response is essential to protect the environment
2. Describe the measures which need to be in place when planning for emergencies.

Chapter Contents

Introduction 152
9.1 Emergency planning to protect the environment 153
9.2 Planning for emergencies 155

INTRODUCTION

Much of the content of the previous chapters has been about avoiding pollution. If an organisation is well run and all the rules and suggestions for controlling operations are followed then the risks of causing pollution are low. However things can go wrong. Most major pollution events have one or more of several common causes and often they could have been avoided or planned for better. An organisation may also be responsible for, or at least involved in or associated with, pollution remote from its own site; for example, an accident involving transport of its potentially hazardous material on a busy motorway in an urban area. Some examples of previous significant global events that attracted widespread publicity were listed in Box 1.1 of which some were on sites and others involved goods being transported. The Environment Agency reported that, for England and Wales in 2015, serious pollution incidents had fallen to the lowest level ever recorded at 499. Under half of these were from companies regulated by them. This implies that over 250 incidents were from those companies that were unregulated, most of which would not normally be considered likely to be a pollution risk. They are unlikely to be large multinational companies conducting high-risk operations. However, although the risks from the larger regulated companies are low, the consequences are probably more serious.

A common link between some incidents is that the environmental damage is often associated with other issues such as accidents and even deaths as a result of poor health and safety practices. This is true of some of the incidents shown in Box 1.1. The investigation reports usually confirm that there may be some common problems with the management and culture of the organisation. Some of the issues in avoiding and planning for pollution emergencies ought to be similar to those involved in avoiding and planning for other accidents. As with other aspects of an EMS, there are opportunities from standardising some of the procedures and paperwork with those for health and safety.

Emergency plans need to take account of the whole range of activities that an organisation is involved in and to identify the risks with the potential to cause an emergency situation. The main risk factors are shown in Box 9.1 and many of these have been mentioned in

Box 9.1 Some common causes of pollution of the environment

- Poor design, manufacture or construction of plant and equipment.
- Failure of power supply causing pumps, monitoring and control equipment or other items to stop working.
- Failure of pipework or containment vessels due to accidental damage, corrosion or high pressure.
- Dangerous or uncontrolled conditions within plant leading to explosion, fire or other damage.
- Overloading of plant or equipment.
- Failure of monitoring and instrumental equipment.
- Failure of automatic systems for emergency shutdown.
- Operators failing to follow procedures.
- Spills and leaks of liquids or gases.
- Illegal activity such as liquid effluent discharges to the wrong point or connection.
- Over-filling of tanks.
- Failure to provide adequate storage and bunding for liquids.
- Disruption by extreme weather such as storm or site flood.
- Interference by vandals or others with ill intent.
- Damage to plant caused by construction, maintenance or other activity on site.
- Traffic accidents on site or involving transport of materials away from the site.
- Mismanagement of waste.
- Inadequate segregation of materials including wastes leading to reactions that emit fumes or cause fire.

- Consequences of an unrelated site emergency such as fire.
- Failures by suppliers or contractors.
- Management and supervisory failures to implement and enforce the necessary measures to prevent or deal with such incidents.
- Inadequate planning to prevent or respond quickly enough to a risk.

previous chapters. Each chapter has been about one medium such as air or water and emergencies can affect several media. Occasionally a minor incident can cause a chain reaction leading to a major emergency. A small fire can result in loss of power to the whole site or fire water can become contaminated with chemicals; an undetected leak of flammable gas can result in a devastating explosion releasing toxic fumes and liquids; a leak of crude oil can cause pollution to water and air. The potential can seem alarming and sometimes it is. The bigger the operation and the more hazardous the materials handled the greater the risk of a major disaster. The experience of BP in the Gulf of Mexico, referred to in Chapter 1, is an example of this. It occurred at a large drilling site 80km offshore; however, small operations can have a significantly damaging local effect if the impact is on a small but sensitive watercourse or near to a housing estate or site of special scientific interest.

The onus is on the potential polluter to assess the risk and to take steps to avoid causes in the first place but to be prepared to respond if something does go wrong. This chapter is about emergency planning for dealing with pollution but it ought to be a part of other business plans to deal with other crises such as risks to health and safety, financial problems, failures of computer and IT systems, loss of consumer confidence, bad publicity and business continuity.

9.1 Emergency planning to protect the environment

The main reasons why emergency planning is important can be summarised under seven headings but they are interlinked.

9.1.1 The need for prompt action to protect people and the environment

The primary reason for any response at all is to protect people and the environment. People on site or nearby may be affected by toxic emissions or other emissions to air that may just be a nuisance. Watercourses or adjacent land may become contaminated and wildlife may be affected. These have already been covered in the relevant chapters. Although some incidents such as an explosion can be dramatic and the impact instantaneous, most develop relatively slowly or at least with enough time to react and respond if adequate preparation is in place. If there is no plan then the response will probably be disjointed, inadequate and too late. The two main reasons for making this a prompt response are:

- To identify the cause as soon as possible and rectify it as far as practicable, thus preventing further deterioration of the situation.
- To react to the immediate consequences and try to contain any pollution to prevent it spreading further.

A prompt response will normally depend on some pre-planning and preparation.

9.1.2 It is part of an environmental management system

The introduction of any EMS, as described in Chapter 2, has avoiding pollution as one of its key elements. A commitment to prevent pollution which has been endorsed by the senior management is a requirement of the adoption of ISO 14001. The background to this can be found in Chapter 2. So any non-compliance with the EMS or pollution that does occur is a breach of the EMS and that undertaking and it could result in loss of accreditation. The existence of an EMS is often a condition of a permit or a contract to supply goods and services, so loss of accreditation could have further consequences.

Section 8.2 of ISO 14001 actually requires organisations to 'establish, implement and maintain procedures to identify potential emergency situations'. The organisation is also required to 'respond to actual emergency situations and accidents and prevent or mitigate associated adverse environmental impacts' and to 'periodically review and, where necessary, revise its emergency preparedness and response procedures, in particular, after the occurrence of accidents or emergency situations'. It must also 'periodically test such procedures where practicable' (ISO 2015). The broader principles of ISO 14001 also apply: roles and responsibilities need to be assigned, training needs to be provided and documentation managed with the necessary control procedures in place.

9.1.3 A general responsibility or duty not to pollute

A responsible organisation does not set out to pollute. Its relationship with the wider community and the tolerance shown by neighbours, customers and other stakeholders is based, at least partially, on the assumption that it will behave responsibly and not cause pollution either directly or indirectly. Direct pollution may arise from the organisation's operations on its own sites or activities under its control. Indirect pollution could arise, for example, as a result of its lack of attention to its supply chain of raw materials, the activities of subcontractors or the disposal of its waste.

Some of this is enshrined in law. Causing pollution is usually an offence against one or more statutes, so avoiding pollution becomes a duty. An example from the UK is the Water Resources Act 1991 in which causing pollution of water is an offence. In England the Duty of Care with waste outlined in Box 6.8 is an example where the duty is a statutory one. So planning to avoid pollution and to deal with any incidents will help to avoid legal action. Box 9.2 in Appendix 4 includes specific examples drawn from UK practice that have been referred to in previous chapters where there is a legal requirement to make plans. Failure to do so is also an offence.

9.1.4 The risk of prosecution and other costs

In Section 9.1.3 we saw that there are legal reasons for not causing pollution. If you break the law you may be prosecuted under criminal law (outlined in Chapter 1). This can result in fines, imprisonment in extreme cases and legal costs. If the incident was a genuine accident (i.e. not due to some form of negligence or a deliberate act) and the response was prompt and effective, then the risk of prosecution is reduced. Even if it does go to court, the fines may be reduced in recognition.

A further risk arises under civil law (also in Chapter 1). The polluter may be required to pay for the response of the regulatory bodies, clean-up of the environment that has been damaged (e.g. mop up any oil), restitution of the environment to its previous condition (e.g. replace fish that have been killed) and pay compensation to any third parties that may have been adversely affected (e.g. a company that had to stop production as it had to cease water abstraction or to evacuate its staff). There could be additional legal costs associated with any civil actions. The civil costs could be substantial and should not be ignored. Remember the case of BP where they are running into billions of dollars.

The legal liabilities extend beyond those above.

- Chapter 6 explained how the polluter will have to pay for remediation of contaminated land.
- Within the EU, the Environmental Liabilities Directive (European Commission 2004) requires polluters to remediate any damage to the environment; recognised in England in the Environmental Damage (Prevention and Remediation) Regulations 2009.
- In some cases, involving protected habitats and sites, Natural England or its equivalents in the rest of the UK may prosecute for causing damage.
- In the event of fire, there are requirements to make reparations on vulnerable sites and liabilities extending beyond the site boundary to those that may be affected.

All of this should be enough to convince any organisation to at least look at their vulnerability before dismissing it out of hand.

9.1.5 Business continuity

The immediate response to an incident may involve shutting down one or more operations and this could entail loss of production. A small incident with quick recovery would not be a problem. However, consider the implications if the incident goes on for several days or a significant part of the plant has been damaged or destroyed. As well as a larger loss of production, the amount of staff and management time deployed in dealing with the incident will also affect productivity. There is also the risk that a regulatory body will force a closure with a notice or revoke or suspend a permit until such time as the cause of the incident has been identified and remedial measures put in place.

The business could be seriously affected by any of these as there is not just the loss of income to consider but the possibility that customers will go elsewhere and not return or decide to source from more than one supplier to reduce their risk of future disruption of supply.

9.1.6 Damage to reputation

If all of these were not enough there is also the reputational risk. BP is a classic case again but by no means alone. Pollution causes bad publicity and damages relationships with a wide range of stakeholders. Neighbours may become less trusting and more aggressive, regulators more persistent and customers may be lost. The whole saga gets another airing later if a prosecution is made. Complaints by third parties can also result in further bad publicity especially if there are disputes about the nature or extent of the damage and civil cases in the courts. The whole lot is compounded if there is any hint of incompetence in the cause or in the response.

The reputational damage is not confined to the location of the incident. We have already seen that damage caused to the environment in other countries where a supplier, a subsidiary or subcontractor is responsible can reflect on the customer or the parent company.

9.1.7 The business case

Putting all of this together, there is potentially a lot at stake: legal, financial and reputational. A well-prepared emergency plan with all the supporting procedures in place should be time and money wisely spent if it avoids any of the above.

9.2 Planning for emergencies

The risks associated with pollution as presented above mean that avoidance makes good sense and the previous chapters have included sections on this as it applies to the specific media such as air or water. An emergency on site may arise from an event remote from that particular part of the operation (such as a power failure) and extend to cover all parts if it gets out of control. So emergency planning has to take a broad view of the site and the operations.

There are three aspects of planning for emergencies:

- Planning for and implementing what needs to be in place to reduce the risk of an emergency arising.
- Preparing a response plan to deal with the full range of likely events should an emergency arise.
- Planning all of the other things that need to be in place for the response plan to work.

In all cases the amount of effort expended and the level of detail and provision made (and hence the cost) need to be consistent with the risks. BP, Shell, Exxon – all the big oil companies – run global operations involving oil and other bulk organic chemicals on a large scale. Their activities extend from the exploration for and extraction of crude oil, refining oil and producing a wide range of products to the transport of all of these by ship, rail, pipeline and truck across the globe. They each rely on thousands of suppliers and contractors, many of them in difficult and remote locations. They have teams of specialists devoted to dealing with emergency planning. Compare that with a small operation whose main environmental aspect in this context is small quantities of non-hazardous waste. There are risks to people and the environment but they can be easily controlled by existing management concentrating mainly on how the waste is stored and managed and the potential failings of the contractor who collects it.

9.2.1 Assessing the risks

The starting point is a risk assessment that brings together:

- The potential sources of environmental damage that could arise from an organisation's activities, products and services.
- The factors that could cause any of these to become the source of the damage.

The former ought to be based on the aspects and impacts identified as part of introducing an EMS as explained in Chapter 2. These should have included all the aspects that could have an impact under normal operating conditions and those that arise under abnormal conditions, which is what concerns us here. The EMS may not have covered the entire site or all operations so this will need to be checked. For example, the EMS is often concerned with the operations in manufacturing but an emergency may be the result of an event in another part of the site such as a fire or a power failure caused in another unit.

The latter is the events, both internal and external, that could be the abnormal conditions. Box 9.1 has some generic causes but it will not cover every eventuality. It needs to be extended to all potential sources of risk for that site and this is where additional work may be needed. The preparation of the EMS may have considered external power supplies as an aspect of the organisation that is responsible for indirect emissions of greenhouse gases. This is different to the failure of that same power supply to operate essential equipment on site. There are other risks such as flooding that could cause an undesirable impact from an operation by causing the release of hazardous materials as well as disruption to the operation.

This assessment is illustrated at the top of Figure 9.1. The risk assessment is based on bringing the two sources of information together to develop the list of the potential scenarios that could result in environmental damage

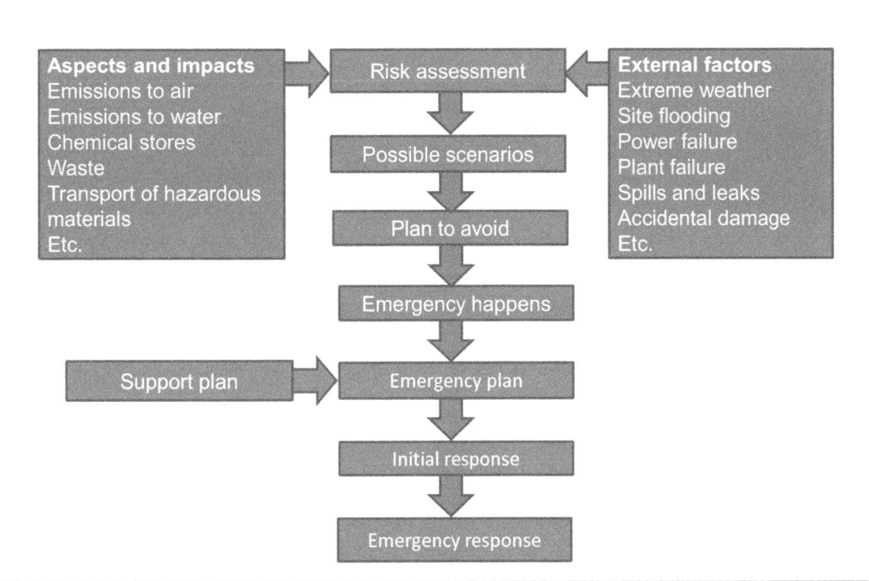

Figure 9.1 Planning for an environmental emergency

or harm directly or contribute by triggering a failure. This could be shown as:

- Environmental damage (impact): dust nuisance to neighbours, pollution of watercourse, plants killed in SSSI, etc.
- Potential source of environmental damage (aspect): emission of gas, discharge of solvent, toxic vapour, etc.
- Trigger for that event: failure of electrostatic separator, pump failure, leak from a pipe, etc.

This would need to be expanded to include all reasonable eventualities. It would include which equipment or pieces of plant are at what risk and how. For example: a discharge of liquid could result from a pump failing to control a level in a tank so that it overflows. This could be due to failure of the pump, failure of the power supply to the pump, failure of the level monitoring device, failure of the control system for the pump, failure of the power to the monitoring and control system, and maybe more.

9.2.2 Avoiding emergencies

The English proverb 'prevention is better than cure' has its equivalents in most countries and is apt for this topic. Developing the pump failure scenario from above, the provision of a bund to contain the contents of the storage tank, as described in Chapter 5, could already have been carried out. There are several other possibilities that may be available to prevent it overflowing in the first place and which may be more appropriate: provision of a standby pump, provision of an alternative power source or standby generation, duplication of essential monitoring and control systems, alarms to alert the operator to any of the potential failures, routine manual inspections to check on performance, etc. depending on the situation. It is not suggested that all of these need to be done; the extent to which any of these are implemented depends on the risk of failure and the potential consequences as well as the balance between cost and risk.

The development of this approach to include all foreseeable causes can be done in a structured way and there will be overlap. A general power failure will affect all pumps and other electrical equipment. The problem is where to draw the line and there may be no simple answer. Japan is subject to frequent and severe earthquakes and they plan and prepare for them. The extent to which the operators of the nuclear reactors at Fukushima (mentioned in Section 7.2.7) should have planned and prepared for a tsunami from an earthquake 70km off shore that flooded and damaged the electrical supply and the cooling system is easier to justify with hindsight. There are earthquakes in the UK and they are not severe but they do happen fairly frequently. They can probably be safely discounted but the risks from flooding from other sources are real in many locations and appear to be increasing. Reflect on the phrase 'can probably be safely discounted'; are you comfortable with that; could you defend it if you applied it to your situation in other circumstances; what are the investment costs and risks associated with that judgement?

In August 2012 India was affected by widespread power failures affecting nearly half the population caused by excess consumer demand and inadequate provision by the generating companies. Similar events over large parts of California and New York have affected a more developed economy like the United States in the recent past. There is concern in the UK that failure to put a strategy in place to meet future energy demands could result in similar problems there. At the time of writing (September 2019) a recent lightning strike knocked out the power supply to a large part of the UK, affecting 1 million people and, a salutary lesson, standby generators at a hospital and a water facility failed to cut in. The implications of these events are that any organisation dependent on electrical power supplies for the continuity of its operations and the protection of health, safety and the environment has to have a plan to respond quickly to a failure of supply.

The results of the above approach should have generated a list of sites, activities, processes, operations, plant and equipment that could be vulnerable and a list of causes that could be responsible. Both of these may need regular review in the light of changes on site or experience. Both should generate some actions to inspect and check that precautions are in place and, if not, an action plan to do so. Precautions can range from installing protection measures such as a bund or emergency power supply, to alarm systems or frequent inspections. They do not have to involve massive expenditure to cover every eventuality if the risks are low but the decisions taken and the reasons behind them should be documented for future reference.

The planning to avoid emergencies in the first place also generates a list of potential emergencies which may arise if subsequent actions are not taken or prove to be unsuccessful. This is where the emergency response plan comes into play.

9.2.3 The emergency response plan

If, despite all the preparation for avoidance, an incident occurs, the risk to the organisation has been explained in Section 9.1 and it has to be dealt with. The emergency response plan lies at the heart of being prepared to deal with all likely eventualities. It can be considered in two parts: information and procedures. The information needs to be in one place, available instantly and expected to be up to date. The procedures outline the actions to be taken in response to different situations and need to be available to those dealing with the incident as well as those managing the response. The emergency plan is a document (or documents) that bring together all of the information and procedures needed to respond in an effective manner and prevent or mitigate any pollution. For a large organisation it may be many documents covering different sites or different processes and, of necessity, including much content. In these circumstances achieving consistency of approach, content and quality of documentation and response will be a challenge that will require specialist teams. For smaller organisations it should be a simpler matter. The contents of the plan need to be tailored to the organisation and the risks but much of it would be built around a common structure. The form it takes may need to reflect how often changes are needed. Some information may need regular updates (such as changes of personnel or telephone numbers) and is best suited to loose-leaf document and/or online. The following sections outline the main contents of the written plan and the actions required to support it. More detailed advice from the UK environment agencies was included in 'Incident response planning: PPG21' and other relevant information on this topic can be found in 'Managing fire water and major spillages: PPG18' and 'Dealing with spills: PPG 22'. All of these were produced by the UK environmental agencies but are currently under review. Progress can be followed at www.netregs.org.uk/: search for pollution prevention guidelines.

The preliminaries to the plan

The front of the plan needs to make several items clear. Bearing in mind that it may be shared with outside bodies and that people and circumstances change, there needs to be no doubt about what the plan refers to and how up to date it is. So the title-page needs to clearly state the name, address and, if appropriate, the site or parts of the plant that it refers to. It should also be dated and show a version number consistent with the document control procedures for an EMS outlined in Chapter 2 and contain the name and contact details of the author or the person to whom enquiries should be made. The document should also be signed off by the person with overall responsibility for the organisation or the site.

The introductory pages need to show which people or organisations have received a copy so that they can be issued with new versions as they are produced. This helps those involved to talk to each other, knowing which information is available to all parties. The circulation list should extend to all those likely to be involved, including external organisations such as the emergency services, if appropriate.

The rest of the plan needs to have the detail which will differ from site to site depending on the circumstances. The order of the contents is not important as long as an index makes them easy to find. The range of possible contents is included below but there may be additional items that a particular site requires. The following sections also include the supporting actions that need to be planned for and in place to ensure that the plan actually works when put to the test.

Emergency control centre

Coordination of response is important so that that actions are taking place as planned (e.g. nothing is missed because someone is on holiday), that effort is not being wasted with duplication (e.g. two people independently liaising with the fire service) and timely responses are made to developments that may require decisions (e.g. to call in contractors to remove spilt liquid). This is normally managed from a control centre.

Box 9.3 Suggested services required in an emergency centre

- Adequate space to accommodate the facility in a secure location.
- Adequate desks, table space (for plans) and chairs.
- Independent phone lines (more than one and not through a switch board or automated call centre).
- Fax machine.
- Internal independent communications equipment such as radios or mobile phones.
- Indicator board for alarms and other important control information.
- Monitors for CCTV.
- Computers with access to the internet and company intranet, e-mail, etc.
- A white board for notices.
- A log book to record events and actions.
- Camera to record events.
- Torch, PPE and any other requirements for safe access.
- Access to the tannoy system.
- Keys to the main parts of the site at risk.
- Adequate stationery to hand.
- Sampling and test equipment as appropriate to the site and operations.
- Kettle and basic supplies.

Consideration should also be given to how failure of the power supply will be dealt with.

Not every site will need a specifically designated control centre: it is absolutely necessary for larger sites where the potential risks are high. A large site may have a dedicated centre permanently set up within a site control centre for normal operations. For a small site with low risks it may be a commandeered office or meeting room. In all circumstances it needs to be ready to swing into action at short notice.

A key requirement for the emergency control centre is that it is in a location that will not be affected by any likely event; so it cannot be the supervisor's office in the middle of a chemical plant or a site vulnerable to flooding. For large sites with high risks it may be off site and/or duplicated. Thought may also have to be given to the details of its design and construction if the site is vulnerable to explosions or similar events. It also needs to be accessible at all times; emergencies have a habit of happening at night or weekends. Some essential services that should be in place are listed in Box 9.3: the sort of key information that needs to be immediately available is listed in Box 9.4. This key information is part of the emergency response plan but maps and site plans are often on large sheets which may not be conveniently bound into a document and, in any event, may be more useful on a wall in the control centre. The list of requirements may seem excessive and not all of them may be relevant. However do consider how you would respond if called out of bed at 2 o'clock in the morning and cannot find a pen or paper to write down details of a telephone call or find the telephone number of the person who holds the key to a part of the site which has not been left with the centre.

Controlling events

The control centre is the hub of response activity and may involve several people. Someone (the controller) needs to be in charge of the centre and events on the ground. The choice of this person, and any deputy to cover absence, is an important decision and it could be a different person for different circumstances. The controller needs to be someone who is familiar with the site and what goes on there, is available on site most of the time,

Box 9.4 Key information required in an emergency centre

Copies of the emergency plan which may also contain the following items.
Large-scale (e.g. Ordnance Survey) and street maps showing local geography.
Maps marked to show neighbours, sensitive adjacent sites, local boreholes, etc.
Site plans showing:

- Locations of accesses, main site features, staff assembly points.
- Locations of process areas, with those especially at risk in more detailed plans.
- Locations of drains, interceptors and any special features (e.g. closure valves).
- Locations of storage for chemicals, waste and other potential risks.
- Locations of fire hydrants and fire-fighting equipment and emergency spill kits, etc.
- Any sensitive areas within the site boundary (e.g. where contaminated water may leak underground, or sensitive ecosystems).
- Locations of any emergency lagoons for diverted contaminated water.

List of key staff and their responsibilities.
Contact details (including out of normal working hours) for:

- Internal staff likely to be needed on site or to consult by telephone, key holders and those that need to be informed.
- External organisations such as emergency services and regulators.
- Neighbours and downstream abstractors that may need to be advised.
- Contractors that may be required to work on site, clean up or remove waste.
- Suppliers of goods and services (e.g. chemicals, catering).
- Local media contacts.

Schedules of chemicals, hazardous substances, etc. stored on site, with quantities.
Safety data sheets, COSHH assessments and other supplier information for chemicals and other materials on site.

- Prepared check lists for any planned actions.
- Copies of procedures to deal with the emergency situations.
- Copies of reports of any previous incidents, exercises or audits.
- Copies of permits or other relevant regulatory documents.

has the necessary level of seniority and authority to take action (e.g. order equipment or services without going through internal purchasing procedures). It is unlikely to be the Chief Executive or Managing Director except in a small company. It is more likely to be a middle manager, the Environmental or Health and Safety Manager or, for a large site, someone specifically recruited to the role.

Boxes 9.3 and 9.4 showed what equipment and what information needs to be available at the control centre or, if there is not one, available to those dealing with an event. Some of this is self-evident but some needs more explanation.

Access to site plans

A range of site plans may be needed to help with the response. General maps showing access roads and the immediate neighbourhood are necessary to discuss safe access with the emergency services and to identify people or sites that need to be advised or protected. If there are emissions to air some access points may have to be avoided depending on wind direction (you do have a wind sock or indicator on site, of course). This also affects the localities at risk down wind. The same map may be used to show safe areas for staff to congregate if evacuating buildings, taking into account the nature of the event and the wind conditions.

A detailed site plan should show the positions of the main locations and the activities carried out along with any particular hazards such as chemical and gas storage or hazardous waste. A plan also needs to show where fire-fighting equipment such as hydrants and extinguishers are located and the controls to any sprinkler systems. Details should be included of the location of stores of materials to deal with spills and equipment and tools such as pumps or shovels, spare drums and plastic sacks, specialised first aid equipment and personal protective equipment not in daily use such as breathing apparatus or chemical suits.

Plans of drainage systems are also required. These should be colour coded in the same way as the drain covers and gullies on the ground to distinguish between the foul and surface water sewers as mentioned in Chapter 5. The plans should indicate the underground routes of the pipes as well as features visible on the surface along with interceptors, valves and access points. Similar plans will be required to show pipelines and the locations of valves conveying gases and liquids around the site as well as services such as water, gas and electricity.

Inventory of materials and locations

This is needed to help with the interpretation of events and predict the potential scale of the incident. The inventory needs to focus on the materials considered a risk such as chemicals, oils and fuels, paints, compressed gases, flammable materials and wastes. The inventory should include the maximum amounts that can be present and their locations. Some materials may be in several locations from stores through production lines to despatch areas. The inventory can be used to build up supporting information about their properties. Health and safety information, environmental effects, fire and explosion risk, incompatible materials (including the effect of water) and any special handling precautions should be included. Most of this should have been provided by the manufacturer or supplier in safety data sheets and COSHH assessments but it may need supplementing from other sources.

Inventory of hazardous activities on site

A site map should show the locations of hazardous activities but it needs to be backed up with more details. The controller or anyone else dealing with an incident may not be familiar with all of the processes but find themselves liaising with the fire service over just such details. There is a balance to be struck over the quantity of information but 'Production area' on a map is not very informative. What is being produced? What is used in the process? Are there any hazardous by-products? Where does the effluent go? Is there a risk assessment available? (If not, why not?)

List of key staff and their responsibilities

Not everyone who works in an organisation knows everyone else and what they do. The list of key staff should include all those likely to be involved in dealing with an incident and their deputies if possible. It should extend beyond the immediately obvious of those with direct responsibilities for operations, environmental management or health and safety. People in purchasing, site security, PR, finance and personnel may be called upon. Administrative services may need to help out with staffing, additional telephone lines or welfare. Some people may need to be told even if they take no active part; it is not wise to exclude the managing director.

Internal contact list

In support of the list of key staff there should be a separate list of contact details: internal and home telephone numbers, mobile phone number and e-mail addresses. This list contains personal and confidential information and should not be freely available. It may also be structured around a cascade principle based on who to contact first and then who to contact if they cannot be raised. It may also have to cope with escalation of an event that a couple of people start with but then realise that more help is needed.

Liaison with regulatory bodies, emergency services and other external services

For sites where emergency planning is required and especially if on a large scale, formal liaison with various organisations will be required. The following arrangements may be needed:

- Police to discuss traffic control or evacuation plans (in the UK the police take overall control in emergencies involving the public or public spaces).

- Fire service, who will inspect and advise on fire risk.
- Ambulance service to review any special issues such as toxic material or potential scale of an accident.
- Environmental regulator to discuss the potential risks, mitigation measures, emergency plans and any permitting restrictions.
- Water and sewerage undertaker(s).
- Contractors who may be required to supply goods or services, especially if at short notice.
- Health and safety regulatory body (HSE in UK) about the H&S issues and special situations (such as COMAH sites in the UK, see Box 9.2).
- Local authority who may also be involved in any response.

The discussions need to ensure that all parties have an understanding of the site and the potential risks involved and the respective responsibilities, roles and procedures and communication details. Liaison meetings need to be on a regular basis as people and procedures change but need not be frequent if agreed methods of communicating small changes such as new personnel are put in place. This requires updates to the plan by the person responsible for it and sharing it with all the listed recipients.

The liaison arrangements need to include contact details again. Most will have 24-hour arrangements involving publicly available telephone lines but they will also have independent lines to deal with such events.

Liaison with neighbours

The neighbours may include not just those who live nearby or occupy adjacent offices or factories that could be affected; there may be water abstractors some distance downstream of any potential pollution of a watercourse and other sensitive sites at risk from air pollution which could also be some distance away. Meeting local businesses is a two-way process; you are their neighbour and just as likely to be affected by activities on their site. Bear in mind that the same principles should apply in sites located overseas. There have been instances where local populations affected by incidents in factories or mines have been ignored in such circumstances to the detriment of their health and safety or their property. Responsible companies will respect their rights just as much as anywhere else.

The purpose of the liaison meetings is to build trust and understanding in circumstances where suspicion could easily develop. The residents will be concerned if they hear talk of explosions or risks from emissions to atmosphere that could cause them to be evacuated or to be confined indoors. The discussion needs to focus on how the company will recognise an emergency situation and advise the residents of what is happening and what to do. Means of communication (in the UK this may be via the police) and practical arrangements for their safety and well-being will need to be discussed.

9.2.4 Recognising risk situations and action to take

The overall effectiveness of dealing with an emergency often depends on the initial responses. If a developing problem can be recognised early and swift action can prevent the situation deteriorating, then an emergency may be avoided. This comes down to good preparation and this can be developed from the work outlined in the previous sections. If the risks areas are known and understood, then actions can be put in place to deal with either the causes or the effects.

Developing the example based on a pump failing to empty a tank, two possible approaches can be suggested assuming the plans to avoid this happening have not been put in place or have failed. These are based on the source, pathway, receptor model. One method may be to tackle the source by stopping whatever is producing the liquid that is fed to the pump. This may involve stopping the process filling the tank or diverting the flow to an emergency tank or lagoon. Another way may be to deal with the pathway and accept that the tank will overflow but the liquid can be trapped and contained in any bund or the drainage system, at least as a short-term measure. Either may avoid having to respond by intervening at the receptor, for example, the local stream, and the embarrassment of disrupting a business downstream.

The planning for the initial response needs some thought. (Consideration of previous problems would be a good starting point as emergency planning is often introduced after an incident has happened.) The scenarios to consider should be the same or derived from those considered to avoid emergencies in the first place, e.g.:

- Fire or explosion.
- Spill or leak.
- Failure of electricity, water or gas supply.
- Failure of plant or containment vessel.
- Damage to plant, pipelines, etc.
- Vehicle accident on site or on the road involving company vehicle or products.
- Extreme weather conditions.
- Actions to take following any alarms being triggered.

Not all of these will apply at all sites and there may be others but the list indicates the scope of the items to be considered. The plan needs to take account of each one of these, consider the consequences and detail the actions to take. These could range from fixing the problem as it happens, through closing valves or switching on a generator to complete site shut down. The plan also needs to take account of the consequences of these actions as they may influence another part of a process. This is one of the reasons for planning; just closing a valve may cause a build-up of pressure, overheating or some other effect which compounds the situation. Some of these things could happen automatically but then the fact that this has happened needs to be recognised and follow-up actions taken. The responses also have to recognise the need for flexibility as there will be occasions when the event has not been planned for or the circumstances are different from the plan (e.g. a temporary process or piece of plant on site or building works blocking access – although the plan should have been adapted to take account of these as they occur).

Initial response on the ground

This initial response is often critical to preventing escalation of an incident and is often influenced by one person; the person who first reported, learned of or observed the problem. This may have come in as a telephone call from outside to the switchboard or been spotted by a worker on his way to the car park. All employees need to be aware of the potential for incidents and how to respond by passing the information on to the responsible person such as the site manager or in the process area involved. Procedures at the switchboard, control centre or other places where information can be received should include how to handle the report. Some basic information needs to be recorded:

- Date and time of report.
- Who complained or reported it with contact details.
- Nature of the incident or alarm.
- What action, if any, have they taken?

Subsequent recipients of this information may want to contact the original source of the information to clarify something or to help plan their response. That response will depend on circumstances but could include informing the controller and activating the emergency response plan immediately if the potential risks are large. On the other hand if the report does not appear serious they may want to quickly check if the report does require action by inspecting the site or source of complaint. Sometimes people report something long after it happened; it may not be from your site (if it is not then it should not be ignored but passed on to the environmental regulator or a suspected neighbouring site); it may be known about already and action is in hand. This should not preclude initiating the emergency response plan if there is any doubt.

Other immediate actions could be to inspect an emission chimney, watercourse, site drains and interceptors or the obvious potential sources such as a delivery point or storage site. Evidence may be found from staining on the ground or smell which could indicate the source. Bearing in mind that the site may be large, this is where coordination comes in and help may need to be organised through the control centre or the controller. It may be appropriate to collect some samples of discharges or from drains or watercourses and arrange analysis.

At some stage further action may be needed and if there is any uncertainty initial preparations to activate the emergency plan should have been started to activate the control centre and inform the necessary people both internally and externally. They may need to be found and have to change their plans. They can be stood down if the event is shown to be containable or has already been dealt with.

Start a log!

If the site is large or a control centre is in operation and several people are involved it is easy to lose track of what is happening, who is doing what, who has been told, etc. Keeping a record in a log at the control centre is a good discipline in dealing with the event but also in later reviews and any possible investigations involving outside organisations. Some of the actions such as notifications may be recorded on pre-prepared tick sheets. There may be more than one log if more than one centre is involved (e.g. the switchboard or someone delegated to carry out a particular task) and individuals should keep their own logs of what they do in diaries or notebooks.

Extending the response

As the position becomes clearer the rest of the emergency plan comes into operation. The procedures in the plan should set out the actions to take for the various potential scenarios: activation of the response team, what to shut down, activation of emergency power supply, who to inform, deployment of specialist kit such as pumps or materials to contain spills and so on. If the plan has been well prepared this should run well but the unexpected can occur and the controller needs to keep a watch for this and be prepared to change plan or initiate additional procedures. For example, the emergency pump may also fail to start (when was it last tested?), a general power failure has not just affected your site but that of someone that you were going to depend on, or one event has triggered another out of the blue (remember Fukushima). Specialist outside help may be needed such as contractors with pumps, generators or containment booms (these should all be part of the plan).

In dealing with the emergency it is important to look outside of the immediate area affected. Site security to keep people away and out of harm may require someone on the ground. Visitors may need to be turned away or directed to a different location. Anyone not directly involved should be kept clear if for no other reason than they get in the way, but their health and safety is at risk and no one wants more casualties to deal with. This means that scheduled deliveries or meetings may need to be changed and other plans altered. Little details like this are important but need to be delegated so that those directly involved with an incident can concentrate on that. Is it in the plan?

Dealing with leaks and spills

Leaks and spills are a common problem but many are small, do not spread far and can be mopped up easily. Larger spills should have been contained by bunds or other design features as described in Chapter 5. However, an emergency indicates that something has gone wrong with this and liquid is spreading out of control. Factors to take into account are the nature of the liquid and its potential impact. A dilute solution of a benign compound or foods could be diverted to the foul sewer after discussion with the sewerage undertaker. Flammable solvents or toxic material need more care, both to prevent environmental damage and risk to people but also to avoid further problems such as fire.

It may be possible to stop leaks by closing a valve or sealing the leak with tape, a clamp or special putties available for different materials but these will need to be available on site. Once a leak has spilled liquid onto the ground, fine sand can be used to contain or direct the flow of liquid by creating channels. Proprietary absorbents can be used to mop up material on the ground or, if floating like oil, on water. These are available in granules or as sheets, pillows or booms. Drain covers, access hatches and gullies should be sealed with rubber sheets unless sealed already. Drainage systems may have valves to be closed or bungs can be inserted into the pipes (sand bags can be used if nothing else is available). Installed interceptors may act as a trap, especially for oil.

The environmental regulator needs to be advised if the material gets into a watercourse or is at risk of doing so; they will have their own plans to deal with this but the more time they are given the better their response. They may deploy floating booms to contain oil and similar products (see Figures 5.10 and 5.11 in Chapter 5) or utilise emergency aeration equipment to protect fish (see Figure 9.2).

Figure 9.2 Surface aerators used to reoxygenate a lake

As the liquid is contained it needs to be removed. Small quantities may be removed by the absorbents. Larger quantities behind booms or trapped in drains and interceptors need to be pumped out. This may be possible with a small pump into drums or it may be better to call in a specialist contractor whose details should be recorded in the plan.

There will be waste material for disposal at the end of the incident: the removed liquid and the materials used to contain it along with any other contaminated material such as soil or protective clothing. All of these may now be hazardous waste and need to be treated as such as described in Chapter 6.

Emissions to air

The opportunities to intercept and deal with emissions to air are limited. The only actions can be taken at the source: the processes or activities that may be responsible or any treatment plant to deal with emissions as described in Chapter 4. The emergency plan should have identified the potential sources and specify what to do. It should also specify what action to take to protect those affected or at risk, taking account of weather conditions, especially wind direction. A nuisance odour will be more of a problem on a sunny day than if it is raining. If it is toxic in any way neighbours may need to be advised to stay indoors with doors and windows shut or else evacuated. At this stage it is beyond the control of the organisation acting on its own and will involve external agencies such as the police, environmental agency and local authority. The plan should have recognised this and these agencies should also be prepared with their own plans. Their plans will be generic to cater for a wider range of events and sites.

Environmental hazards associated with fire.

Fire is a risk in itself as property and materials are destroyed and lives are threatened. Dealing with fires is the responsibility of the fire service and they should be advised as soon as any fire or risk of fire has been identified. The response usually involves spraying large volumes of water unless special equipment such as CO_2 suppression is used. The water will spread onto the ground and will contain any contaminants present from the property, the fire combustion products or picked up on the way from the fire to its destination. It could soak into the ground or drain to a watercourse or get into the foul sewer and disrupt a sewage treatment works with the consequences outlined in Chapter 5. The water could well be toxic and so means of dealing with the volume need to be considered. It may be acceptable for it to go to foul sewer with the permission of the sewage treatment works operator. More likely it will not be acceptable or that route is not available anyway. In that case it needs to be diverted to a suitable location and captured. In the confusion of a fire this will not be easy to organise on the go: it needs preplanning based on the areas at risk, the drainage characteristics and suitable sites for containment. Large volumes will need a lagoon of sufficient capacity with impermeable base and sides. Sometimes it may be possible for the fire service to reuse stored water but this will need to be designed into the storage facility. At a push a car parking area could be designed with ramps and kerbs to act as a containment area. It may also be necessary to tanker contaminated water away during or after the event and arrangements need to be in place to deal with that. Fire also produces smoke and gaseous emissions that may be toxic, especially if plastics or organic chemicals are involved. These need to be responded to in the same way as other emissions to air referred to above.

Close liaison with the environmental agency and fire emergency services should take place at the planning stage and be reviewed at subsequent liaison meetings.

Road traffic accidents

If the company delivers products by road (or by rail, ship or air) then there is always the possibility of an incident involving the mode of transport. If the vehicle is directly owned and operated then the company is certainly involved; if it is through a contractor, then the involvement is less but not totally absent. Emergency procedures need to take account of this. The accident will be dealt with by the emergency services but they will get in contact to check information about the contents and how to clean up or remove them from the vehicle if necessary. The extent to which this is important will depend on the nature and properties of the contents. The company emergency control centre or someone in a position of responsibility will need to liaise over these issues and may have to organise recovery of the materials and the vehicle.

Clean-up and disposal

There have been several references to clean-up operations. This may be material on site or in the surrounding environment including from neighbouring premises. Clearing the mess may involve direct employees, contractors, the environmental agencies or a mixture of all three. It will include such items as escaped liquids or solids, contaminated soil and debris from the incident or a fire. These materials will have to be sent for disposal along with other materials used in the response and clean-up operations such as sand, absorbents and contaminated protective clothing. All of this is likely to be treated as hazardous waste. It is the responsibility of the site owner that the disposal of waste is handled properly as described in Chapter 6 and the normal waste disposal contractor may not be suitable.

Handling the press and media

Only a local reporter for press or radio may learn of and turn up at a small incident. Large incidents can attract all of the media, including television, from across the world. The organisation's reputation is at stake on three grounds:

- The cause of the incident.
- The response to the incident.
- How these come across to the public through the media.

The media are in search of a story and they will see it as their role to uncover any signs of mismanagement or incompetence. An incident managed well could just enhance the organisation's reputation if it was caused by an external factor outside of their control but to which the response was timely and effective. However well managed though, if an interviewee comes across as evasive or not in control, this can cause just as much harm. For this reason it is not wise to let just anyone deal with the press but to use only trained people or those who have already demonstrated competence. Large organisations employ press officers to deal with the media or at least arrange the interviews with other employees. Such luxuries are not available to small companies. Training is available to help the novice and the emergency plan should limit media contact to those trained or at least those with the authority and competence to speak on behalf of the organisation. The spokesperson needs to be kept informed with what is happening and what is proposed by the emergency control centre and to be able to respond to concerns raised by the public or the media however outlandish they may seem. If the information is not to hand they should say so and find it out. The plan may need to take account of the interviewees being away from the site; they could be at another office, possibly in a newsroom or at another public space where the impact has occurred. Statements made in public need to be truthful and any promises made capable of being followed through.

Health and safety

It is beyond the scope of this book to go into the details of the health and safety requirements of dealing with incidents. There are plenty of texts available to help including the companion volume to this series (Hughes and Ferrett 2011). The organisation should have a Health and Safety Manager or Adviser who should be involved in the planning. The points to bear in mind though are that, in the emergencies that are the subject of this chapter, substances that are a risk to the environment are, by definition, a risk to people and the closer they are to the incident the higher the risk. This means that in planning for emergencies the health and safety implications have to be taken into account. This includes selection and training of the staff involved, how the response is organised and managed, the provision of protection equipment and the risks to others on the site or neighbours.

Included in this is the need to keep people away and the possibility of evacuation of buildings or parts of or the entire site. Suitable mustering points need to be available out of harm's way and directions clearly signed. Responsibilities need to be assigned for raising the alarm and organising the evacuation and checking that everyone is accounted for. This ought to be part of other plans to deal with emergency situations that have no environmental implications. It does not help to have conflicting plans.

Welfare

The planning needs to take account of the welfare of those involved in responding to an incident. In many cases this may not be a big issue but people get tired, thirsty and hungry, especially if they are under stress and the action is hectic and prolonged. The attention of the controller and the rest of the team should be on the event, not on where to find a coffee or a sandwich. People also need breaks and, if an event goes on for many hours or days, they will need to be covered by alternative staff. This all needs to be included as part of the plan and such information as where to get food after midnight may be helpful.

Maintenance of the plan

Reference has been made several times to ensuring that the plan is kept up to date. Sites and people change, temporary works may run for weeks or months, old processes change and new processes are introduced, contractors go out of business, the inventory of materials needs to reflect new supplies or products. The potential list is long and many of these details will be stable but procedures need to be in place to monitor and report the changes and make any alterations to the plan. Bear in mind that some events such as the contractor going out of business may escape this process – have more than one.

The planning documents are important but they are useless if everything else is not in place. Someone needs to be responsible for ensuring that all of the assumptions in the plan about availability of people, equipment and facilities are justified. This requires regular inspection and testing of the control room and its facilities, the stocks of materials to deal with an incident and the fire protection arrangements. The various elements should not be being used frequently, indeed it is important to ensure that they are not used for routine tasks (radios and computers may get taken to cover for some logistical problem but that is not an emergency).

The inspections need to be based on a checklist of what should be available and it must indicate what additional checks may be required: is it clean and working, do motors or engines start, are batteries charged, is the PPE available and serviceable, are the spill kits complete, is anything past its use by date, has someone blocked access to equipment, etc. Some of this may be conducted by supervisors or managers as part of their routine but it could be part of the routine of another person who is also doing the detailed testing. These inspections need to be scheduled to ensure that they happen and the findings recorded even on something as simple as a tick list.

Training and exercises

Training is required in support of all of the aspects of planning. It needs to start with the introduction of an emergency plan and regular refreshers will be needed to ensure that the employees retain their competence and are brought up to date with any changes. A programme of induction for new employees also needs to be in place. Different programmes will be required for various groups such as those conducting the emergency planning, those manning the control room, those dealing with problems on the ground, those dealing with the media and the remainder who may not be directly involved in the response but still have to take action such as reporting unusual events or evacuation of a building. The training needs to explain the site and the risks, the alarms or other warning arrangements that they need to be aware of, how the organisation has prepared to respond and the roles of the individuals, even if it is just to keep out of the way.

Training needs to be complemented by regular exercises. These should involve the external bodies and neighbours as far as practicable and be made as realistic as possible by using smoke bombs and any other means to simulate potential events. Exercises may need to simulate different types of emergency such as fire, a leak of liquid or an emission of toxic gas and could just focus on one part of the plan such as evacuation of a building or sealing the drainage system. The purpose is to test the systems such as alarms and the responses such as evacuation times, the deployment of people and equipment and communications. Simulations should be recorded and actions logged so that subsequent analysis can identify where improvements are needed. Planning a good exercise does require some time and effort but it does provide an opportunity to check that all the details are correct and up to date. A surprise exercise may find that some of the simple things like the telephone numbers of contacts have changed or the computer software has not been updated and e-mail does not work.

The training records should contain details of training events and exercises and the participants. They may be consulted by enforcement agencies in the event of a real emergency.

Follow-up to an event

As soon as possible any equipment used needs to be cleaned and checked and materials consumed need to be replaced. Another incident could occur at any time and if there is a fault somewhere on the site that caused the first one that has not been found or fully sorted out, the risks are high.

Debrief

The ending of the response and closure of the control centre should not be the end of the event. An investigation needs to follow to establish what went wrong, how the response coped and if there are any lessons to be learned (the fact that an incident has happened means that there must be). This will require interviewing staff, assessing the information in the logs, reviewing analytical information, receiving reports from outside agencies involved, testing equipment ... the list could be long. Outside agencies such as the environmental regulator may be conducting their own investigation and the results of your investigation may be needed in court. The debrief needs to be as soon as possible after the incident whist memories are fresh and it should be well documented. There may be disciplinary issues or issues with suppliers or contractors to follow up as well. The debrief findings are fed back into future reviews of the plan but may also mean that normal work procedures need to change or training is required or additional protection measures need to be implemented.

Audit

Regular audits are part of an EMS and the audit process is described in Chapter 2. The audit of the emergency plan itself and the supporting facilities and services is just as valid as for other parts of the EMS. The way it is conducted and reported will be similar but it is important that audits are carried out with a frequency and depth that reflect the risks and any recent events. The results of audits will feed into the regular reviews of the plan.

Reviews of emergency plans

The plans need regular review to ensure that they are kept up to date. These may be planned on a regular basis such as annually but a review should also be triggered by some events. An actual incident or inadequacies uncovered during an exercise should cause a review as a matter of course. Other reasons could include an incident in another organisation from which lessons could be applied, new processes or materials on site, a change in legal requirements, organisational changes, new key personnel or the availability of new equipment. The external bodies may need to be involved in the review process. The aim is to ensure that the plans remain fit for purpose and are subject to continuous improvement.

Reviews need to be documented and, along with any revised plan, subject to the document control procedures and issued to all those with an interest.

CHAPTER

10

The examination for the NEBOSH certificate in environmental management

After this chapter you should be able to:

1. Understand how the NEBOSH certificate examination works
2. Decide if you wish to enter the examination
3. Prepare yourself by studying the sample questions and looking at other guidance.

Chapter Contents

Introduction **168**
10.1 The written examination **168**
10.2 The practical application **168**
10.3 Course providers **169**
10.4 Sample questions **170**

INTRODUCTION

The general introduction to this book explained that it is a guide to the NEBOSH Certificate in Environmental Management (CEM) and it is structured around the syllabus. The syllabus is available through the NEBOSH web site www.nebosh.org.uk. It contains general background information about the qualification, how the standard fits into level 3 of the National Qualifications Framework for England, Wales and Northern Ireland, assessment and criteria for the qualification, access to examiners' reports, the detailed syllabus and information about the practical application. This chapter will only concern itself with the last two items. There is also an example examination paper in the syllabus.

10.1 The written examination

The examination consists of 11 questions and all are compulsory. The first scores 20 points and the rest score 8 each: the maximum total mark is 100. The first question is longer and candidates would be expected to spend more time on the answer. The exam paper recommends about half an hour on question 1 of the two hours available in total. Candidates should expect to write about two sides of A4 paper for question 1 and between a half and a full side for the others. If you are writing a lot more than that you are probably getting carried away and risk running out of time for the other questions. If it is a lot less then you may have missed something in the question.

Each question may have more than one part to it. The question shows how many marks are awarded for each part as an indication of where the effort is to be expected.

10.1.1 Command words

Questions use standard language to guide candidates in their answers using command or action words. These are to show candidates the depth of detail required in an answer. The five command words that you may come across and their meaning are as follows.

Give the meaning of

This question is seeking a simple definition, such as, 'give the meaning of sustainability'. The best way to answer this is to use a standard definition (such as the Brundtland one in this case) if such exists. Other questions may require you to make up your own definition, the key being to capture all of the elements of the word or phrase in a few sentences.

Identify

This is asking for a list. For example, 'identify two greenhouse gases' should elicit the answer: 'carbon dioxide and methane' or any other valid combination. The answer to some questions may be more than a three words: such an example would only be part of a whole question. How would you answer 'identify the control measures available to reduce the concentration of solids in an industrial emission to atmosphere'? The answer would contain a list of the various types of filters, gravity separators, electrostatic separator and water-based methods.

Outline

This is seeking more information; an outline is not a list. It requires more information enlarging on the topic of the question. For example, 'outline two control measures available to reduce the concentration of solids in an industrial emission to atmosphere' requires the candidate to identify each one and then explain how they work. This is not to elicit a full technical description but to set down the main features, again in a few sentences (typically half a page). An answer to this question might identify the bag filter and give information about the principles on which it works (particle size and pore size), how pressure builds up and how the bag can be cleaned. The second example chosen could be the electrostatic precipitator with the principles of using static electricity to attract the particles to a plate and how the plate is eventually cleaned.

Explain

This is requiring more detail again and would be expected to give more information about how a concept works or some reasons or justification for the information provided in the question. For example, 'explain the concept of source, pathway and receptor when assessing environmental risk'. Here the candidate would be expected to show that they understood how a pollutant at the source is linked to the receptor by the pathway; that the concept is used to clarify how the risk to the receptor may arise; and how it could be managed by intervention at the source or the pathway.

Describe

This is similar to 'explain' in that it requires more detail but the subject matter of the question will be different. For example, an answer to the question 'describe two outlets available for waste' would be expected to identify two of them such as landfill and incineration and put into words how they work and operate. So the description of an incinerator would include the types of waste for which it is used and the principles on which it works with a little about the limitations.

These are the main command words that you will be asked for in this examination. You will not be asked to draw anything although you can provide a drawing if it helps you with the answer. It is important to read the command words and shape your answers accordingly. Candidate fail to gain marks for an answer that is too short for the command word.

Access to past exam papers and examiners' reports

Past exam papers give the complete questions set at one sitting. Following an examination NEBOSH publish an examiner's report which comments on how each question was answered with comments on misunderstandings or common mistakes. The reports do not give the answers.

These can be downloaded from NEBOSH via their web site shop.

10.1.2 Examination technique

Everyone has their own methods of answering examinations. Some just start at the beginning and plough on to the end. Some pick the one that they are most comfortable with to start and then answer them in the order of their perceived difficulty. Others alternate between those that they find easy and those that they find difficult. You can answer them in any order. It always pays to take a few minutes to read all of the questions at the start and make up your mind which ones you want to tackle first. If one looks difficult, perhaps because you do not understand it or you do not know the answer, leave it and come back towards the end. Do not spend ages trying to formulate an answer leaving inadequate time to do justice to the rest.

Do read the question carefully. Bearing in mind the time allowed and the expected length, you will not be asked to write down everything you know about a topic. Questions are focused on just a part of the syllabus and there are no marks awarded for content that is not relevant to the question. Many candidates find that it helps to jot down a few words at the start to help them formulate the answer and get the order right. You can write that on the answer paper and just put a line through if you want to. Before you launch into the answer, recheck that you are answering the question set, not one that you would have liked!

Keep an eye on the clock. It is better to answer all of the questions if you can rather than leave any out: the first few marks are usually the easiest for candidates to find. Leave a few minutes in hand to read over your answers again to see if you have missed anything out that has suddenly come into your mind.

Revision is important. Your learning may have been spread over several weeks or months and refreshing the memory will help recall in an examination setting. Preparing your own notes as you go is one good way of helping the revision process. There is also a published Revision Guide available (Blackhouse 2018).

Sample questions

Some sample questions from past papers are given at the end of this chapter. Two have been selected from each element of the syllabus which is represented by a chapter in this book. The format is as set with bold type being used to draw attention to specific points expected in the answers. Note that some questions may cover more than one element. All of them are short (8 mark) questions.

10.2 The practical application

The aim of the practical application is to demonstrate that you can put the knowledge learned to use in a work environment by conducting an environmental review. It is expected to be carried out in the candidate's work place unless prior agreement has been reached with NEBOSH beforehand. It can be done at any time but must be handed in no more than 14 days after the written examination. It requires the candidate to complete a proforma questionnaire about the work place or part of the work place and write the report in about three hours. The proforma covers most of the elements of the syllabus so you are not expected to take on a major complete site. In this case a unit of operations would be more appropriate. It recognises that not everything will apply to sites.

The assignment is completed with a written report containing an introduction to the area and the activities reviewed, an executive summary, the main findings, the conclusions and the recommendations. It will of necessity be brief and could not constitute a full formal assessment for any other purpose.

The report is marked by an assessor who will be looking to see that they can identify:

- The nature and location of the environmental issue;
- The degree of risk associated with the environmental issue;
- Preventative and protective environmental measures already in place;
- The remedial actions, where appropriate, with relevant prioritisation.

10.3 Course providers

The training for this examination (both parts) can be bought from a range of providers. Some offer part-time courses, some full-time and others online. There are providers across the UK and some offer courses in other countries. The list of approved providers can be found on the NEBOSH web site.

10.4 Sample questions

These are drawn from past papers with the kind permission of NEBOSH.

Chapter 1

1. **Outline** the benefits to business of good environmental management. (8)

2. **Outline** why deforestation is seen as an environmental concern. (8)

Chapter 2

3. A company is implementing a formal environmental management system.
   Identify:
   (a) the possible *internal*; (4)
   (b) the possible *external* (4)
   sources of information which should be reviewed during the initial review.

4. **Describe** the main features of an ISO 14001:2015 environmental management system. (8)

Chapter 3

5. **Explain** what is meant by the term *'pollution pathway'*. (6)
   Give TWO examples of typical pollution pathways. (2)

6. **(a)** **Give** the meaning of the following terms:
   (i) environmental aspect; (2)
   (ii) environmental impact. (2)
   (b) **Identify FOUR** examples of environmental *impacts*. (4)

Chapter 4

7. **Outline** the control hierarchy for air pollution control **AND give TWO** suitable examples for **EACH.** (8)

8. **(a)** **Identify FOUR** atmospheric pollutants that arise from the combustion of coal. (4)
   (b) **Outline TWO** methods that may be used to reduce emissions of air pollution from a coal-fired power station. (4)

Chapter 5

9. During a routine inspection of a manufacturing site an external storage tank containing fuel oil was found not to have a bund,
   Outline the features that should be included in the design of a suitable bund for the tank. (8)

10. **Outline** the reasons for practicing water conservation. (8)

Chapter 6

11. **Outline** the main requirements applying to the storage and consignment for disposal of hazardous waste. (8)

12. **Outline** the potential environmental effects that may arise from contaminated land. (8)

Chapter 7

13. **(a)** **Identify FOUR** options for the generation of energy from sources that do not rely on fossil fuels. (4)
    (b) **Outline ONE** possible limitation associated with each option. (4)

14. **(a)** **Outline** the benefits associated with hydropower. (4)
    (b) **Outline** the limitations associated with hydropower. (4)

Chapter 8

15. A large dust extraction fan and collector unit is to be installed against the outside wall of a factory building.
    (a) **Identify TWO** possible sources of noise from this equipment. (2)
    (b) **Outline** the issues to be considered so that the equipment does not constitute a noise nuisance when in operation. (6)

16. **(a)** **Outline** the potential sources of noise that would be associated with the construction of a new airport. (6)
    (b) **Identify** the potential environmental effects of the noise. (2)

Chapter 9

17. A company has an emergency plan to deal with a range of emergencies including fire, explosion, chemical spillage and flooding.
    Outline the practical measures which should be taken to ensure these plans will work effectively when required. (8)

18. **Outline** the key actions that need to be taken to protect the environment in the event of a serious fire breaking out on industrial premises. (8)

Appendix 1

Units of measurement used in environmental management

The principal units used are based on the International System (SI) of units used in science. It does not follow it exactly though due to convention. For example, the SI unit of time is the second but for practical reasons we have used hours, days and years.

The units used in this text are:

Parameter	Unit	Symbol
Length	metre	m
Volume	litre	l
	cubic metres	m^3
Mass	gram	g

The units of metre, litre and gram are familiar from everyday use in most countries of the world. To express larger and smaller volumes a prefix is used. For larger measures:

kilo	1000 (10^3)	k
mega	1,000,000 (10^6)	M
giga	1,000,000,000 (10^9)	G
terra	10^{12}	T

Giga and terra prefixes are mainly found in energy measures. One gigawatt (GW) is one thousand million Watts (a billion Watts in the US).

For smaller measures:

milli	one thousandth (10^{-3})	m
micro	one millionth (10^{-6})	μ
nano	one billionth (10^{-9})	n

Thus one megalitre (ML) is a million litres. One kilogram (kg) is a thousand grams. One microgram (μg) is a millionth of a gram.

Large volumes of air and water have been measured in two ways by convention. A litre is the familiar unit and 1ML, at one million litres, is quite large. However, for very large volumes m^3 (metre cubed) may be used. One m^3 is the same volume as 1,000 litres. This unit is used in measuring the volumes of water reservoirs and the concentration of substances in air as described in Chapter 4.

Parts per million (ppm) is also used frequently. It has to be used for recording in the same units, e.g. ppm by weight means one gram in a million grams. Ppm by volume is one litre in a million litres. Historically ppm was used for concentrations in water e.g. one gram in a million grams of water. Because one litre of water weighs 1,000 grams, a concentration in ppm is numerically the same as mg/l: 5ppm is the same concentration in water as 5mg/l. This does not work in air as one litre of air weighs a lot less than one kg.

The range of prefixes is much larger than shown here; they can be found easily in reference books or on the internet. Conversion factors for imperial and US measures can be found in the same way.

Appendix 2

Background briefing to some of the scientific terminology used in the text

INTRODUCTION

There are several terms used in the text that come from chemistry, biology and other sciences. This appendix is aimed at helping those readers who are unfamiliar with them or are looking to refresh their memories. It is not a textbook on these subjects, rather a series of definitions with somewhat simplified explanations. There are many good textbooks available to get more detail if you are interested; anything aimed at general science at secondary school level ought to suffice.

Elements and compounds

An **element** is the purest form of a substance; it cannot be subdivided and still retain the same properties. There are 98 naturally occurring elements on the earth of which common examples are hydrogen, carbon, iron, cadmium and uranium. The smallest part of an element is the **atom**; if it is subdivided it is broken down into the building blocks of neutrons, protons and electrons. All elements are made up from different combinations of these subatomic particles as outlined in Box 7.2 on nuclear fission.

Elements combine together chemically to form **molecules.** A molecule is the smallest part of a **compound.** Examples of compounds are water, calcium carbonate (chalk) and sulphur dioxide.

Symbols for elements and compounds

Each atom has a symbol. The symbols are used as shorthand for the names of elements and compounds. Some examples of symbols for elements in the text are as follows:

Hydrogen	H
Carbon	C
Oxygen	O
Nitrogen	N
Sulphur	S
Uranium	U

The atoms combine together to form molecules of the same compound in fixed ratios. The same atoms can combine to form more than one compound in different ratios. Some examples of molecules to illustrate this are:

Water	H_2O
Ammonia	NH_3
Carbon dioxide	CO_2
Carbon monoxide	CO
Methane	CH_4
Nitrogen dioxide	NO_2
Sulphur dioxide	SO_2
Sulphur trioxide	SO_3

The small numbers show how many atoms of each element are present in the molecules that make up the compound. If there is no number shown it is one by default. A water molecule is a combination of two hydrogen atoms and one oxygen atom; a methane molecule is one carbon and four hydrogen atoms.

Most elements exist in nature as single atoms: C, S, Fe, U. The common gases – hydrogen, oxygen, nitrogen – exist in nature as molecules of two atoms: H_2, O_2 and N_2 for example. Oxygen is unusual in that in can exist as two forms: the oxygen (O_2) that we breathe and an unstable form ozone (O_3) which is harmful if breathed in but protects the earth from harmful UV rays from the sun.

All substances in the universe are made up from molecules of increasing complexity. Nitric acid is HNO_3; one hydrogen, one nitrogen and three oxygen atoms. Sucrose (sugar) is $C_{12}H_{22}O_{11}$. Polymers such as polyethylene are long chains of molecules made up from thousands of atoms.

Halogens

The term **halogen** is used to describe any of the elements fluorine, chlorine, bromine or iodine. They show similar chemical properties.

Organic and inorganic compounds

Organic compounds are the building blocks for all animal and plant cells: proteins, fats, carbohydrates are all organic compounds. The term organic is applied to compounds which have mainly carbon atoms in their structure. Organic compounds may also have hydrogen and oxygen present along with small amounts of nitrogen, phosphorus and sulphur in some examples. Inorganic refers to the rest of the compounds such as iron oxides or water.

Volatile organic compounds (VOCs) are compounds containing carbon which are volatile at normal temperatures and pressures. Methane and chlorofluorohydrocarbons (CFCs) are examples.

Solutions in water

When soluble inorganic compounds dissolve in water the atoms that they are made from dissociate into **ions**. As an example, common salt is sodium chloride NaCl (one atom each of sodium Na and chlorine Cl). In water it splits into Na^+ and Cl^-, positive and negative ions due to the exchange of an electron between the two atoms. Acids also dissociate in the same way: hydrochloric acid HCl dissociates into H^+ and Cl^- ions. Sulphur trioxide reacts with water to form sulphuric acid H_2SO_4. It dissociates into two hydrogen ions, each H^+, and one sulphate ion SO_4^{2-} (the charges have to balance out).

Pure water also dissociates itself into to one hydrogen ion and one hydroxide ion – H_2O forms one H^+ and one OH^- ions. The numbers of each are the same.

Acids, alkalis and pH

The hydrogen ions in solutions of **acids** are responsible for the acidity. They attack and dissolve stonework or kill plant tissue in the fallout from acid rain. The acid fumes in air where sulphur dioxide and trioxide are dissolving in water vapour or the water around your eyes and in your lungs causes irritation and damage to those cells. It does not matter which acid is responsible (from sulphur oxides forming sulphuric acid or nitrogen oxides forming nitric acid). The strength of a solution of sulphuric acid (illustrated by its corrosiveness) depends on how many hydrogen ions are present and this is determined by how much sulphur oxide dissolved in the water. Common everyday acids are acetic acid (vinegar) and citric acid (in lemon juice). They will dissolve chalk in the same way that acids dissolve some building stones.

In contrast sodium hydroxide (NaOH) dissociates in water to form a sodium ion Na^+ and a hydroxide ion OH^-. Compounds that dissociate to form hydroxide ions are known as **alkalis** (or bases). Solutions with hydroxide ions are **alkaline**.

The strength of an acid solution is measured by its **pH** on a scale that runs from 0 to 14. A pH value of 7 is neutral when the concentrations of hydrogen ions and hydroxide ions are equal (as found in pure water) with lower numbers meaning the solution is more acidic and higher numbers more alkaline. Most natural water bodies have a pH of between 7 and 8, i.e. slightly alkaline. The scale is logarithmic which means that a change of one unit in pH value is a tenfold change in the hydrogen ion concentration. An acid solution with a pH of 3 is one hundred (10×10) times more acidic than one with a pH of 5: one with a pH of 11 is a thousand ($10 \times 10 \times 10$) times more alkaline than a solution with a pH of 8. The mathematics behind the pH scale is a bit complicated if you are not familiar with logarithms but the main thing is to recognise the scale of 1 to 14. Below pH 7 hydrogen ions predominate and above pH 7 (i.e. alkaline) hydroxide ions predominate.

Acids and alkalis react together and if the concentrations of each are the same they will neutralise each other to end up with a solution of pH 7. This is the principle behind neutralisation of acids. Sodium or calcium carbonate can be used to neutralise acids following a spill. Sulphur trioxide formed by burning coal containing sulphur reacts with moisture in the flue gases and the atmosphere to form sulphuric acid. Calcium hydroxide (lime) is used to neutralise and remove the sulphur oxides in a flue gas desulphurisation plant. Note that heat can also be produced in these reactions as is common with many chemical reactions. The amount of heat depends on the strength of the acids and alkalis; even water added to concentrated sulphuric acid will react violently and is dangerous.

The term **salt** is used to describe a compound that is formed from the reaction between an acid and an alkali. Hydrochloric acid and sodium hydroxide react to form sodium chloride (also known as common salt or table salt). Sodium hydroxide reacts with sulphuric acid to form the salt sodium sulphate.

Oxidation

Oxidation refers to the chemical process by which oxygen combines with other elements or compounds to form new compounds. Plant life, bacteria and animals, including humans, get their energy by oxidising organic substances in their food during a process called respiration. In doing so the carbon in the organic molecules is oxidised to carbon dioxide and the other main product is water from the oxidation of the hydrogen atoms in the compounds. This is a biochemical process that occurs in cells; energy is released to enable the cells to function.

Inorganic elements and compounds can also be oxidised in other ways. Iron is oxidised to rust in a damp atmosphere and carbon is oxidised to carbon dioxide when it burns. Coal is virtually pure carbon but any sulphur present is also oxidised to sulphur dioxide and sulphur trioxide. Other organic compounds such as methane are oxidised to carbon dioxide and water (from the carbon and hydrogen atoms in the methane molecule). Oxidation of the carbon in fossil fuels is the source of the increasing carbon dioxide in the atmosphere.

Photosynthesis

Plants have another exclusive feature. In the presence of sunlight they are able to absorb carbon dioxide from the atmosphere and use it to form organic compounds to build their cell structure. In the process they release oxygen. This is the process known as **photosynthesis**. Animals get their energy by eating plants or other animals that have fed on plants and oxidising the carbon content: all life is ultimately dependent on photosynthesis for survival. Photosynthesis provides food and reduces some of the carbon dioxide in the atmosphere to release oxygen to replenish that which was used by all living matter as they respire. Unfortunately, plants cannot cope with the increasing amounts of carbon dioxide being released from burning fossil fuels. Deforestation and loss of other plants by desertification or other means reduces the capacity for the reduction of carbon dioxide through photosynthesis. The cycle of oxygen and carbon dioxide through photosynthesis and respiration is part of what is known as the carbon cycle.

Aerobic and anaerobic processes

The biological process of oxidation is known as an **aerobic** process meaning that is in the presence of air. The term is most often applied to bacterial processes whereby they oxidise organic matter as their source of food to obtain their energy. This is the basis for the decay of dead cells in the presence of oxygen in the atmosphere. It also is the means used in sewage treatment works to oxidise organic matter in a biological filter or activated sludge plant. The use of **BOD** as a measure of biochemical oxygen demand (described in Chapter 5) is simulating what happens naturally in rivers where bacteria oxidise the organic matter present. The bacteria consume the oxygen from the water and reduce its concentration.

Some bacteria can use organic substances as a source of food and energy in the absence of air or oxygen. This is known as an **anaerobic** process. The bacteria get their oxygen from other molecules containing it and the waste products produced include carbon dioxide and methane. Anaerobic processes occur in waste organic matter after all of the available oxygen from the air has been used up; this is what happens in a landfill site and in an anaerobic digester.

Appendix 3

Some examples of environmental management in practice

The first edition of this book included some specific examples taken from several sources to illustrate the different opportunities to reduce environmental impact and save money. They included large multinational companies tackling their water and energy use and waste production in their operations, through the building design and construction of retail stores to smaller companies looking at their water use in the washrooms. A lot has been published since then and a web search on any well-known company should deliver examples of their environmental or sustainability reports. These should include objectives and targets as well as current performance against these. The first four of the earlier reports have been updated below and the sources quoted are from the companies' own web sites (other previous examples have disappeared from public view). Others are linked to sites which contain other case studies from different business sectors which may be of interest.

Pepsico

Pepsico is an international company producing drinks and food products such as potato snacks. In the first edition of this book it was reported that the Company had gained the Stockholm Industry Water Award for sustainable water management. The water efficiency achievements were a 20 per cent efficiency per unit of production, conserving nearly 16 billion litres of water in 2011, from a 2006 baseline, through the application of water-saving equipment and technologies, creative recycling and reuse, and by deploying a water management system throughout its manufacturing facilities. Since then the company reports that they have not sent any manufacturing site waste to landfill for more than six years and that the anaerobic digestion plant in Leicester UK generates 40 per cent of the site electricity. Potato waste is offered to local farmers for crop fertilisation. The longer term targets include achieving 100 per cent recyclable, compostable or biodegradable packaging by 2015. Part of this will be by reducing the amount of packaging used but also by using different packaging materials, including those made from biological material.

Source: www.pepsico.co.uk/what-we-believe/Planet

Coca-Cola Enterprises Ltd and recycled PET

Coca-Cola Enterprises Ltd is part of Coca-Cola Enterprises Inc., independent bottlers of Coca-Cola drinks. Plastic drinks bottles are made from PET (polyethylene terephthalate) which is traditionally manufactured from fossil fuels. The company introduced bottles made from a combination of up to 25 per cent recycled plastic and of up to 22.5 per cent plant-based material as reported in the first edition of this book. The target in 2018 is to move to 40 per cent recyclable PET in new bottles.

Source: www.cokecce.co.uk/sustainability

Marks and Spencer plc Plan A

Marks and Spencer plc (M&S) is a UK multiple retailer with stores across the country. The company has a plan (Plan A) with the objective of making the company the world's most sustainable major retailer. This covers more than just the environment. It has focused on a wide range of targets including reducing food waste and energy use and reducing and recycling packaging. Much wider coverage can be found at https://corporate.marksandspencer.com/annual-report-2018/mands_plana_planet.pdf

Fujitsu

Fujitsu is a Japanese technology company with a worldwide market in IT products and services. It has a focus on reducing greenhouse gas emissions, water consumption and reducing releases of chemical pollutants. The most recent progress report for 2018 can be found at www.fujitsu.com/global/documents/about/resources/reports/sustainabilityreport/2018-report/fujitsureport2018-040201-e.pdf

Energy saving in business

The Carbon Trust is a UK based advisory body helping businesses to reduce their carbon emissions through implementing energy efficiency and using alternative sources of energy. It has produced a series of publications for a range of business sectors. These include agriculture; ceramics, glass and cement; chemicals; construction; engineering; food and drink; health care; hospitality; manufacturing, metals and metal products; mining and quarrying; plastics and rubber; retail and distribution; sport and leisure as well as types of educational establishments and central government departments. There is too much information to repeat here but you may wish to look at any topics that you are interested in.

Source: www.carbontrust.com/resources/reports/advice/sector-specific-publications.

Food and drink

The Food and Drink Federation provides guidance to that sector on environmental management and sustainability including on energy, packaging and sustainable resourcing of palm oil. Some of this is for members only but much is open access. More can be found at www.fdf.org.uk/toolkit_environment.aspx.

Automotive industry

The UK-based Society of Motor Manufacturers and Traders (SMMT) produced a report in 2018 on 20 years of Sustainable Progress. It claims that carbon dioxide emissions from new cars are down by over 50%, relative waste to landfill down by over 95 per cent and this largely due to end of life vehicle reuse and recovery which increased to 95%. The average age of cars scrapped has increased as well. The report includes data on reductions in nitrogen oxide emissions from vehicles and the performance of the sector on energy, VOCs, etc. in the manufacture of vehicles and the supply chain.

Source: www.smmt.co.uk/industry-topics/sustainability/.

Waste and Resources Action Programme (WRAP)

WRAP has been mentioned in the main text as a source of information on resource efficiency based on the three Rs: reduce, reuse and recycle. Its latest plan is closely allied to the Circular Economy promoting how design, production, selling and consumption allied to the three Rs can accelerate the move to a sustainable economy. WRAP has concentrated on food and drink, clothing and textiles and electricals and electronics for further work. These collectively contribute significantly to the UK's carbon footprint, water footprint and household waste production referred to in the main text. More detail can be found at www.wrap.org.uk/about-us/our-plan.

The Ellen Macarthur Foundation

The Circular Economy has also been referred to in the main text and the concept has been promoted by the Ellen Macarthur Foundation. Its aims are to: design out waste and pollution; keep products and materials in use; and regenerate natural systems. It is supported by a range of global partners and offers various services aimed at encouraging the adoption of its aims. These include a learning hub, education and training, conferences and a wide range of case studies across various sectors and across the world. See www.ellenmacarthurfoundation.org/circular-economy/concept.

Edie

A web site that is worth signing up to (free) in order to see what other businesses are currently doing for environmental management is Edie at www.edie.net/.

Appendix 4

Further details on chapter references to UK law

Chapter 1

Box 1.2 Environmental law in England

The law is quite complex in all its aspects and environmental law is no exception. This brief summary is intended to give only an outline and does not cover all the details or nuances of the legislation. It does not constitute legal advice! If you find yourself involved in a legal dispute or threatened with prosecution then it is sensible to seek professional advice at the earliest opportunity.

The UK has a long history of developing environmental legislation going back many centuries. There are two primary sources of law: common law and statute law and the legal processes by which cases are determined are known as civil law and criminal law respectively.

Common law

Common law (also known as case law) is based on precedents that have been set by judges deciding cases that have been brought to settle previous disputes. New cases are determined by civil law whereby one person, the claimant, has sued another, the defendant, for doing some wrong or sought protection for some infringement that is affecting their life or well-being. The court (usually a judge on their own) makes a decision based on precedents or may, if referred to a higher court, set a new precedent. Decisions may be appealed to a higher court but, once finally decided, the case becomes a reference for future similar events. Civil cases involving small amounts up to £5,000 may be decided in the Small Claims Court; between £5,000 and £50,000 the case goes to a County Court. Above £50,000 they go to the High Court. Appeals go initially to the next court up and then to the Court of Appeal and, ultimately the Supreme Court in the UK and finally to the European Court of Justice within the EU.

These cases are not set down in statutes but only in the records of the courts. The main purpose of civil law is to settle disputes. These can cover many areas such as interpretation of contracts or settlement of debts but in environmental matters they are usually to rule on 'torts' – a word used to describe an act which affects another in a way that impinges on their rights. The commonest types are nuisance and negligence (known as 'derelict' in Scotland). Nuisance means interference with someone's enjoyment of their land and could cover such things as loud or excessive noise, odour or dust particles. To be successful in court the nuisance has to be sufficiently significant to merit a case and the actions have to be considered unreasonable. A small amount of smoke for a few minutes at night would not qualify but thick black smoke for several hours a day when many people would be in their gardens or forced to close their windows would. It does not have to be the result of a deliberate act and the court would consider the practicality of avoiding the problem in reaching a decision.

Where an individual is suing for nuisance this would be a private case. If many people are affected then the case becomes a public nuisance which is also a crime that can result in prosecution by, for example, a local authority. Excessive dust depositing on property and causing damage affecting many households and also public areas would be an example.

Negligence goes beyond nuisance in that it requires a breach of duty of care. Duty of care means that an individual or a business has a duty not to cause damage or injury to a neighbour and negligence is the result of acts or omissions that could reasonably be foreseen to be likely to cause harm. An example would be failing to take reasonable precautions to secure a reservoir outlet valve against vandalism which, on being opened, caused flooding onto a neighbour's land. In order to succeed, a case has to show that a duty of care was owed, that there was a breach of that duty, that harm was caused and that it must be foreseeable by the defendant that such harm was likely to be the result of their actions or lack of them. Employers are liable for the negligence of their employees even if they have not personally been negligent. This reinforces the need for training of employees so that they understand what the risks are and how to avoid them and supervision to ensure that procedures and instructions are followed.

Decisions in civil law are based on the balance of probabilities, i.e. it does not have to be proved beyond reasonable doubt. The consequences of a successful case may be an injunction to stop the defendant continuing the actions with, if appropriate, damages to compensate for the harm caused.

Criminal law

In contrast, criminal or statute law is based on decisions by Parliament. In recent decades new legislation generally starts with Directives from the European Union. These are agreed by the Environmental Ministers of the member states but still have to go through a process of becoming national law within each state. The approach in the UK is to develop laws or statutes which are passed as Acts of Parliament. These set up a framework which allows the Secretary of State for the Environment to issue Regulations which contain all of the details. The Regulations can be issued later and changed without having to refer back to Parliament. An example is the Environmental Permitting (England and Wales) Regulations 2010 made under powers set out in the Pollution Prevention and Control Act 1999. The latter was passed to implement Council Directive 96/61/EC concerning integrated pollution prevention and control.

Breaches of statutes are a criminal offence and are dealt with under criminal law. This exists to enforce the acts and regulations and failure to comply can result in prosecution by the enforcing authority (the Environment Agency in England, Natural Resources Wales, SEPA in Scotland or the relevant local authority). Cases are brought before a magistrates' court that may decide directly or, if the case is sufficiently serious, refer it to the Crown Court for decision by a judge and jury. If there is a point of law at stake the case would be referred to the High Court. In either court the case has to be proven beyond reasonable doubt meaning, if there is any uncertainty in the interpretation of the evidence presented, the defendant is likely to be found 'not guilty' (in Scotland there is also a finding of 'not proven'). Higher courts can also require that remedial action is taken. Appeals follow a hierarchy from magistrates' court to Crown Court to High Court to Court of Appeal and then above as before.

Criminal courts punish offenders for their misdemeanours, which means that they may be liable to pay a fine or face imprisonment or both. Magistrates' courts have a limit on the levels of penalty that they can impose on those found guilty (called summary conviction). These are set out in the legislation and can be unlimited or up to two years imprisonment. If the magistrate thinks that the seriousness of the offence is such that a higher fine or term of imprisonment (known as an indictable offence) is warranted then they refer to the Crown Court. There, the level of fine is unlimited and prison terms may be sentenced up to five years.

The accused may be able to claim that they took reasonable care and that it was not reasonably practical (for example due to the scale of the event) to avoid the offence. However, for some offences such as causing water pollution, this is not possible – in these cases it is known as 'strict liability'.

Statutory nuisance is also a criminal offence as defined, for example, in the Environment Protection Act 1990. It includes the state of premises, emissions of smoke, gases fumes, dust and odours, the accumulation of waste, noise and various other causes that could be prejudicial to health or a nuisance. The definition of nuisance is as in civil cases but the case is taken up by a local authority due to the scale or extent of the problem and the potential effects on health. The local authority may issue an abatement notice requiring the offending party to take steps to stop or restrict the nuisance or other measures as required. Failure to comply with an abatement notice is also a criminal offence. Noise is probably the most frequent cause of statutory nuisance and the local authority will take account of the nature of the noise, its location, the time of day, the duration and the activity causing it. A noisy pub with loud music late at night could result in an abatement notice in a residential area. Noisy road works to carry out an emergency repair to a burst water main under similar circumstances would not.

Other sanctions

The regulatory agencies have other legal sanctions available to them which are discussed in section 1.4.5 and Box 1.11.

Box 1.6 Environmental enforcement agencies in the UK

England

Legislature: UK Parliament www.parliament.uk/
Environment: Environment Agency www.gov.uk/government/organisations/environment-agency
Nature conservation: Natural England www.gov.uk/government/organisations/natural-england

Wales

Legislature: Welsh Assembly http://wales.gov.uk/
Natural Resources Wales https://naturalresources.wales/?lang=en

Scotland

Legislature: Scottish Office http://home.scotland.gov.uk/home
Environment: Scottish Environment Protection Agency (SEPA) http://sepa.org.uk/
Nature conservation: Scottish Natural Heritage www.snh.gov.uk/

Northern Ireland

Legislature: Northern Ireland Executive www.northernireland.gov.uk/
Environment: Northern Ireland Environment Agency (a part of Department of Environment, Agriculture and Rural Affairs) www.daera-ni.gov.uk/northern-ireland-environment-agency
Nature conservation: Department of Environment Northern Ireland www.daera-ni.gov.uk/articles/council-nature-conservation-and-countryside

Box 1.7 Role of the Environment Agency in England

The areas regulated by the Agency are:

- Integrated Pollution Prevention and Control (IPPC).
- Radioactive substances.
- Waste management.
- Water quality.
- Land quality.
- Water resources.
- Flood risk management.
- Fisheries.
- Navigation, conservation and recreation – mainly on rivers.

Incidents reportable to the agency include:

- Pollution to air, water or land.
- Damage or threat to the natural environment.
- Poaching, illegal fishing or distressed and dead fish.
- Blockages in watercourses.
- Illegal dumping of hazardous wastes or large amounts of industrial waste.
- Incidents at waste sites regulated by the Agency.
- Illegal abstraction from watercourses.
- Collapsed or badly damaged river banks.
- Flooding.

The details of the above relevant to environmental management are expanded in Chapter 1 and later chapters. Air quality is largely the preserve of local authorities except for some types of installation regulated by the Agency under IPPC. Local authorities are also involved in dealing with fly tipping.

Box 1.8 Facilities requiring a permit

- An installation regulated under the integrated pollution prevention and control regime (IPPC, see main text and Box 1.9).
- Mobile plant regulated under IPPC or used for waste operations.
- A waste operation.
- A mining waste operation.
- An activity involving radioactive substances.
- A water discharge activity.
- A groundwater activity.

Water impoundment and abstraction are still governed by an abstraction licence but are expected to require a permit in due course.

Box 1.9 The roles of the Agency and local authorities in permitting under IPPC

The overall regime for regulating pollution derives from an integrated approach to reducing pollution from all sources and was applied initially to large-scale operations which could cause emissions to all media. This was known as Integrated Pollution Prevention and Control (IPPC) and was in response to an earlier EU Directive which has recently been updated (European Commission 2010). Premises regulated by the Agency or the local authorities are known as 'installations'. They are divided into classes: Part A1 activities, Part A2 activities and Part B activities. In all cases they include both fixed sites and mobile plant. The classification of an activity is based on the Schedules in The Environmental Permitting (England and Wales) Regulations 2010. Broadly, Part A activities are those most potentially polluting or emitting substances for which limits are set under EU directives. The Agency regulates Part A1 and most waste activities. Part A1 includes higher risk installations based on their size or potential to cause pollution such as large chemical plants, power stations and cement works.

Local authorities regulate Part A2 activities which include the remaining industrial premises such as glassworks, sawmills and foundries and Part B installations which are the smaller operations often carrying out a limited range of activities but potentially a cause of air pollution. This is under a regime known as Local Air Pollution Prevention and Control (LAPPC). Examples include paint spraying or dry cleaning.

District or borough councils are normally the regulator. In areas with only one council (a Unitary Council), it is the regulator. The port health authority may be the regulator in port areas.

Box 1.10 Powers available to inspectors from the Environment Agency

The following list of powers is a summary of those in the Environment Act 1995. It has been paraphrased but does indicate the extensive range of powers available. Similar powers are available to the other UK environmental regulatory bodies. The details vary according to which legislation is being enforced but the principles are similar.

- To enter any premises at any reasonable time or, in an emergency where in their opinion there is immediate risk of serious pollution to the environment, at any time and, if need be, by force.
- To take any other person duly authorised by the enforcing authority including a constable.
- To take any equipment or materials required for any purpose for which the power of entry is being exercised.
- To make such examination and investigation as may in any circumstances be necessary.
- To direct that premises or any part of them, or anything in them, shall be left undisturbed.
- To take measurements and photographs, make recordings and inspect and photocopy documents.
- To take samples, or cause samples to be taken, of any articles or substances found in or on any premises and of the air, water or land in, on, or in the vicinity of, the premises.
- To cause an article or substance which appears to have caused or to be likely to cause pollution of the environment or harm to human health to be dismantled or subjected to any process or test (but not so as to damage or destroy it, unless that is necessary).
- To take possession of any article or substance and detain it for so long as is necessary in order to examine it, or cause it to be examined, to ensure that it is not tampered with before examination of it is completed and to ensure that it is available for use as evidence in any proceedings.
- To require any person to answer questions and to sign a declaration of the truth of the answers.
- To require the production of information and records and to inspect and take copies of, or of any entry in, the records.
- To require any person to afford such facilities and assistance as may be required.
- Any other duty required for the discharge of their function.
- To take enforcement action such as the issuing of notices or initiate proceedings for prosecution.

Box 1.11 Enforcement powers available to the Environment Agency

The Agency has a range of powers available many of which are specified in the Acts or Regulations that relate to the control of pollution and the issuing of permits going back several years. To understand the full range would require those details and any subsequent amendments. This box sets out a summary of the range available and the circumstances under which they are likely to be used.

Civil sanctions

Recognising that many offences against statute may be relatively minor but require some proportional action short of prosecution in court, The Regulatory Enforcement and Sanctions Act 2008 enabled civil sanctions for the first time. For environmental offences The Environmental Civil Actions Order 2010 enables the regulatory agencies to issue notices or apply penalties directly. For the Environment Agency these came into force in January 2011 and were limited to certain aspects of the environmental legislation (regulations associated with packaging, oil storage, water resources, dangerous substances, nitrate pollution, hazardous waste, fisheries and sludge) but are being extended in application. The current status is summarised in www.gov.uk/government/publications/environment-agency-enforcement-and-sanctions-policy. For minor offences (such as those involving paperwork) there are fixed penalties

of £100 for individuals or £300 for companies. For more serious offences including negligence or which result in financial gain, there is a variable maximum penalty up to £250,000.
The basic approach is the issuing of a:

Warning or formal caution

- **Stop notice** which requires an immediate stop to an unlawful activity that is causing serious harm or has the potential to cause harm to human health or the environment;
- **Compliance notice** which requires actions to comply or to return to compliance, within a specified period (including failure to complete returns and similar offences);
- **Restoration notice** which requires steps to be taken, within a stated period, to restore harm caused by non-compliance.
- **A fixed penalty notice or monetary penalty notice** as above.
- A further possibility is an **Enforcement undertaking** which is based on an offer to take steps to comply and make amends for non-compliance and its effects. This can result in improvements to the local environment as compensation for damage caused elsewhere.

These notices can also be used where the existing legislation is inadequate in providing the necessary powers.

Statutory sanctions under previous legislation

For offences not subject to civil sanctions or where the circumstances suggest that they are inadequate, the Agency can issue various notices or take other actions depending on the Act that applies. The terminology used has changed over time as a result of amendments to various regulations and similar provisions still apply in the devolved administrations. The current sanctions available are:

- **Revoke, suspend or vary** a permit to stop an activity or bring about changes in operations to protect the environment.
- Issue an **enforcement notice** requiring compliance with the conditions of a permit.
- Issue a **groundwater prohibition notice** to stop an activity that may cause pollution of groundwater.
- Issue a **notice to remove waste.**
- Issue an **anti-pollution works notice** where activities on a site cause or threaten to cause pollution of controlled waters.
- The Agency also has powers to **prevent or remedy pollution and recover costs.**

The ultimate sanctions involve **prosecution** in the courts for criminal offences as described in Box 2.1. The offences are included in the relevant legislation and vary but the principles are similar. The decision whether to prosecute (e.g. for causing pollution, operating without a permit or failure to comply with permit conditions or notices and for any other breaches of legislation) is based on the seriousness of the offence, the history and attitude of the offender.

Various other sanctions are available directly or through the courts but are not of general interest. Details are available through several links on the Agency web site.

Box 1.12 Designated statuses by Natural England for protected sites

- Sites of Special Scientific Interest (SSSIs) for the best wildlife and geological sites.
- Special areas of conservation (SAC) designated under the EU Habitats Directive.
- Special Protection Areas (SPAs) designated under the EU Birds Directive.
- RAMSAR sites designated under the RAMSAR Convention (see Box 1.3).
- Areas of Outstanding Natural Beauty (AONBs).

Chapter 3

Box 3.1 Projects that require an environmental impact assessment within the EU

Below is a simplified list from Annex 1 of the Directive. The full list contains details of sizes or capacities that need to be exceeded before the Directive applies.

- Crude-oil refineries;
- Thermal power stations and other combustion installations;
- Nuclear power stations and other nuclear reactors including the dismantling or decommissioning of such power stations or reactors;
- Installations for the production, reprocessing and disposal of nuclear fuel and the storage and disposal of radioactive waste;
- Integrated works for the initial smelting of cast iron and steel;
- Installations for the production of non-ferrous crude metals from ore, concentrates or secondary raw materials by metallurgical, chemical or electrolytic processes;
- Installations for the extraction and processing of asbestos and products containing asbestos;
- Integrated chemical installations for the production of a specified range of products;
- Construction of lines for long-distance railway traffic, airports, motorways and express roads;
- Waste disposal installations for the incineration, chemical treatment or landfill of hazardous waste;
- Waste disposal installations for the incineration or chemical treatment of non-hazardous waste;
- Groundwater abstraction or artificial groundwater recharge schemes;
- Works for the transfer of water resources between river basins;
- Waste water treatment plants with a capacity exceeding 150,000 population equivalent;
- Extraction of petroleum and natural gas for commercial purposes;
- Dams and other installations designed for the holding back or permanent storage of water;
- Pipelines for the transport of gas, oil, chemicals or for the transport of carbon dioxide streams for the purposes of geological storage;
- Installations for the intensive rearing of poultry or pigs;
- Industrial plants for the production of pulp, paper and board;
- Quarries and open-cast mining;
- Construction of overhead high voltage electrical power lines;
- Installations for storage of petroleum, petrochemical, or chemical products;
- Installations for the capture of CO_2 streams and storage sites for the geological storage of carbon dioxide.

Source: European Commission (2014)

Box 3.2 UK requirements for environmental impact assessment

The UK legal requirements arise from the implementation of EU Directives described in the main text. The devolution of some aspects of government to the regions means that the regulatory regime is becoming more disparate and subject to change. If up to date information is required it is necessary to check on the latest versions of regulations and guidance through the relevant web sites for the devolved regional governments or the regulatory agencies.

The implementation of Directive 2011/92/EU on the assessment of the effects of certain public and private projects on the environment (European Commission 2014a) is through the Town and Country Planning (Environmental Impact Assessment) Regulations 2011. These only apply in England (with minor exceptions) with similar regulations applying in other parts of the UK (e.g. the Environmental Impact Assessment (Scotland) Regulations 2011). Schedule 1 of the Regulations list projects for which an EIA is mandatory and is similar to the list in Box 2.1. Schedule 2 is similar to the discretionary list in directive for the projects likely to cause a significant environmental impact due to the characteristics of the development, the location of the development or the characteristics of the impact. The selection criteria are specified in greater detail in Schedule 3. The 2011 Regulations are based on the 1999 Regulations with subsequent amendments consolidated into the text.

The implementation of European Directive 2001/42/EC on the assessment of the effects of certain plans and programmes on the environment (known as the SEA Directive; European Commission 2001a) is through the Environmental Assessment of Plans and Programmes Regulations 2004 for England with comparable regulations for the devolved regions. These are relevant to Structure and Local Plans and those concerned with minerals, waste, air quality, National Parks, transport, river basin management and the licensing of energy developments as examples. They apply, for example, to strategies and plans produced by the Environment Agency. Guidance for the UK is available in *A Practical Guide to the Strategic Environmental Assessment Directive* (Office of the Deputy Prime Minister 2005).

The Control of Major Hazards Regulations 1999 (with amendments in 2005) implements the Seveso II Directive (European Commission 1996). These are enforced by the Health and Safety Executive and the Environment Agency (and their devolved counterparts). The Regulations apply mainly to the chemical and oil industries and those that use or store quantities above the limits for the specified substances listed in Schedule 1. The majority of installations are 'lower tier' sites but they become 'upper tier' if the quantities exceed higher limits.

Chapter 4

Box 4.2 Air pollution control in the UK

Current air quality standards in the UK are derived from European Directives as described in Chapter 1 and are set out in the Air Quality Standards Regulations 2010. These are mainly derived from the most recent EU Directive (European Commission 2008e) and the Gothenberg Protocol to Abate Acidification, Eutrophication and Ground-level Ozone (UNECE 1999) which set standards for emissions to be achieved by 2010. Responsibility for implementing the regulations is devolved to the national administrations outside of England. The combined administrations produced an Air Quality Strategy in 2007 (DEFRA 2007) which contained air quality objectives and targets for a range of pollutants. The Strategy recognised that action would be needed at national, regional and local levels. Following legal challenges driven principally by concerns over emissions from vehicles, there have been further changes. A strategy dated 2019 can be found on the UK Government web site which explains the action required at various sources but the detailed standards are not included. The long-term future for the standards will be determined as the UK withdraws from the EU.

Local Authorities are required to review air quality in their areas and designate air quality management areas if improvements are necessary. Monitoring is conducted in areas known to be at risk of pollution using sampling and continuous monitoring with instruments. An air quality action plan is put in place to meet the targets where they are not being achieved. Some of this can be done utilising their powers for permitting and regulating local air pollution and their roles under IPPC as described in Chapter 1 and Box 1.9. The environmental agencies also play a role where they issues permits to sites under IPPC.

Where local action is insufficient or not appropriate, the problem becomes a national one and is achieved by national policies on issues such as energy production and vehicle emissions.

Chapter 5

Box 5.1 The regulation of water in the UK

In the UK water quality and abstraction have been regulated for many years. The main current legislation is the Water Resources Act, 1991 and subsequent Regulations although some aspects are covered by other legislation. The details of the regulations vary in the different devolved regions. The Act applies to 'controlled waters' which are surface fresh waters (such as rivers and streams, canals, lakes and ponds); groundwater; estuaries above the freshwater limit; coastal waters (essentially above high tide level or below the freshwater limit in an estuary) and territorial waters (out to sea for three nautical miles).

The UK environmental agencies monitor and report on water quality in the environment and establish water quality objectives for water quality improvement. They also manage the catchment to protect the flow regimes in watercourses so that the requirements of all users (including the environmental requirements) can be met as far as possible. The quality objectives and flow requirements are used to establish the conditions for licences for water abstractions and impoundments and permits for discharges. The balance between abstractions and residual flow affects the dilution available for polluting substances in discharges. The lower the flow, the tighter the permit conditions will be for a discharge.

Information about water quality and other aspects of their regulatory activities, including information about abstraction licences, permits and compliance are in the public domain. Details can be accessed through their web sites and at their offices.

Drinking water quality is regulated by the Drinking Water Inspectorate in England and Wales, an arm of DEFRA. The standards are defined in the Water Supply (Water Quality) Regulations 2016. Water that meets these regulations is accepted as 'wholesome'. In Scotland there is a Drinking Water Quality Regulator and in Northern Ireland the Drinking Water Inspectorate is part of the Northern Ireland Environment Agency: both agencies have their own versions of the Regulations.

Water supply, sewerage and sewage treatment in England and Wales is provided by water companies. In the other administrations the services are in the public sector but at arm's length from the regulators. There are price controls in place to balance the income requirements of the utility suppliers and the interests of the consumers. Companies wishing to make a discharge to their sewers require a formal document called a consent. This will contain limits on the flow and quality conditions in order to protect their infrastructure and the final quality of their own effluents.

Box 5.3 Pollution Prevention Guidelines issued by the environmental agencies in the UK

- PPG 1 General guide to the prevention of pollution.
- PPG 2 Above ground oil storage tanks.
- PPG 3 Use and design of oil separators in surface water drainage systems.
- PPG 4 Treatment and disposal of sewage where no foul sewer is available.
- PPG 5 Works and maintenance in or near water.
- PPG 6 Working at construction and demolition sites.
- PPG 7 Refuelling facilities.
- DEFRA Groundwater Protection Code: Petrol stations and other fuel dispensing facilities involving underground storage tanks.
- SEPA: Underground storage tanks for liquid hydrocarbons: code of practice for the owners and operators of underground storage tanks (and pipelines), 2006.

- PPG 8 Safe storage and disposal of used oils.
- DEFRA Code of Practice for Using Plant Protection Products.
- DEFRA Groundwater Protection Code: Use and disposal of sheep dip.
- Sheep Dipping Code of Practice for Scottish Farmers, Crofters and Contractors.
- PPG 13 Vehicle washing and cleaning.
- British Marine Federation: Environmental Code of Practice for marine businesses, sailing clubs and training centres; and The Green Blue: Self Assessment Environmental Toolkit
- Environment Agency: Environmental management toolkit for the food and drink industry
- PPG 18 Managing fire water and major spillages.
- PPG 20 Dewatering underground ducts and chambers.
- PPG 21 Pollution incident response planning.
- PPG 22 Incident response - dealing with spills.
- PPG 26 Drums and intermediate bulk containers.
- PPG 27 Installation, decommissioning and removal of underground storage tanks.
- PPG 28 Controlled burn.
- There is also a simple checklist 'Is your site right?'

The gaps in the numbering are for withdrawn documents, some of which have been replaced by non-PPG items listed. All are available by searching for a PPG on the web site www.gov.uk/guidance/pollution-prevention-for-businesses or www.netregs.org.uk/environmental-topics/pollution-prevention-guidelines-ppgs-and-replacement-series/guidance-for-pollution-prevention-gpps-full-list/

Chapter 6

Box 6.4 Waste definitions in the UK

In the UK you will come across the following classifications for waste types:

- **Inert:** as the EU directive. This does not degrade and so can be landfilled. Examples are bricks, concrete and sub-soil.
- **Hazardous:** as the EU directive. These wastes need to be kept separate and can only be disposed of as hazardous waste. Examples are toxic metals such as mercury, acids and asbestos.
- **Non-hazardous:** waste that is neither hazardous nor inert. This can also be disposed of to landfill. Examples include paper, cardboard, food waste and wood.
- **Special waste:** this was the term previously used for hazardous waste across the UK until the introduction of the classification as 'hazardous' under EU legislation. The term 'special' is still used in Scotland.

For legal and regulatory purposes the following definitions are also used:

- **Controlled waste:** this is household, commercial, industrial, agricultural and mining waste depending on source. This is a term used in the regulations and it includes waste in the type categories above which can come from any source.
- **Clinical and radioactive wastes:** these are hazardous waste as described in the main text.

Note that some wastes such as packaging and electrical and electronic wastes are also covered by other EU directives and UK regulations as described in the text. Packaging will be non-hazardous waste; electrical and electronic goods will be mainly hazardous waste. Their sources can be from any listed under controlled waste.

Box 6.5 UK waste production

Waste statistics are slow to be compiled and subject to later adjustment. The latest compiled data for the whole of the UK published in 2019 is for 2016 and is compared with 2011 from the first edition of this book.

	2011	2016
Waste source	Weight Mt	Weight Mt
Construction	101.0	136.2
Mining and quarrying	86.0	81.1
Soils	58.7	–
Commercial and industrial	63.7	41.7
Household	31.5	27.3
Other	2.7	17.7
Total	**288.6**	**222.9**

In 2011, the amount of waste recovered, including energy recovery, was 143Mt; just under 50 per cent. Most of the remainder was deposited on land. In 2016 the comparable figures were 117Mt and about

53 per cent. Waste production has declined by about 23 per cent and recycling rates have increased marginally.

Sources: Waste data overview (2011b), and UK Statistics on waste, 7 March 2019, available at www.defra.gov.uk

Box 6.8 Regulation of waste in the UK

The main requirements are listed below. They are broadly similar in all the jurisdictions. Reference is made to exemptions which apply in certain circumstances such as low volumes or particular types of waste or activities and for charities. The details are beyond the scope of this book and are subject to change. More information can be found on the DEFRA and the regulatory agencies' web sites. Standard permits are available for some types of operation.

- Producers of waste need an environmental permit if waste is stored for more than 12 months.
- Producers of hazardous waste must register with the relevant Regulator.
- Businesses that treat their own waste on site (e.g. by composting) need to register with the EA.
- If waste is received from another site then the activity may need to be registered or a permit may be required.
- Carriers of waste must be registered unless they are a waste collection authority or otherwise exempt.
- Facilities that store, treat or receive waste must have an environmental permit appropriate to the activities carried out unless exempt.

There are other relevant issues in Box 6.9.

Box 6.9 Duty of care as respects waste

The following is a summary of the main points as set out in the Environment Protection Act 1990, section 34, and subsequent regulations and guidance.

- The duty applies to anyone who imports, produces, carries, keeps, treats or disposes of controlled waste or, as a dealer or broker, has control of such waste.
- Waste must not be treated, kept or disposed of in a manner likely to cause pollution to the environment or harm to health.
- Different waste types to be segregated.
- Waste may only be deposited at a site with a permit.
- The producer of the waste must only pass it on to a person with the appropriate permit (a registered waste carrier).
- The producer should ensure that the ultimate disposal is at a site with the appropriate environmental permit.
- The producer should prevent the escape of the waste and keep it secure and protected from the weather.
- Waste containers should be suitably labelled.
- On transfer the waste should be accompanied by a transfer note with a written description of the waste and the EWC code.
- Hazardous waste to be accompanied by a consignment note with a written description of the waste and the EWC code.
- Records to be kept for two years: three years in the case of hazardous waste.

Chapter 8

Box 8.1 Noise in the work place

In the UK the Control of Noise Regulations sets limits for noise intensity in the work environment. The limits and action required vary with duration of exposure and whether the noise is continuous or peak but the upper peak sound pressure limit is 140dB. More information on sound measurement and the health and safety aspects of sound at work can be found in the companion book in this series (Hughes and Ferrett 2011).

Chapter 9

Box 9.2 Legal requirements for emergency plans in the UK

The Control of Major Accident Hazards (COMAH 2015) applies mainly in the chemical industry but there are others affected. The requirements are to 'Take all necessary measures to prevent major accidents involving dangerous substances; limit the consequences to people and the environment of any major accidents which do occur' (from the Health and Safety Executive (HSE) web site www.hse.gov.uk). They apply to hazardous chemicals above specified quantity thresholds. The requirement is for plans to prevent or mitigate serious accidents which could cause harm to people or the environment. The regulations are enforced by the HSE and the environmental regulators who can stop operations if they believe that the plans are deficient. The details can be found on the HSE web site and in the regulations.

Permits issued under the Environmental Permitting Regulations for discharges or operations subject to IPPC may also require operators to plan for emergencies that threaten the environment.

Bibliography

Baxter PJ, Kapila M, and Mfonfu D. (1989) 'Lake Nyos disaster, Cameroon, 1986: The medical effects of large scale emission of carbon dioxide?' *British Medical Journal*, 298: 1437–1441.

Blackhouse J (2018) *Environmental management revision guide for the NEBOSH certificate in environmental management*, Routledge, Abingdon, Oxon.

BP (2012) *BP statistical review of world energy June 2012*, available at www.bp.com/statisticalreview.

British Standards Institution (2003) *British standard 8555: Guide to the phased implementation of an environmental management system including the use of environmental performance evaluation*, BSI, London.

British Standards Institution (2019) *British standard 4142: Method for rating industrial noise affecting mixed residential and industrial areas*, BSI, London.

Brundtland, G (1987) *Report of the world commission on environment and development: Our common future*, United Nations General Assembly document A/42/427, New York.

COMAH (2015) *Control of major accident hazards (amendment) regulations*, available at www.legislation.gov.uk/uksi/2015/483/contents/made.

Convention on Climate Change, see United Nations 1998.

DEFRA (2001) *Guidance note for the control of pollution (oil storage) (England) regulations 2001*, Department for Environment, Food and Rural Affairs, London.

DEFRA (2007) *The air quality strategy for England, Scotland, Wales and Northern Ireland*, Department for Environment, Food and Rural Affairs, London.

DEFRA (2009a) *Protecting our water soil and air, a code of good agricultural practice for farmers, growers and land managers*, Department for Environment, Food and Rural Affairs, London, available at www. defra.gov.uk.

DEFRA (2009b) *Environmental permitting guidance: The landfill directive for the Environmental Permitting (England and Wales), Regulations 2007 Updated October 2009*, available at www.defra.gov.uk

DEFRA (2010a) *Environmental permitting guidance, core guidance for the environmental permitting (England and Wales) regulations 2010, version 3.1*, updated March 2010, available at www.defra.gov.uk.

DEFRA (2010b) *Environmental permitting guidance the waste incineration directive for the environmental permitting (England and Wales) regulations 2010, version 3.1*, updated March 2010, available at www.defra.gov.uk.

DEFRA (2011a) *Government review of waste policy in England 2011*, available at www.defra.gov.uk/publications/2011/06/14/pb13540-waste-review/.

DEFRA (2011b) *Waste data overview*, available at www.defra.gov.uk.

DRAX Group plc (2011) *Annual report and accounts*, p. 5, available at www.draxpower.com/investor.

Environment Agency (2005) *Hazardous waste, interpretation of the definition and classification of hazardous waste*, Technical guidance WM2, available at www.environment-agency.gov.uk.

Energy Institute (2015) *Guidelines on environmental management for facilities storing bulk quantities of petroleum, petroleum products and other fuels*, 3rd edition, Energy Institute, London.

Environment Agency (2018) www.gov.uk/government/publications/electrical-and-electronic-equipment-eee-covered-by-the-weee-regulations/electrical-and-electronic-equipment-eee-covered-by-the-weee-regulations.

Environment Agency (2019) *Land contamination: Risk management*, available at www.gov.uk/guidance/land-contamination-how-to-manage-the-risks.

European Commission (1994) *Directive 94/62/EC of 20 December 1994 on packaging and packaging waste*, Official Journal of the European Union L365/10, 31 December 1994.

European Commission (1998) *Directive 98/83/EC of 3 November 1998 on the quality of water intended for human consumption*, Official Journal of the European Union L330/32, 5 December 1998. This is under review, available at https://ec.europa.eu/environment/water/water-drink/review_en.html.

European Commission (1999) *Directive 1999/31/EC of 26 April 1999 on the landfill of waste*, Official Journal of the European Union L182/1, 16 July 1999.

European Commission (2000a) *Directive 2000/60/EC of the European Parliament and of the Council of 23 October 2000 establishing a framework for community action in the field of water policy*, Official Journal of the European Union L327/1, 22 December 2000. Later amendments are available at https://eur-lex.europa.eu/legal-content/EN/TXT/?uri=CELEX:02000L0060-20141120

European Commission (2000b) *Directive 2000/76/EC of the European Parliament and of the Council of 4 December 2000 on the incineration of waste*, Official Journal of the European Union L332/91, 28 December 2000.

European Commission (2001a) *Directive 2001/42/EC on the assessment of the effects of certain plans and programmes on the environment*, Official Journal of the European Union L197/30, 21 July 2001.

European Commission (2001b) *Commission Decision of 16th January 2001 amending Decision 2000/532/EC as regards the list of wastes*, Official Journal of the European Union L47/1, 16 February 2001.

European Commission (2003) *Directive 2002/96/EC on waste electrical and electronic equipment*, Official Journal of the European Union L37/24, 13 February 2003.

European Commission (2004) *Directive 2004/35/EC on environmental liability with regard to the prevention and remedying of environmental damage*, Official Journal of the European Union L143/56, 30 April 2004.

European Commission (2006a) *Regulation (EC) No 1907/2006 of the European Parliament and of the Council of 18 December 2006 concerning The Registration, Evaluation, Authorisation And Restriction Of Chemicals (REACH)*, Official Journal of the European Union L136/3, 29 May 2007 (corrected version).

European Commission (2006b) https://eur-lex.europa.eu/eli/dir/1976/464/2000-12-22. Details with later amendments are available at https://ec.europa.eu/environment/water/water-dangersub/lib_dang_substances.htm.

European Commission (2006c) *Directive 2006/118/EC of the European Parliament and of the Council of 12 December 2006 on the protection of groundwater against pollution and deterioration*, Official Journal of the European Union L372/19, 27 December 2006.

European Commission (2008a) *EMAS factsheet*, 3rd edition, EMAS and ISO/ENISO 14001 differences and complementarities.

European Commission (2008b) *Directive 2008/1/EC concerning integrated pollution prevention and control*, Official Journal of the European Union L24/8, 29 January 2008. The latest proposal can be found at https://eur-lex.europa.eu/legal-content/EN/TXT/PDF/?uri=CELEX:52007PC0844&from=EN.

European Commission (2008c) *Regulation (EC) No 1272/2008 of the European Parliament and of the Council of 16 December 2008 on classification, labelling and packaging of substances and mixtures, amending and repealing Directives 67/548/EEC and 1999/45/EC, and amending Regulation (EC) No 1907/2006*, Official Journal of the European Union L353/1, 31 December 2008.

European Commission (2008d) *Directive 2008/98/EC on waste and repealing certain directives*, Official Journal of the European Union L312/3, 22 November 2008.

European Commission (2008e). *Directive 2008/50/EC of 21 May 2008 on ambient air quality and cleaner air for Europe*, Official Journal of the European Union L151/2, 11 June 2008.

European Commission (2010) *Directive 2010/75/EU on industrial emissions (integrated pollution prevention and control) (recast)*, Official Journal of the European Union L 334/17, 17 December 2010.

European Commission (2012) *Directive 2012/18/EU of the European Parliament and of the Council of 4 July 2012 on the control of major-accident hazards involving dangerous substances, amending and subsequently repealing Council Directive 96/82/EC*, Official Journal of the European Union L197/1, 24 July 2012.

European Commission (2014a) *Directive 2014/52/EU amending Directive 2011/92/EU on the assessment of the effects of certain public and private projects on the environment*, Official Journal of the European Union L 124/1, 16 April 2014.

European Commission (2014b) *Commission Decision of 18 December 2014 amending Decision 2000/532/EC on the list of waste pursuant to Directive 2008/98/EC of the European Parliament and of the Council*, Official Journal of the European Union, L 370/44,30 December 2014.

European Commission (2018) *Directive (EU) 2018/849 of the European Parliament and of the Council of 30 May 2018 amending Directives 2000/53/EC on end-of-life vehicles, 2006/66/EC on batteries and accumulators and waste batteries and accumulators, and 2012/19/EU on waste electrical and electronic equipment,* available at Official Journal of the European Union, L150/93, 14 June 2018.

European Commission (2019) *Proposal for a directive of the European Parliament and of the Council amending directive 1999/31/EC on the landfill of waste*, available at https://eur-lex.europa.eu/legal-content/EN/TXT/PDF/?uri=CELEX:52015PC0594&from=EN.

Houghton J (2009) *Global warming: The complete briefing*, 4th edition, Cambridge University Press, Cambridge, 18–23.

Hughes P and Ferrett E (2011) *Introduction to health and safety at work, the handbook for the NEBOSH national general certificate*, 5th edition, Butterworth-Heinemann, London, 394.

International Association for Impact Assessment (1999) *Principles of environmental impact assessment best practice*, available to download at www.iaia.org.

International Association for Impact Assessment (2009) *What is impact assessment?* available to download at www.iaia.org.

IPPC (2018) *Global Warming of 1.5°C, an IPCC special report on the impacts of global warming of 1.5°C above pre-industrial levels and related global greenhouse gas emission pathways, in the context of strengthening the global response to the threat of climate change, sustainable development, and efforts to eradicate poverty*, available at www.ipcc.ch/2018/10/08/summary-for-policymakers-of-ipcc-special-report-on-global-warming-of-1-5c-approved-by-governments/

ISO (2015) *ISO 14001 Environmental management systems – Requirements with guidance for use*, International Organisation for Standardisation, Geneva.

ISO (2004) *ISO 14004 Environmental management systems – General guidelines on principles, systems and support techniques*, International Organisation for Standardisation, Geneva.

ISO (2006a) *ISO 14040 Environmental management – Life cycle assessment – Principles and framework*, International Organisation for Standardisation, Geneva.

ISO (2006b) *ISO 14044 Environmental management – Life cycle assessment – Requirements and guidelines*, International Organisation for Standardisation, Geneva.

ISO (2010) *ISO 14005 Environmental management systems – Guidelines for the phased implementation of an environmental management system, including the use of environmental performance evaluation*, International Organisation for Standardisation, Geneva.

ISO (2011) *ISO 14001: Environmental management systems – An easy-to-use checklist for small business – Are you ready?* International Organisation for Standardisation, Geneva.

ISO (2017) *ISO 14001: Environmental management systems – A practical guide for SMEs*, International Organisation for Standardisation, Geneva.

Mauna Loa Observatory (2019) The observatory publishes a record that is continuously updated at http://co2now.org/

NASA (2019) https://climate.nasa.gov/evidence/

Office of the Deputy Prime Minister (2005) *A practical guide to the strategic environmental assessment directive*, available at www.communities.gov.uk.

Our world in data (2019) https://ourworldindata.org/world-population-growth

Royal Commission on Environmental Pollution (1988) *The best practicable environmental option*, 12th Report, FEB 1988, Cm 310, HMSO, London.

Smithsonian Institute (2019) https://ocean.si.edu/through-time/ancient-seas/sea-level-rise

Sustainable Build (2019) www.sustainablebuild.co.uk/ReducingManagingWaste.html.

UNECE (1999) *Gothenberg protocol to abate acidification, eutrophication and ground-level ozone*, available at www.unece.org/env/lrtap/multi_h1.html.

UNEP (1998) *Basel convention for the control of trans-boundary movement of hazardous wastes and their disposal*, United Nations Environment Programme, available at http://basel.int/TheConvention/Overview/TextoftheConvention/tabid/1275/Default.aspx.

UNEP (2008) *Sustainable societies in Africa, modules on education for sustainable development*, available at www.unep.org/training/programmes/Instructor%20Version/Part_2/Activities/Human_Societies/Population/index.html

UNEP (2009) *The Montreal protocol on substances that deplete the ozone layer*, Nairobi, Secretariat for the Vienna Convention for the Protection of the Ozone Layer and the Montreal Protocol on Substances that Deplete the Ozone Layer.

UNFAO (2019) www.fao.org/sustainable-development-goals/indicators/211/en/

United Nations (1992) *Report of the United Nations conference on environment and development, Rio de Janeiro, 3–14 June 1992*, UN, New York.

United Nations (1998) Kyoto Protocol to the United Nations Framework,. http://unfccc.int/kyoto_protocol/items/2830.php

United Nations (2019) www.un.org/sustainabledevelopment/development-agenda/

United Nations Economic Commission for Europe (1992) *Convention on Environmental Impact Assessment in a Transboundary Context*, available on the UNECE web site www.unece.org.

Waste and resources action programme (2019) www.wrap.org.uk/food-drink.

WHO (2006) *Copenhagen, Air quality guidelines global update 2005*, WHO Regional Office for Europe, WHO, Geneva.

WHO (2017a) News release, 14 June 2017, available at www.who.int/en/news-room/fact-sheets/detail/drinking-water

WHO (2017b) *Guidelines for drinking water quality*, 4th edition, WHO, Geneva, available at www.who.int/water_sanitation_health/publications/drinking-water-quality-guidelines-4-including-1st-addendum/en/

WRAP (undated) www.wrap.org.uk/sites/files/wrap/Reducing%20your%20construction%20waste%20-%20a%20pocket%20guide%20for%20SME%20contractors.pdf

Index

Page numbers in **bold** indicate a box or table, *italics* indicate a figure.

acid rain, effects of 54, 56, **61**, 69, 174
agriculture and horticulture: monoculture and biodiversity **56**; water pollution 82; water usage 79
air pollution: atmospheric inversion **67**; control methods 71–76; damage to ozone layer 7; effects of poor air quality 66–67; emergency response 160; sources and impacts 66–71; waste incineration 132
air quality standards 66, **187**; units of measurement 66, 171
anaerobic digestion 96, 105, 117, 151
atmospheric inversion 67, *67*

bacterial disease 78, **80**
Basel Convention (1989) **15**, 117
Best Available Technique (BAT) 17, **17**
Best Practical Environmental Option (BPEO) 17
biodiversity, loss of 9, 54, **56**, **107**
biofuels 131–132
BP, Gulf of Mexico spill 10–11, 153, 154
British Standards: BS 4142 145; BS 8555 **24**, 25
Brundtland Report 12

carbon dioxide (CO_2) emissions: capture and storage 75; climate change 5–7, 68; reduction measures 122; reduction targets 6–7
carbon footprint 7, 46, 51
cars: choice of fuel 138; greener manufacture 14–15, 68, 72
CFCs (chlorofluorocarbons): damage to ozone layer 7, 54, 69; emissions, control of 67; emissions impact 7
climate change 5–7; cause of drought 8; effects on biodiversity 9; emissions causing 69, 115, 116; fossil fuels as cause 124; impacts 28, **32**; international action 16; life cycle analysis 50
COMAH regulations 23, 58, 89, **193**
companies: corporate social responsibility 14–15, 24; environmental credibility 23, 153; press and media 163; unethical practice, exposure of 9–10
complaints, response to 37, 64, 144–147
consent (sewage discharge) 85–86, 88, **188**
contaminated land: environmental impact 54, 58, *182*; pollutants within 118; regulation of 18, 19; remediation techniques 119; risks from 118, 120; water pollution 84
corporate social responsibility 14, 23; *see also* Environmental, Social and Governance reporting
COSHH 89, 92, **158**

DECC (Department for Energy and Climate Change) 18
deforestation 8–9, 55, **56**, 175
DEFRA (Department for Environment, Food and Rural Affairs) 18, 19, 104, 119, **187**
desalination 79
developing countries: drinking water 82; energy supplies 122, 124, 135; solar energy 126; standards enforcement, lack of 11
district energy systems 132, 133
document control emergency response plan 153, 157, 165; environmental management system 27, 34–35, 43; operational risks and impacts 63; waste management 19, 111
drinking water 78–79
Drinking Water Inspectorate **188**

ecosystems **28**, 54, **56**, **107**, 118, **158**
embodied carbon and water 51
emergency planning: reasons for 153–154; response plan 155–159; risk assessment 155, 160
emergency response plan: debrief and review 161–165; emergency control centre **157**, 157–159, 161–164; external services, liaison 159–160; fire hazards **189**, 162; initial response *155*, 160–161; key information **158**; key staff **158**, 159; leaks and spillages 93–94, 161; maintenance of 164; preliminaries 157; press and media handling 163; transport incident 163; waste disposal 163
emissions: atmospheric pollutants 54, 68–71; elimination at source 71–72; emergency response *153*, 162; minimising 72; monitoring data 39, 57; natural sources 68; removal processes 71–76; standards for 67–68
emissions removal: activated carbon 74–75, *75*; bag filter 73–74, *74*; electrostatic precipitators 73–74, *74*; solids by gravity 73; wet scrubbers 74, *75*
employee involvement: emergency planning 161, 164; energy efficiency 138, 140; environmental auditing 42; environmental management system 31–33, **33**, 35, 39; job and task procedures 58; noise control 147; recycling, training in 112; travel efficiency 138, 140; waste management training 106–108
energy efficiency: building layout and design 137–138; electricity usage 137; cost reduction 136–137; heat generation and use 136–137; human behaviour 138; incentive schemes 135; on site production 135; transport 138–140, *139*
energy recovery from waste 114–115, 132–133
energy supplies: alternative sources 124–135; biofuels 131–132; combined heat and power 133, *133*; fossil fuels 122–124, *122*; fuel cells and hydrogen 134; geothermal power 128–129, *129*; heat pumps 133; hydropower 127–128, *128*; methane recovery 132; nuclear power 129–131; solar energy 124–126, *125*, *126*; waste incineration 132–133; wave and tidal power 128; wind power 126–127, *126*, *127*
environment, definition of 2, *3*
Environment Agency: carbon trading schemes 18, **182**; discharge and disposal permits 18, **182**; emergency planning 154; enforcement powers **183–184**; enforcement procedures 19; information source 30, 59; inspection and sampling 19; inspectors' powers **183**; pollution incidents 152; public information 20; regulatory and advisory functions 18–19, 47, **181–182**; water regulation 188
Environment Protection UK 149
environmental aspects: definition of 22; direct and indirect 27; examples of **28–29**; identifying 27
environmental auditing 41–42, **42**

environmental damage: major incidents **10**; UK regulation 19, 154
environmental impact: aquatic 54; archaeological and historical sites 56; atmospheric 54; categories of 28, 53; definition of 22; direct and indirect 27, 53; ecosystems 55; examples of **28–29**; land contamination 54; raw material extraction 53; significance of 62–63; socio-economic and community effects 55; transport 56; types of 53–56
environmental impact assessment (EIA): aims and objectives 49; definition of 48; external information sources 58–60, **59**; impacts, significance of 62–63; internal information sources 57–58; legal requirement 48–49, **185**; life cycle analysis 49–53, *50*, **52–53**; operational risks, potential and impact evaluation 60–61, **61**; receptors at risk 61–62, **61**; review, reasons for 64; UK legislation **185**; voluntary action 49
environmental legislation 10–11; **187–188**; common law (UK) **179–180**; criminal law 11, **145**, 154, **180–181**; EIA Directive (EU) 48–49; Environment Act (1995, UK) 18, **183**; Environment Protection Act (1990, UK) 111, 119, **180**, **191**; frameworks and standards 11, 16, **180**; information sources 58; international agreements 6–7, 14–15, **15**, 49–50; local compliance 16–17; Radioactive Substances Act (1993, UK) 18, 103, **181**, **182**; Regulatory Enforcement and Sanctions Act (2008, UK) **183–184**; terminology used 17–18; Water Resources Act (1991, UK) 154, **188**
environmental management: commercial examples 177–178; incident costs 11; reasons for 3–4, 9; scale of problem 4; scientific terminology 173–175; units of measurement 171
environmental management systems (EMS): active monitoring 38–39; auditing of 41–43; benefits and limitations 45–46; checking and monitoring 37–38; development process 24–26, *25*; development through BS 8553:2016 **24**; document control 34; emergency planning 34, 36–37, 152, 153; environmental policy **24**, 26–27, **27**; financial benefits 23; implementation responsibilities 35, **36**; improvement, continuous 25–26, 31, 45, 135, 138, 165; initial review 30; ISO standards compatibility 23–24, 46; key performance indicators 40, **40**; management commitment 23; management review 44; non-compliance with law 41; operational integration 34–35; performance review 43–44; planning objectives and targets 30–31, **31**, **36**; reactive monitoring 39, 41, 49, 58; training programme 31–33, **33**
environmental noise: control of 146–148; effects of 145; measurement of 145, nuisance 145; operational procedures 148–149; sources of 146; workplace noise **192**
environmental organisations 60
Environmental Permitting Guidance for waste 114
Environmental, Social and Governance reporting 15
European Union legislation: CO_2 emission reduction targets 7; Directives, application of 11, 16, **180**; EIA Directive, projects affected 48–49, **185**; EMAS (Eco-Management and Audit Scheme) 24, 26, 34, 46; Environmental Liability Directive 19, 119, 154; European Waste Catalogue 101, **102**; Industrial Emissions Directive 115; Landfill Directive 102, 103, **114**; legislative institutions **16**; Packaging and Packaging Waste Directive 112–113; Waste Electrical and Electronic Equipment Directive 113; Waste Framework Directive 101; Waste Incineration Directive 115; Water Framework Directive 85

fossil fuels: benefits and limitations of 122–124; sources and uses 122–123
Friends of the Earth 9, 60

geothermal power 128–129, *129*
global warming 6; *see also* climate change
Gothenberg Protocol **187**
Greenpeace 9, 60
Groundwater: abstraction licence **182**; agricultural and industrial use 79; natural sources 78–9; over abstraction 79; pollution, effects of 54, 61, 80–81; pollution remedial regulation 89, **184**; source protection 114, **188–189**

halogens 7, 69, 174
hazardous liquids container construction 92, **92**; drainage, contamination prevention 90–91, *90*, *91*; fire water 83, 94, 157, **189**; siting and storage 89–90, *90*, operational risk management 92–93; spillages 93–94, *93*; UK regulation **182**
health and safety: emergency planning 152, 153, 156, 159; hazardous substances 94; ISO standard 23; operational procedures 58; supplier data 58, 90
Health and Safety Executive: COSSH **186**, **193**; legionella 80; noise 146
hydropower 124, 127–128, **128**

incidents: financial costs 11–12, 154; nuclear power 131, 156; press and media handling 163; reactive monitoring 39, 41
indigenous peoples 5, 9, 13, 62, 132
information, sources of: environmental management system 30; external to organisation 58–60, **59**; internal documents 57–58; *see also* document control
instrumentation and sampling 39
Integrated Pollution Prevention and Control (IPPC) 17, 18, 19, 49, 92, 114–115, 119, **180–182**, **187**, **193**
international agreements 7, 8, 15–16, 15, 49, 67, 70
International Association for Impact Assessment (IAIA) 48
International Standards: compatibility 23, 46; PDCA model 25, *25*; information source 59; ISO 14001 22, 23–7, 153; life cycle analysis 51
ISO 14001; compliance with 42, continual improvement 45; document control 34; environment, definition of 2; implementation model 25, *25*; standards compatibility 23, 46

key performance indicators 22, **22**
Kyoto Protocol 6, 16

land issues 8
land pollution *see* contaminated land
lead 28, 69, 71, 83, **87**, 90, 105, 113
life cycle analysis 14, 35, 49–53, *50*, **52**, 135
local exhaust ventilation 72

maintenance: need for 37, 72, 92, 97, 136–137, 140, 146–147; records 37, 57
mercury 56, 69, 70, 83, **87**, 113, **190**
methane: atmospheric pollutant 54, **61**, 68; energy recovery 96, 105, 116, 132, 134–136; greenhouse effect 6, **52**; risk management 54, 70, 72, 115, 119
mineral and ore extraction and processing 3–5, 8, 11, **28**, 51, **51**, 69, 83, 95, **102**, 117, **185**
mining and quarrying: environmental impact **28**, 54, 72, 82–83, 103, **107**, 118, 135; waste disposal permit **182**; water pollution 82
monitoring: active monitoring 38–39; reactive monitoring 39
Montreal Protocol 7, **15**, 67

Natural England 19, 20, 49, 60, 154, **181**, **184**
NEBOSH Certificate examination: practical application 169; sample questions 170; written examination 168–170
negligence: legal action 11, **179–180**; penalties and fines 154, **184**
neighbours: complaints 37, 58, 73, 119; consultation 30, 148; corporate social responsibility 15; emergency planning 154–164; noise nuisance 148–149
nitrogen, oxides of 34, 66, 68, 69, 124, 132, 135, 174

nuclear power **10**, **185**, 103, 129
nuisance: legal action 10, **179–181**; noise 27, 41–42, 52, 144–148, **145**; odours 67, 162; wind turbines 127

oil storage: container construction **85**, 89–91, **92**, **188**; drainage interceptors *91*; spillages 93; UK regulation **186**
ore extraction 51, **51**
OSPAR Convention **15**
ozone layer 7, 52, 61–62, 66, 67, 69, 71; and OSPAR Convention **15**

particulate matter 66–71, 11–16, 132; removal from diesel engines 75
penalties and fines 11, 23, 46, 63, 120, 145, 154; in UK **180**, **183–184**
permits: authorisations 18–19; discharges to atmosphere 68; discharges to watercourses 85–88; hazardous liquids 102; incineration 115; UK regulation **187**, **188**, **191**, **193**; waste management 112, **191**; water abstraction 81
persistent organic pollutants (POPS) 70
personal behaviour 12
pollution: common causes **152–153**; definition of 3–4; direct and indirect 53–54; Pollution Prevention Guidelines (PPG) **188–189**; rapid industrialisation 9
pollution, air: control methods 71–76; emergency response 153, 155, 159, 162; ozone layer, damage to 7; poor quality, effects of 66–67; sources and impacts 68–71; standards 67–69; waste incineration 132
pollution, land *see* contaminated land
pollution, water: agricultural and industrial use 79; caused by contaminated land 84; natural sources 84; reduction by control measures 84–85; regulation 85–86; risk reduction 84–85, 89–93, 152–153; sources and impacts 82–84; spillages 161–162; water quality, effects of 80–81
population growth 4
pressure groups 9–10; EIA consultation 60
product comparison 51, **52–53**; life cycle analysis 51
professional organisations 60

RAMSAR Convention **15**, 16, **184**
raw materials impact of extraction 53, 100; sourcing of 48; supply records 27, 30, **40**, 42
recycling barriers to and solutions 106–109; benefits **107**; limitations **107**; types of 105
regulations in UK: Air Quality Standards Regulations 2010, **187**; COMAH Regulations 89, **193**; Contaminated Land (England) Regulations 2012 119; Control of Major Hazards Regulations 1999 186; Control of Pollution (Oil Storage) Regulations 2001 91, **188**; COSSH regulations 89, 92; Environmental Assessment of Plans and Programmes Regulations, 2004 48–49, **186**; Environmental Damage Prevention and Remediation Regulations 2009 19, 154; Environmental Permitting Regulations (England and Wales) 2010 18, **193**; The List of Wastes (England) Regulations 2005 101; Site Waste Management Plans Regulations 2008 113; The Waste (England and Wales) Regulations 2011 104, 111–112; Waste Electrical and Electronic Regulations 2006 113, **190**; Water Supply (Water Quality) Regulations 2000 **188**
regulatory bodies: CO_2 emissions, reduction targets 7; emergency planning 159–160, 162; information source 59, **59**; pollution prevention guidelines 84, **188–189**; role of 18–19; UK enforcement **181**, **187–188**
resource efficiency 13, 40–41, 178
resources, consumption rise 4–5, 8
Rio Summit (1992) 13
risk management: emissions monitoring 57; environmental management system 23; hazardous liquids *88*, 88–89; receptors at risk 61–62, 146–148; source, pathway, receptor model 61, **61**, 63, 89, 119, 146, 160

Seveso II Directive 49, **186**
site inspections 19, 30, 38–39, 43, **85**, 93
site safety: contamination history 58; documentation and training 94; hazardous liquids, siting and storage 89–90; hazardous liquid containers 91–92; operational risk management 92–93, 155–156; pipe and connective networks 92; spillages, dealing with 93–94, 93, 161–162, *162*; waste storage 110–111, *110*
SMART objectives and targets **31**, 45
smog 7, 54, 69–71, 124; inversion as a factor **67**
solar energy 124–126, *125*, *126*
Strategic Environmental Assessment (SEA) 49, **185**
sulphur, oxides of 39, 54, 67–69, 74, 124, 132
suppliers: EMS accreditation, benefits of 63–64; material data safety sheets (MSDS) 58, 90, 93, **158**
surface water: pollution, effects of 54, **69**, 80–81; pollution threats 82; sources of 78; usage 79–80
sustainability: balancing aspects of 12–14, *12*; biofuels debate 132; definition of 12; UN Goals 12

tidal power 128
trade associations 29, 30, **52**, 58–59
transport: energy efficiency 138–140, *139*; environmental impact 54, 55, 56; incident response 163; noise reduction 148; suppliers' environmental credentials 64–65
UN Environment Programme 103
UN World Commission on Environment and Development 12

virtual water 8, 51
volatile organic compounds (VOC) 54, 70, 72, 123

waste: causes of 3–4, 100–101, **101**, **190**; clinical 103; controlled 103; hazardous and non-hazardous 101–103, **102**, **190**; inert 103; radioactive 103, 130, **185**; special 190; UK classifications **190**
waste disposal: documentation 57, 104, 111; illegal export 8, 117; incineration of 106, 111, 113, 115–116, 132–133; landfill sites 111; operational competency 111–112; plastic 8, 118; residual waste 106, 116–117; rising costs 117–118; trans-boundary movement 117; wastewater treatment of 96
waste management: construction materials 113; containers and storage *110*, 110–11; domestic 113; electrical and electronic 113; good practice (duty of care) 111, 117, **191**; hierarchy 104; minimising production 104; nuclear **130**; packaging 112–113; recycling and reuse 100–101, 105–106, 116–117; regulation **191**; use of BPEO 17; zero waste 26, 104, 106
wastewater sewers, consenting for discharges 94, 96, 162
wastewater treatment 94–96, *94–96*; measures of organic carbon **86–87**
water pollution: agricultural and industrial sources 84; contaminated land 84; natural sources 84; risk reduction measures 84–85, *85*, 89–93, *90*, *91–93*; sources and impacts 82–83; spillages, dealing with 93–94, *93*, 161–162, *161*; water quality, effects on 82
water resources: abstraction 79, 81; agricultural and industrial use 79; conservation measures 81; goods production usage **8**, 77–78; hydropower 124, 127–128, *128*; safe drinking water 79; UK regulation **188**; unequal consumption 7; water borne disease 71, **80**; water cycle 78–79, *79*; water quality and regulatory control 85–88, **87–88**, 88–89, **188**
wave power 128
wildlife: acid rain, effects of 69, 81; contaminated land 118; global warming and climate change 2, 6, 9; noise, effects on 145; over-abstraction of water 81; organisations to protect 20, 60, **184**; water pollution, effects on 54, 80, 82, 85, 153
wind power 126–127, *126*, *127*
World Health Organisation 67, 78